MOUNTAIN ENVIRONMENTS
IN CHANGING CLIMATES

Will there still be glaciers in the Alps at the end of the twenty-first century? Is mountain-based hydro-electric power supply likely to be disrupted by the enhanced greenhouse effect? Can one predict climate change in mountain regions?

Once regarded as hostile and economically non-viable regions, mountains have in the latter part of the century attracted major economic investments for tourism, hydro-power, and communication routes. The uncertainties of climate change will inevitably have an effect on the economy of mountainous regions. Global change will exacerbate the conflictual situation between Environment and Economy.

This interdisciplinary book summarizes some aspects of current work on the physical and human ecology of mountain environments. International experts in different fields are brought together to address topics related to the impacts of climate change on mountain environments and socio-economic systems, as well as means of understanding the fundamental processes involved, their observation and their prediction.

Martin Beniston is Senior Scientist in the Department of Geography at ETH-Zurich, and Co-Chair of the Intergovernmental Panel on Climate Change, Working Group II/C. As director of the Swiss National Climate Program from 1990–92, he organized the 1992 International Conference in Davos whose themes are reflected in this book.

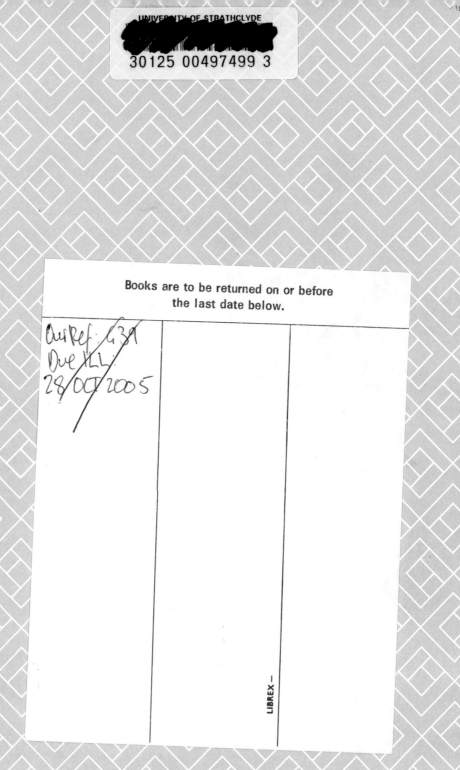

Books are to be returned on or before
the last date below.

AuRef G31
Due ILL
28/OCT 2005

LIBREX—

MOUNTAIN ENVIRONMENTS IN CHANGING CLIMATES

Edited by Martin Beniston

London and New York

First published 1994
by Routledge
11 New Fetter Lane, London EC4P 4EE

Simultaneously published in the USA and Canada
by Routledge
29 West 35th Street, New York, NY 10001

© 1994; collection © Martin Beniston; chapters © contributors

Typeset in Garamond by Solidus (Bristol) Limited

Printed and bound in Great Britain by
Biddles Ltd, Guildford and King's Lynn

British Library Cataloguing in Publication Data
A catalogue record for this book is available from the British Library

Library of Congress Cataloging in Publication Data
Mountain environments in changing climates / edited by Martin
Beniston.
Includes bibliographical references and index.
1. Mountain climate. 2. Weather, Influence of mountains on.
3. Climatic changes. I. Beniston, Martin.
QC993.6.M66 1994
574.5'264—dc20
93–45308

ISBN 0–415–10224–3

CONTENTS

Part I Climate change in mountain regions: fundamental processes, historical and contemporary observations, modelling techniques

Part II Impacts of climate change on vegetation: observations, modelling, networks

Part IV Conclusion

FIGURES

TABLES

CONTRIBUTORS

Bruno Abegg Department of Geography, University of Zürich-Irchel, Switzerland

Gabriela Apfl Department of Geography, University of Berne, Switzerland

Alan Apling Global Atmosphere Division, Department of the Environment, London, UK

Antoine S. Bailly Department of Geography, University of Geneva, Switzerland

J.S. Baron Natural Resource Ecology Laboratory, and National Park Service, Water Resources Division, Colorado State University, Fort Collins, USA

Roger G. Barry CIRES, University of Colorado, Boulder, USA

Michael F. Baumgartner Department of Geography, University of Berne, Switzerland

Martin Beniston Convenor of the Davos Conference, Department of Geography, Swiss Federal Institute of Technology (ETH), Zürich, Switzerland

Meinhard Breiling International Institute for Applied Systems Analysis (IIASA), Laxenburg, Austria

Felix Bucher Department of Geography, University of Zürich-Irchel, Zürich, Switzerland

Harald Bugmann Systems Ecology, Institute of Terrestrial Ecology, Swiss Federal Institute of Technology (ETH), Zürich, Switzerland

Tim R. Carter Agricultural Research Centre, Joikioinen, Finland

S.R. Chalise Mountain Environment Management Division, International Centre for Integrated Mountain Development (ICIMOD), Kathmandu, Nepal

Pavel Charamza Department of Statistics, Charles University, Prague, Czech Republic

T.N. Chase Department of Atmospheric Science, Colorado State University, Fort Collins, USA

J.M. Cram Cooperative Institute for Research in the Atmosphere, Colorado State University, Fort Collins, USA

Bernard Doche Laboratoire de Biologie Alpine, Université Joseph Fourier, Grenoble, France

G. Dürrenberger Human Ecology, Swiss Federal Institute for Environmental Science and Technology (EAWAG), and Swiss Federal Institute of Technology (ETH), Zürich, Switzerland

Andreas Fischlin Systems Ecology, Institute of Terrestrial Ecology, Swiss Federal Institute of Technology (ETH), Zürich, Switzerland

B.B. Fitzharris Department of Geography, University of Otago, Dunedin, New Zealand

Rainer Froesch Department of Geography, University of Zürich-Irchel, Switzerland

C.E. Garr Department of Geography, University of Otago, Dunedin, New Zealand

Hartmut Grassl Max-Planck-Institut für Meteorologie, and Meteorology Institute, University of Hamburg, Germany

Lisa J. Graumlich Laboratory of Tree-Ring Research, University of Arizona, Tucson, USA

Richard Guyette School of Natural Resources, University of Missouri, Columbia, USA

Wilfried Haeberli Laboratory of Hydraulics, Hydrology, and Glaciology, Swiss Federal Institute of Technology (ETH), Zürich, Switzerland

Pat N. Halpin Global Systems Analysis Program, Department of Environmental Sciences, University of Virginia, Charlotteville, USA

Friedrich-Karl Holtmeier Department of Geography, Landscape Ecology Division, University of Münster, Germany

John L. Innes Swiss Federal Institute for Forest, Snow and Landscape Research, Birmensdorf, Switzerland

François Jeanneret Geographical Institute, Universities of Berne and Neuchâtel, Switzerland

H. Kastenholz Institute for Behavioral Sciences, Swiss Federal Institute of Technology (ETH), Zürich, Switzerland

T.G.F. Kittel Natural Resource Ecology Laboratory, Colorado State University, Fort Collins, and Climate System Modeling Program, University Corporation for Atmospheric Research, Boulder, USA

Christian Körner Department of Botany, University of Basel, Switzerland

T.J. Lee Department of Atmospheric Science, Colorado State University, Fort Collins, USA

Brian H. Luckman Department of Geography, University of Western Ontario, London, Canada

Clifford J. Martinka Glacier National Park, West Glacier, Montana, USA

Martin L. Parry Environmental Change Unit, University of Oxford, UK

J.D. Peine National Park Service, University of Tennessee, Knoxville, USA

David L. Peterson National Park Service and Cooperative Park Studies Unit, University of Washington, Seattle, USA

Christian Pfister Institute of History, University of Berne, Switzerland

R.A. Pielke Department of Atmospheric Science, Colorado State University, Fort Collins, USA

Martin F. Price Environmental Change Unit, University of Oxford, UK

R. Rudel Institute for Research in Economics (IRE), Bellinzona, Switzerland

André Pornon Laboratoire de Biologie Alpine, Université Joseph Fourier, Grenoble, France

PREFACE

A. Apling

Much has happened since 1991, when an international conference entitled 'Mountain Environments in Changing Climates' which eventually resulted in this book, was proposed. In particular, the United Nations Conference on Environment and Development (the so-called 'Earth Summit') took place in June 1992 in Rio de Janeiro, during which two major international conventions relevant to the topics treated in this book were each signed by more than 150 countries: the Conventions on Climate Change and on Biological Diversity. Not only are these conventions relevant to mountain environments but, in the case of the Climate Convention, better understanding of the impact of climate on the natural and socio-economic environments of mountains could have a direct influence on the future development of the Convention itself.

The Climate Convention's major objective is stated in its Article 2, an extract of which reads:

> The ultimate objective of this Convention and any related legal instruments that the Conference of the Parties may adopt is to achieve, in accordance with the relevant provisions of the Convention, stabilization of greenhouse gas concentrations in the atmosphere at a level that would prevent dangerous anthropogenic interference with the climate system. Such a level should be achieved within a time frame sufficient to allow ecosystems to adapt naturally to climate change, to ensure that food production is not threatened, and to enable economic development to proceed in a sustainable manner.

This raises the question of how we define 'dangerous anthropogenic interference with the climate system'. The answer must lie in research to understand in detail the interaction of climate change with natural and human-managed ecosystems, and with human settlements, activities, and infrastructure. The level of detailed understanding needs to be sufficient for identifying quantitatively how systems respond to climate change, both in terms of rate of change and absolute level, so that 'thresholds' can be identified corresponding to a 'dangerous anthropogenic interference with the climate system'.

This is an extremely challenging task, but a vital one. It is possible that the threat of climate change, unlike the other classic air pollution problems of urban smog, acid deposition, and ozone depletion, can be tackled before effects begin to be felt. The Climate Convention is explicitly a precautionary instrument. However, until observable impacts are detected, the basis for action depends solely on the modelled and predicted future effects. Identifying the limits to tolerable climate change quantitatively and, through climate models, interpreting them in terms of limiting the atmospheric concentrations of greenhouse gases, will provide the key information to guide governments and develop the Climate Convention.

Amongst impact study areas, mountain regions appear to be reasonably well advanced in the level of mechanistic understanding of the ecosystems they contain and, for a number of regions, in terms of the detailed geographic information available to calibrate and initialize climate impact assessments. Most importantly, the very nature of mountain regions means that results are regionally specific, which is an essential characteristic for climate change impact thresholds if they are to command attention.

Of course, much of the work described in this book is not immediately concerned with climate change threshold identification, although virtually all of it could contribute. The relationship of climate and mountain systems will remain an important area of basic ecological research of interest in its own right, but the possible contribution that it can make to our appreciation of the threat of climate change and the measures needed to combat it will add considerably to the urgency with which the work is pursued.

ACRONYMS

ALPEX	Alpine Experiment
ASCAS	Alpine Snow Cover Analysis System
AVHRR	Advanced Very High Resolution Radiometer
BATS	Biosphere-Atmosphere Transfer Scheme
BP	before present
BRIM	Biosphere Reserves Integrated Monitoring Programme
CAPE	Convective Available Potential Energy
CCM	Community Climate Model
CFCs	Chloro-Fluoro-Carbons
CLIMAP	Climates of the Past
COHMAP	Commission for Oceanography, Hydrology, Meteorology, and Atmospheric Physics
CPU	Central Processor Unit
DBMS	Data Base Management System
DEM	Digital Elevation Model
ECHAM	European Centre/Hamburg General Circulation Climate Model
ECMWF	European Centre for Medium-range Weather Forecasting
ELA	Equilibrium Line Altitude
ENSO	El Niño-Southern Oscillation
EOF	Empirical Orthogonal Function
EPA	Environmental Protection Agency
EPOCH	European Program on Climate and Natural Hazards
ESF	European Science Foundation
ETH	Eidgenössische Technische Hochschule (Swiss Federal Institute of Technology)
FIFE	First ISLSCP Field Experiment
ForClim	Forest-Climate Model
FORECE	FOREst succession model for central Europe
FUTURALP	FUTURe ecosystems in the ALPine region project
GARP	Global Atmospheric Research Programme
GCM	General Circulation Model
GFDL	Geophysical Fluid Dynamics Laboratory

GHG	Greenhouse Gas
GIS	Geographic Information System
GISS	Goddard Institute for Space Studies
GOES	Geostationary Orbiting Earth Satellite
HKH	Hindu Kush-Himalayas
IIASA	International Institute of Applied Systems Analysis
ICIMOD	International Centre for Integrated Mountain Development
ICSU	International Council of Scientific Unions
IGBP	International Geosphere Biosphere Programme
IHD	International Hydrology Decade
IPCC	Intergovernmental Panel on Climate Change
ITCZ	Inter Tropical Convergence Zone
IUCN	International Union for the Conservation of Nature
LAI	Leaf Area Index
LEAF	Land Ecosystem-Atmosphere Feedback
LST	Local Standard Time
MPI	Max Planck Institute
NCAR	National Center for Atmospheric Research
NDVI	Normalized Difference Vegetation Index
NOAA	National Oceanic and Atmospheric Administration
OSU	Oregon State University
RAMS	Regional Atmospheric Modeling System
RAMSES	Regional Application Modeling System for Ecosystem Studies
RHESSys	Regional HydroEcological Simulation System
SiB	Simulation of the Biosphere model
SMA	Swiss Meteorological Agency
SRM	Snowmelt Runoff Model
UKMO	UK Meteorological Office
UNCED	United Nations Conference on Environment and Development
UNECE	United Nations Economic Commission for Europe
UNEP	United Nations Environment Programme
UNESCO	United Nations Educational, Scientific and Cultural Organization
USEPA	United States Environmental Protection Agency
USGS	United States Geological Survey
UTC	Universal Time Change
WCMC	World Conservation Monitoring Centre
WCP	World Climate Programme
WMO	World Meteorological Organization

INTRODUCTION

M. Beniston

MOUNTAINS AND THE CLIMATE SYSTEM

Global climate change has attracted much scientific and public attention in recent years, as a result of fears that man's economic activities are leading to an uncontrolled increase in Greenhouse Gas (GHG) concentrations. These are expected to result in a global rise in temperature due to the radiative properties of these gases. Predictions on the basis of atmospheric General Circulation Models (GCMs) presented in the Intergovernmental Panel on Climate Change (IPCC) report (Houghton *et al.*, 1990) indicate that the lower atmosphere could warm on average by 1.5–4.5°C by the end of the next century. Among the possible consequences of this unprecedented rate of warming are increased desertification, sea-level rise, reduced water resources, collapse of intensive agriculture in the producing countries, irreversible ecosystem damage and loss of biodiversity, and drop in energy availability resulting from dwindling hydro-power. In the context of a rapidly increasing world population, global climate change will need to be prepared for by adaptation (which only the economically strong nations can afford) or prevention (which has not yet been accepted widely enough by political decision makers).

Of major, but not single, importance in the climate system is the atmosphere, which is probably the most rapidly reacting element of the climate system to internal and external forcing. The atmosphere is made up principally of three gases – oxygen, nitrogen, and argon – but also of minor quantities of other gases, such as carbon dioxide (CO_2), methane (CH_4), nitrous oxide (N_2O), and totally artificial molecules such as the chloro-fluoro-carbons (CFCs). Despite their very small concentrations – CO_2 represents about 300 parts per million by volume – their radiative properties and their role in the thermal structure of the atmosphere is of prime importance. Indeed, in the absence of GHGs, the average temperature of the atmosphere would be 35°C less than today, thus preventing any form of life on our planet. However, since the beginning of the industrial era, and because of the fact that the world population has increased quasi-exponentially since that time, GHGs have increased at a very high rate compared to the natural fluctuations of these gases. It is the increase of GHGs, and the imbalance in the natural uptake or absorption of the trace gases, that lead climate specialists to believe that the atmosphere will undergo a period of accelerated warming.

A second key element of the climate system is the ocean, which acts as a regulator of atmospheric temperature and gas concentrations. The exchange of heat and moisture between the ocean and the atmosphere is a key process which is still incompletely understood. The complex surface and deep ocean circulations transport and store enormous quantities of heat and GHGs, and a more complete understanding of the climatic role of the ocean through modelling and observations is required. Recent results indicate that carbon uptake by the ocean is perhaps less important than previously estimated, so that probably a vital climate control is in the biosphere (the 'missing carbon sink'). Living organisms, particularly forests and plants, as well as soils, play a key role in atmospheric heat, moisture, and energy budgets close to the surface. However, knowledge of two-way interactions between the biosphere and the atmosphere is too poor for these effects to be adequately included in present-day climate modelling systems.

Other components of the climate system include the cryosphere – continental ice caps and floating sea ice – whose presence generates instabilities in the general circulation of the atmosphere because of the temperature difference between the Poles and the Equator. In addition, the surface energy balance is significantly influenced by the high reflectivity, or albedo, of the ice. The seasonal fluctuations of sea ice, and its possible disappearence in the Arctic Ocean in a warmer atmosphere, could have significant repercussions on accelerated global warming.

Even if major mountain systems account for a relatively minor surface area of the globe, they are none the less an integral part of the climate system. As a physical barrier to atmospheric flows, they perturb synoptic patterns and are considered to be one of the mechanisms of cyclogenesis in middle latitudes. Because of significant altitudinal differences, mountains such as the Alps, the Rockies, the Andes, and the Himalayas incorporate many elements of the climate system such as ice-fields and glaciers, albeit at a smaller scale than in polar regions; they also feature considerable biodiversity more often associated with latitudinal differences. Abrupt changes in the climatic parameters which have led to the present distribution of vegetation, ice, snow, and permafrost zones, as well as precipitation patterns would impact heavily on these features.

M. BENISTON

POTENTIAL IMPACTS OF ABRUPT CLIMATE CHANGE IN MOUNTAINS

Mountain regions are characterized by sensitive ecosystems, enhanced occurrences of extreme weather events and natural catastrophes; they are also regions of conflicting interests between economic development and environmental conservation. Once regarded as hostile and economically non-viable regions, mountains have in the latter part of the century attracted major economic investments for tourism, hydro-power, and communication routes. In the context of climate change, significant perturbations can be expected to natural systems and these will inevitably have an influence on the economy of mountainous regions. Global change will probably exacerbate the present conflictual situation between environment and economy.

Examples of systems in mountain areas which could be significantly perturbed by abrupt climate change, in particular hypothesized global warming consecutive to GHG increases of anthropogenic origin, are given below. These topics will be developed in greater detail in the various sections of this book.

Impacts on water resources

The sharing of water resources poses immense political problems when major hydrological basins are shared by several different countries, and it is exceedingly difficult to manage the waters of a river which crosses several international boundaries. For example, the Swiss Alps are the source of many of Europe's major rivers, in particular the Rhône, the Rhine, the Inn which feeds into the Danube, and the Ticino which flows into the Pô. Rivers originating in Switzerland flow into the North Sea, the Mediterranean, the Adriatic, and the Black Sea. Any uncoordinated control of these river basins by any one country, due to decreased precipitation and increased demand would lead to serious conflictual situations downstream. Such problems would become exacerbated by climate change in sensitive regions. Simulating precipitation patterns in mountain areas is today exceedingly difficult, because many precipitation events are local or mesoscale in nature and are poorly resolved by forecast models. Predicting the future trends of precipitation over areas as small as the Alps, for example, is practically impossible with present-day GCMs, so that little can be inferred for the impact of global warming on precipitation patterns. However, this is one key parameter which urgently requires quantification, as any significant change in rainfall or snowfall will have wide-ranging effects on natural and economic environments, both within the mountains themselves and far downstream in the river basins which they control.

Impacts on snow, glaciers, and permafrost

The IPCC has attempted to assess some of the potential effects of a warmer atmosphere on the mountain cryosphere. According to a report by the National Research Council (1985), a 1°C increase in mean temperature implies that the present snowline could rise by about 200 m above present levels in the European Alps. Kuhn (1989) has estimated that, for a 3°C warming by the middle of the next century, ice would no longer occur below the 2,500 m level in the Austrian Alps, with the consequence that 50 per cent of the remaining ice-cover would disappear. It is, however, difficult to estimate the exact response of glaciers to global warming, because glacier dynamics are influenced by numerous factors other than temperature, even though temperature, precipitation, and cloudiness may be the dominant controlling factors. According to the size, exposure, and altitude of glaciers, different response times can be expected for the same climatic forcing. For example in Switzerland, a number of glaciers are still advancing or are stationary despite a decade of mild winters, warm summers, and lower than average precipitation.

A rise in the snowline would be accompanied by ablation of present permafrost regions. Zimmermann and Haeberli (1989) show that permafrost melting as well as the projected decrease in glacial coverage would result in higher slope instability, which in turn would lead to a greater number of debris and mud slides, and increased sediment loads in rivers.

Impacts on forests and natural ecosystems

Species will react in various manners to global climate change; increases can be expected for some species while others will undergo marked decreases. Under conditions of rapid warming, it is quite likely that certain ecosystems may not adapt quickly enough to respond to abrupt climate changes; this could be the case of forests which may not be able to migrate fast enough, or of species with very local ranges which have nowhere to migrate to. The possibilities of migration of species to higher altitudes, in order to find similar conditions to those of today, will probably be limited by other factors such as soil types, water availability, and the human barriers to migration such as settlements and highways. Analogies with migrations of the past are hazardous at best because of human perturbations to the environment which were not present millennia ago, and because the expected climatic change will be at a rate unprecedented in the last 10,000 years. The disappearance of certain types of protective vegetation cover, particularly forests, from mountain slopes would render these more prone to erosion, mud and rock slides, as well as avalanches.

Impacts on mountain economies

The environmental impacts of a warmer atmosphere would have a number of consequences for the regional economy of a mountain region. Increases in debris flows and avalanches would be responsible for greater incidence of damage to buildings and traffic routes, vegetation, and the obstruction of rivers with corresponding flooding. In the severe flooding episode of August 1987 in Central Switzerland, road and rail traffic on one of the densest north–south communication routes across the Alps was severely disrupted. Increases in such episodes in future would have severe economic consequences not only for the region itself, but also for all users of such vital communication links. The additional risks due to climate change would sharply raise insurance premiums to cover the higher frequency of damage to property, as well as the basic costs of civil engineering works necessary for the protection of settlements and communication routes. Such costs would weigh heavily upon regional mountain economies.

Tourism might also suffer from climate change in mountain regions; in particular, uncertain snow cover during peak winter sports seasons might discourage tourists to come to the mountains. However, in recent years, numerous winter sports resorts have been facing financial difficulties even during favourable winters. Many are now reorganizing their sports and cultural infrastructure in order to attract vacationers whatever the climatic conditions.

Hydro-electric power is a key element of mountain-based economy in the Alpine regions, as well as in the Rockies and in the Southern Alps of New Zealand. In this latter country, over 70 per cent of electricity is generated by hydro-power (Fitzharris, 1989). Accelerated glacier melting would result in an increase in electricity production and a reduction in water storage requirements, at least until the glaciers find a new equilibrium level. However, if the glacier melting is intense, then river flooding and damage to dam structures may occur, offsetting the temporary 'benefits' of increased storage and production capacities.

RATIONALE FOR THIS BOOK

It is on the basis of some of the considerations just described that the International Conference on Mountain Environments in Changing Climates was organized in Davos, Switzerland, from 11–16 October 1992. The meeting was co-sponsored by a number of Swiss and international organizations, including the World Meteorological Organization, the American Meteorological Society, the Swiss National Science Foundation, and the Swiss Academy of Sciences. It was the first meeting on climate-relevant topics to be held in Switzerland since the signing of the Climate Convention at the United Nations Conference on Environment and Development in Rio de Janeiro (UNCED, June 1992).

The objective of the Davos Conference was to bring together scientists from a variety of disciplines, such as climatology, hydrology, biology, ecology, and economics, to exchange views and ideas on climate processes and impacts in mountain regions of the globe. Over 100 scientists from fourteen countries were present at what was truly an interdisciplinary event.

The book has been divided into three distinct sections, namely:

- climate change in mountain regions: fundamental processes, historical and contemporary observations, modelling techniques;
- impacts of climate change on mountain vegetation: observations, modelling, networks;
- socio-economic aspects of climate change in mountain regions.

Each section includes individual chapters contributed by a number of authors, which address many of the issues which mountain regions need to confront in the face of climate change.

In Section 1, the introductory chapter by Barry gives a comprehensive overview of issues related to past and future climate change in mountain regions; it is followed by a detailed review of climate and environmental stress factors in the Alps by Grassl. Palaeoclimatic records help provide an insight into natural climatic fluctuations, as described by Luckman. Historical evidence of climatic conditions in Europe at the end of the seventeenth century provided by Pfister illustrates the synoptic conditions which prevailed at that time. The extreme cold events of the so-called Maunder Minimum period was felt not only in the more populated lowlands of western and central Europe, but also on mountain ecosystems and the sparse agricultural communities already established in European mountains at that time. This was a period of significant advance of Alpine glaciers, which have been retreating for over a century, as described in the chapter by Haeberli. The section returns to the present day with a contribution by Baumgartner and Apfl on snow-cover monitoring techniques, and a modelling study by Pielke and others on factors other than greenhouse gas increases on regional climates. The section ends with an overview of possible approaches to climate scenario generation for mountain regions by Beniston.

Section 2 is introduced by Körner with an overview of the principal controlling climatic factors which can influence mountain vegetation. Graumlich analyses how palaeoecological studies can shed light on contemporary vegetation dynamics. Several papers address various aspects of climate impacts on mountain forests, through modelling techniques (Halpin, Bugmann and Fischlin), and observations (Holtmeier, Petersen, and Pornon and Doche). The section then moves on to describe links between climate factors, soil chemistry, and tree growth (Guyette), a methodology to analyse the sensitivity of certain plants to climate change by Bucher and Jeanneret, and ends with a description of an intensive monitoring network for mountain forests in Switzerland by Innes.

The third section is devoted to socio-economic aspects of climate change in mountain regions, including risk analysis, human perception, and socio-economic models. Risk perception concepts are introduced by Dürrenberger and others and are applied to the problem of tourism in mountains by Bailly. A specific example of the manner in which lack of snow during the winter season affects ski-lift and cable-car companies in Switzerland is given by Abegg and Froesch. A modelling approach to socio-economic impacts is developed by Breiling and Charamza. The potential impacts of changing climatic conditions in mountains on energy supply are introduced by Garr and Fitzharris, while Chalise provides an overview of some of the major socio-economic pressures on the Himalayan region which are likely to be exacerbated by abrupt climate change. A paper by Carter and Parry provides an insight into the sensitivity of mountain agriculture to present and possible future climate. The section ends with a suggested outline for a network of mountain protected sites worldwide which would enable research on climate and environmental factors to be carried out in a comparable manner from site to site (Martinka and others).

The conclusion to the book is provided by Price, who summarizes many of the topics covered in more detail by the different contributing authors. 'Should mountain communities be concerned about climate change?' is also a view to the future on the numerous and different problems which can be expected in the various mountain regions of the globe under conditions of abrupt climate change.

The editor of this book would like to acknowledge the work of a number of scientific experts who contributed to this book through the peer review of papers which were submitted to this volume.

CONCLUDING REMARKS

It is intended that this book be a reference document for all persons interested in the issues of global climate change and their potential impacts on mountainous regions. It will certainly be cited in the 1995 Assessment Reports currently being prepared by the Intergovernmental Panel on Climate Change where, in particular, one chapter will be devoted specifically to mountain regions in the IPCC Second Assessment Report on the Impacts of Climate Change. The papers in the present volume therefore represent one of the most up-to-date sources of information for the IPCC and others working in the field of climatic change in mountain regions.

REFERENCES

Fitzharris, B.B. (1989) 'Impact of climate change on the terrestrial cryosphere in New Zealand', *Summary Paper*, Department of Geography, University of Otago.
Houghton, J.T., Jenkins, G.J. and Ephraums, J.J. (eds) (1990) *Climate Change – The*

IPCC Scientific Assessment, Cambridge University Press, Cambridge.

Kuhn, M. (1989) 'The effects of long-term warming on alpine snow and ice', in J. Rupke and M.M. Boer (eds), *Landscape Ecological Impact of Climate Change on Alpine Regions*, Lunteren, The Netherlands.

National Research Council (1985) *Glaciers, Ice Sheets, and Sea Level: Effect of a Carbon-Induced Climatic Change*, National Academy Press, Washington, DC.

Zimmermann, M. and Haeberli, W. (1989) 'Climatic change and debris flow activity in high mountain areas', in J. Rupke and M.M. Boer (eds), *Landscape Ecological Impact of Climate Change on Alpine Regions*, Lunteren, The Netherlands.

Part I

CLIMATE CHANGE IN MOUNTAIN REGIONS

Fundamental processes, historical and contemporary observations, modelling techniques

1

PAST AND POTENTIAL FUTURE CHANGES IN MOUNTAIN ENVIRONMENTS

A review

R.G. Barry

INTRODUCTION

Scope of the problem

Our knowledge of the climatic characteristics of mountain regions is limited by both paucity of observations – short records that seldom span 100 years and a sparse station network – and insufficient theoretical attention given to the complex interaction of spatial scales in weather and climate phenomena in mountains. Meteorological research has tended to focus on the upstream and downstream influences of barriers to flow and on orographic effects on weather systems (Smith, 1979), rather than on conditions in the mountain environment. Atmospheric phenomena themselves display a hierarchy of spatial and temporal scales as illustrated in Table 1.1. Microscale features of the atmosphere – wind gusts, for example – are superimposed on larger scales of motion and are not fixed in place, whereas small-scale elements of the landscape surface, such as vegetation canopy, large rocks and hollows, can create microclimatic contrasts in surface healing, soil moisture or snow-cover duration that have longer-lasting significance. The role of sunny (adret) and shaded (ubac) slopes in the Alps in influencing land use and settlement location has been long recognized (Garnett, 1935). Local wind systems, generated by radiational and thermal contrasts in complex terrain, are best developed when synoptic-scale pressure gradients are weak and there is little cloud cover; hence, they are intermittent components of the climate. Synoptic-scale systems – midlatitude cyclones, for instance – are extensively modified as they move across a mountain barrier. Modifications occur on a large scale in terms of changes in frontal structure and cloud systems, and on a local scale in terms of wind systems, such as Föhn (Chinook) occurrence, or precipitation intensity. Isolating these processes in order to understand

3

Table 1.1 Scales in mountain weather and climate

	Microscale	*Local scale*	*Regional scale*
Meteorological phenomena	Turbulent motion (gusts)	Slope and valley winds; fall winds	Thunderstorm cluster; synoptic system
Landscape elements	Rocks; vegetation clumps	Terrain elements (slopes; valleys)	Mountain range; plateau
Climatic features	Snow patches	Adret/ubac radiational contrasts; thermal belts	Plateau; monsoon

Table 1.2 Climatic effects of the basic controls of mountain climate

Factors	*Primary effects*	*Secondary effects*
Altitude	Reduced air density, vapour pressure; increased solar radiation receipts; lower temperatures	Increased wind velocity and precipitation (mid-latitudes); reduced evaporation; physiological stress
Continentality	Annual/diurnal temperature range increased; cloud and precipitation regimes modified	Snowline altitude increases
Latitude	Daylength and solar radiation totals vary seasonally	Snowfall proportion increases; annual temperatures decrease
Topography	Spatial contrasts in solar radiation and temperature regimes; precipitation as a result of slope and aspect	Diurnal wind regimes; snow cover related to topography

their operation is vastly complicated by the inadequate data base for most mountain areas of the world.

Mountain influences on climate and related environmental features are a result of four basic factors: altitude, continentality, latitude and topography – each of which affects several important meteorological variables. The role of these factors, detailed in Barry (1992), are summarized schematically in Table 1.2 where the effects refer to responses to an increase in the factor listed. In turn, these climatic differences are expressed in vegetation type and cover, geomorphic, and hydrologic features. The particular influences of topography are complex and worthy of further discussion. At a primary level, we need to distinguish between isolated peaks, mountain ranges that are large enough to modify the upstream and downstream airflow substantially, and extensive high plateaux that create major barriers to air motion and generate their own climates. Tabony (1985) notes that high peaks experience similar temperatures to the free air of the same altitude. However, elevated plateaux are heated in summer and cooled in winter by radiative processes. Valleys within uplands have 'enclosed' atmospheres that are diurnally modified by nocturnal cooling, especially in winter, and enhanced daytime heating.

Table 1.2 identifies one area important for human activities in high mountains, namely, the physiological stresses imposed by climate. The primary factors are reduced air pressure, which limits oxygen intake and working capacity, and low temperatures, which necessitate adequate clothing and shelter and place limits on crop production potential. However, these primary factors differ in their significance. Low air pressure is a permanent unavoidable condition, whereas low temperatures can be counteracted. It is also worth noting that for a given altitude, say 4,000 m (neglecting the small numerical difference between geopotential and geometric height), the corresponding standard atmosphere value of pressure is approximately 630 mb at latitudes equatorward of 30°, but at 60°N latitude ranges between 593 mb in January and 616 mb in July (Sissenwine, 1969). Thus, at an equivalent height, the effect of reduced air pressure is substantially greater in middle and high latitudes, particularly in winter. The passage of a cyclonic system could further lower the pressure by some 20–30 mb, equivalent at the given altitude to a height difference of 250–375 m. Such latitudinal and seasonal effects may be important for visitors to high mountains, as well as in terms of high-altitude settlements and work enterprises in mid-latitudes.

Specific mountain climate data

The European Alps are by far the best known mountain area of the world in terms of weather and climate and related environmental characteristics. Several summit observatories have records spanning 100 years and there is a dense network of regular observing stations and precipitation gauges. Detailed climatic atlases exist for the Tirol (Fliri, 1982) and Swiss Alps (Kirchhofer, 1982). Complementary information exists on Alpine glaciers (Haeberli et al., 1989), hydrologic regimes (Baumgartner et al., 1983), forest growth (Aulitsky et al., 1982) and other more specific topics. The Alps have also been the locale for the meteorological research programme ALPEX (Alpine Experiment) of the Global Atmospheric Research Programme (GARP) of the World Meteorological Organization, conducted between September 1981 and November 1982 (Smith, 1986), and many university research projects. Extensive studies also exist for the Carpathian Mountains (Obrebska-Starkel, 1990).

The island of Hawaii is well known climatically through observations made at Mauna Loa Observatory since 1957 and various research programmes conducted on the windward and leeward slopes of the mountains in Hawaii and other islands. Similar data exist for Mount Fuji, Japan which also has a summit observatory. However, most of the detailed

5

studies on Mount Fuji's climate are published in Japanese.

The climatic features of the Rocky Mountains and Coast Ranges in North America and the mountains of Scandinavia are known primarily through specialized university and other agency research programmes. Experiments on winter cloud-seeding and mountain lee-wave phenomena have been of particular importance in North America. Mountain-valley winds and orographic precipitation have been investigated in both Europe and North America. The station networks in these countries are moderately good but permanent mountain observatories are lacking. Detailed glacio-hydrological records are also available for Norway, in particular. The Colorado Rocky Mountains and Scandinavian Mountains have also been sites of ecological research under the International Biological Programme's Tundra Biome studies in the 1970s, and in the former area related work has continued through Long-Term Ecological Research Program activities (Barry, 1986). A similar level of information is available for the Greater Caucasus in terms of climate and glacio-hydrological topics, but almost all of it is in Russian-language articles and monographs.

A moderate level of information on the climate of very high mountains is available in the case of the northern Andes and Nepal Himalayas. The former area has a reasonable station network, including some high-altitude stations; microclimatic studies, related to ecological conditions, have provided valuable supplementary information (Monasterio, 1980). The major lack is aerological sounding data for information on free-air conditions. The large-scale climatic controls of this area are still poorly defined. In the case of Nepal, most stations are in the Himalayan valleys, but short-term measurements, particularly by recent Japanese glaciological expeditions, have provided additional information (Barry, 1992, chapter 5). For the adjacent Tibetan (Qinghai-Xizang) Plateau there is now a growing meteorological literature, although several key studies are available only in Chinese.

Mountains where there have been discontinuous research programmes and only a sparse station network is available include such diverse areas as New Guinea, Ethiopia (Hurni and Stähli, 1982), Mount Kenya, the Hoggar and the Mount Saint Elias Range. In New Guinea, where there is only one permanent station above 2,000 m, information on mountain environments (Barry, 1980) has focused on ecological conditions (Smith, 1975) and on the glaciers of Mount Carstenz in Irian Jaya (Hope et al., 1976). More recently, frost occurrence and agro-climatic variability in the New Guinea Highlands have been analysed intensively (Brookfield and Allen, 1989).

Selected issues

This section provides a brief review of some selected topics of particular importance in mountain climatology. They are issues in the sense that incomplete knowledge of them may present problems in attempting to mitigate their consequences for the mountain environment and its inhabitants.

Orographic precipitation

Precipitation forecasting in mountain regions is nowadays performed using a combination of model prediction and near-real time radar measurement of precipitation falling, wherever such technology is available. Various numerical models are available to calculate precipitation rates for air ascending adiabatically over a mountain slope (Sarker, 1966) and incorporating 'spill-over' effects (Colton, 1976). The concept can also be extended for snowfall (Rhea, 1978) and for convective precipitation (Georgakos and Bras, 1984). Hydro-climatological research has identified four factors that determine precipitation amount: rate of ascent, water vapour supply, wind direction, and wind speed (Browning, 1980). Topographic variables that interact with the meteorological factors are: elevation, relative relief, aspect (orientation), and slope angle.

Calculations of the influence of mountainous terrain in England and Wales on air streams from different directions, with specified wind speeds and moisture contents, are illustrated by Hill (1983). Similar calculations of orographically induced vertical motion for different wind directions have been made for Kyushu, Japan, using a three-dimensional numerical model with a 10 × 10 km topographic grid. The patterns of vertical motion compared satisfactorily with storm event precipitation similarly stratified by wind directions (Oki et al., 1991).

In the western United States there have been a number of pilot studies and operational programmes aimed at increasing winter snowfall by orographic cloud seeding. These include work in the Rocky Mountains of Colorado and in the Sierra Nevada. The seeding is carried out using silver iodide either from aircraft or, more cost-effectively, ground generators. Experiments near Climax, Colorado, with careful statistical control, indicated that a 25 per cent precipitation increase was achieved during southwesterly flow at 3 km with relatively high temperatures at 500 mb (Mielke et al., 1981). However the effect on seasonal totals is usually much less pronounced. Nevertheless, related studies in the San Juan Mountains, Colorado (Steinhoff and Ives, 1976) and the Sierra Nevada, California, have sought to evaluate possible ecological impacts of increased winter snowpack and of silver iodide on flora, fauna, hydrological and sedimentation processes. The studies in Colorado included intensive plot studies where snowpack was increased several fold by the use of snow fences.

One of the major issues relating to precipitation is the recurrence of extreme events. An intense storm in July 1976 was responsible for the 'Big Thompson' flood in the Colorado Front Range that caused at least 139 deaths and $35 million property damage (Maddox et al., 1978). Jarrett (1990)

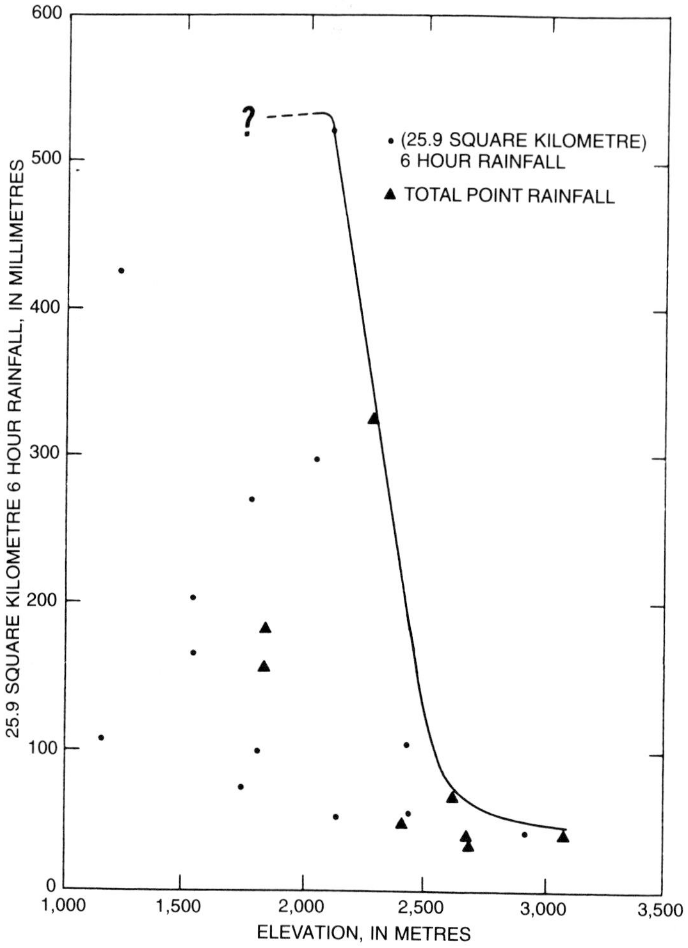

Figure 1.1 Relation between precipitation and elevation, including envelope curve, for large storms in Colorado (from Jarrett, 1990).

shows that flooding in basins above about 2,400 m altitude in Colorado is related to snowmelt, whereas at lower elevations floods result from intense rainstorms. These appear to be both slow-moving summer thunderstorm systems and synoptic-scale cold lows, mainly in spring and autumn. At the higher elevations, summer storm precipitation is about an order of magnitude less than in the foothills (Figure 1.1). Myers and Morris (1991) confirm this conclusion using data from SNOTEL (Snow Telemetry) and National Weather Service stations. A frequency analysis of maximum *daily* precipitation values for June–September at 49 SNOTEL stations above 2,400 m in Colorado shows a 0.01 per cent probability value less than 100 mm. A similar analysis using summer maximum rainfalls at 37 climatological stations above 2,400 m gave a similar result. These results contradict earlier ideas derived from simple transposition of storm models over the higher terrain. It has importance for structures; thus, reservoirs and culverts at higher elevations may have been over-designed.

Extreme precipitation events have geomorphological significance in the Himalayas where they may cause widespread slope failure. Froehlich *et al.* (1990) show that storm events of 60–100 cm magnitude recur 2–4 times/ 100 yr at Darjeeling, where the annual totals average 250–400 cm with extremes of 600 cm. In October 1968, at the end of the monsoon season, such an extreme storm event resulted in modifications to about one quarter of the cultivated area around Darjeeling (Starkel, 1972). The issue of the response of hydrological systems, erosion processes and sedimentation to climate change has been examined for mountains of western Canada by Slaymaker (1990). He outlines the type of impacts likely to result from climate warming, but notes that shifts in the limit of periglacial erosion, permafrost and forest vegetation will be slow.

Air quality and pollution deposition

Many mountain valleys are increasingly subject to the effects of air pollution originating from local settlements, industrial plants and automobiles, as well as from the regional transport of pollutants into the mountains from industrial conurbations. The diurnal wind regime in mountain valleys, during intervals of weak regional-scale pressure gradients, gives rise to upvalley and upslope motion by day that may transport pollutants into the mountains. At night the circulations are reversed but, typically, inversion conditions in valley bottoms and basins can lead to high concentrations of pollutants at low levels (Hanna and Strimaitis, 1990). In winter, when solar heating is insufficient to remove them, such inversions can persist for most of the day.

Pollutants reach the surface by any deposition and washout. In mountain areas, the orographic enhancement of precipitation rates can lead to increased washout. Forland and Gjessing (1975) report that washout rates (relative to dry deposition) for Mg, SO_4 and Ca were 5–15 times greater on the 600 m high mountains around Bergen, Norway, than in the city itself.

Afforestation/deforestation of slopes

Studies of the effects of forest cover on local climate have received considerable attention, often showing conflicting results (Penman, 1963). The effects of differing vegetation cover on catchment hydrology have been studied for many years in Europe, North America and elsewhere (Bosch and Hewlett, 1982). Studies of upland moorland areas transformed by coniferous plantations have also been made in Great Britain (Calder, 1990). In a survey of 94 paired catchment experiments, where trees were either removed by cutting or burning, or killed by application of herbicides, or the area was planted, Bosch and Hewlett (1982) conclude that the direction and magnitude of changes in water yield can be predicted with fair accuracy. Many of the experiments were at elevations of 1,000 m or more in middle mountain terrain. Pine and eucalypti forest types give a change in annual water yield of about 4 cm per 10 per cent change in forest cover, while deciduous hardwoods cause a 2.5 cm change in yield per 10 per cent change in cover. The changes are largest in areas of higher rainfall, but increases in water yield following forest cover reduction are longer-lived in drier climates as the vegetation recovers more slowly. A change in vegetation cover from forest to grassland increases surface albedo, but reduces aerodynamic roughness and precipitation interception, all of which affect evaporative losses. Increased albedo lessens the absorbed solar radiation, but surface temperatures are higher at the surface of a grassland. Reduced roughness lowers the turbulent flux of moisture from the surface to the atmosphere. A multilayer simulation of interception by a pine forest, compared with grass, indicates that the latter intercepts only one quarter of the amount in a pine forest (Sellers and Lockwood, 1981). Using meteorological data from Oxfordshire as input, they calculate total annual evapotranspiration losses of 555 mm from pine versus 398 mm from grass (a 28 per cent reduction) and corresponding annual runoff values of 152 mm and 303 mm, respectively. Simulated peak flows were also 33 per cent greater for the grass cover. In mountain areas, a further consideration is that winter snowpacks are generally greater in forested areas than in clearings and they persist longer in the spring.

In some tropical and Mediterranean climates, forested slopes can increase the available moisture through fog drip as a result of the vegetation removing cloud droplets from the airflow. This is an important determinant of the distribution of montane cloud forest, for example (Cavelier and Goldstein, 1989). However, the possible management of this moisture source appears to have received little attention.

PAST CLIMATE

Sources of evidence

The major types of evidence for past climatic conditions are briefly summarized. Details of the techniques used to extract such information are available in the references cited.

Glacial geomorphology and glaciology

Numerous studies are available in the various mountain areas of the world on glacier variations and past glacier extent (Röthlisberger, 1986; Karlén, 1988; Haeberli *et al.*, 1989). The lower altitudinal limits of ice-cover can be mapped and dated by carbon-14 techniques for the last glaciation and by lichenometric methods for the last few centuries. For the more recent glacial advances or still stands, ice thickness, and therefore volume, may be estimated from trimlines on the valley sides. These changes in ice extent and inferred changes in snowline altitude can be interpreted in terms of likely changes in ablation-season temperature and winter accumulation, although unique solutions cannot be derived. Some attempts have been made to model ice volume changes in terms of climatic forcings, such as the work of Allison and Kruss (1977) for Irian Jaya (New Guinea). More recent studies on glacier–climate modelling for the Alps have been carried out by Kuhn (1989), Oerlemans and Hoogendorn (1990) and others (see Oerlemans, 1989). Statistical analyses of glacier data indicate that in western North America cumulative deviations of mass balance from the mean value for each glacier are strongly correlated up to about 500 km distance, implying that useful regional climate information can be inferred from glaciological observations (Letréguilly and Reynaud, 1989).

Ice cores

The extraction of climatic and other atmospheric signals from ice-cores has considerable potential in high-altitude ice bodies above the zone of melting (Thompson *et al.*, 1984a). Falling snow traps air that is sealed off when ice is formed at about 50–100 m depth. The ice preserves records of atmospheric gas concentrations, aerosols, pollen and volcanic dust, as well as isotopic composition of the snow (related to crystallization temperature of the atmospheric water vapour) and net accumulation rate. In zones of high accumulation rate, annual layers can be distinguished for up to about 10,000 years, but in alpine glaciers the records span at best a few thousand years. Ice-core studies in mountain areas include those on Quelccaya Ice Cap, Peru; Colle Gnifetti and Mont Blanc, Switzerland; and Dunde Ice Cap, western China (Wagenbach, 1989). Drawbacks are the limited number of suitable cold firn sites and the rather short records available from alpine ice bodies. The information is representative of a regional-scale area and this may be valuable in assessing regional anthropogenic effects, especially for pollution by heavy metals such as lead.

Palaeoecology

Evidence of past climates is also obtainable from several biological sources of which pollen and dendrochronological records are the most important.

11

Pollen records extracted from lake sediments and peat bogs can provide information on ecological changes at the genus, or occasionally the species, level for the last 10,000 to 30,000 years and in some instances longer. However, pollen may be transported long distances, the vegetational composition is not always fully or proportionately represented, and the ecological record must be converted to climatic information by some form of transfer function (Webb and Clark, 1977). For Pleistocene materials, an additional problem is caused when former plant assemblages lack a modern analogue. In mountain regions, local wind systems complicate the problem of interpreting the local versus regional pollen rain (Markgraf, 1980). In particular, Solomon and Silkworth (1986) draw attention to the 'mixing lid' formed by valley inversions that may act as a strong barrier to upslope transport of lower-montane pollen. Thus, pollen embedded within the valley circulations may be isolated from that in the airflows above the ridge line. Hence, altitudinal shifts of species are hard to isolate. In this respect, pack-rat middens are useful, in the western United States for example, in providing specific plant macrofossil information (Betancourt *et al.*, 1990). Further difficulties in the utilization of pollen data are the necessity to date the sediments by carbon-14 and the relatively coarse temporal resolution that is usually available, given the intervals between such dates, their standard errors, and variable sedimentation rates. Fifty to 200 years is the typical resolution over time intervals of 1,000 to 10,000 years, respectively.

Dendrochronology has several advantages. It provides annual or even seasonal, resolution over hundreds to a few thousands of years. Two distinct techniques provide climatic information. The first, widely employed, dendroecological method is to measure annual ring-width and the second is to analyze wood density (Fritts, 1976; Schweingruber *et al.*, 1979). Ring-width, when corrected for age trend, varies strongly in response to summer temperature towards the polar and upper altitudinal treelines, and to summer moisture near the lower altitudinal treeline in arid regions. Detailed quantitative interpretation of the ring-width or wood density information involves the development of numerical transfer functions, analogous to those for pollen studies, based on climatic records at nearby stations. Tree-ring studies can also be used to determine fire histories and volcanic eruptions, as well as streamflow condition using ring-width – runoff calibrations.

Historical records

Direct reports of weather-related conditions are contained in personal diaries and official records in European countries, China, Japan, and elsewhere from the sixteenth to seventeenth centuries, as interest in natural science developed, or even earlier in association with documentation of agricultural production for taxation purposes. In the Alpine countries, Britain, and Scandinavia, in particular, some of these sources provide information on the climatic deterioration in mountain and upland areas during the Little Ice Age, AD 1550–1850, compared with the Medieval Warm Period (Lamb, 1977; Parry, 1978; Pfister, 1985 a and b; Grove, 1988; Zumbühl, 1988). Extraction of quantitative material from such sources requires skilled knowledge of historical document interpretation (the necessity for primary sources, content analysis, and consistency checking) and means of standardizing incomplete or short-period records from a particular location with modern data.

Instrumental observations

The origins of mountain meteorology can be dated to the work of Horace-Benedict de Saussure in the Alps (Barry, 1978). In August 1789, he and his son maintained hourly weather observations for two weeks on the Col du Géant, Mont Blanc. Systematic observations at mountain observatories, however, did not begin until the late nineteenth century (Barry, 1992, Table 1.1). H.F. Diaz and R.S. Bradley (personal communication, 1992) identify some 22 stations in Europe, 8 in Asia, and 37 in the USA above 1,500 m with long-term (≥ 30 yr) records. However, not all of these are mountain stations – some are on high plateaux and others in mountains valleys. Second-order and auxiliary stations with shorter records are perhaps ten times more numerous.

Remote sensing

Remote sensing, by aerial photography beginning in the 1930s and satellite sensing beginning in the late 1960s, provides spatially extensive views of surface conditions at specific time intervals. Examples of useful records include air photographs of glacier extent in remote alpine mountain areas (Allison and Peterson, 1976) and global satellite mapping of snow cover since about 1970 (Wiesnet et al., 1987). However, the latter data are suitable only for analyses of large regions. Operational snow mapping of 28 river basins in western North America is performed by NOAA/NESDIS using daily-coverage GOES 1 km-resolution data. Daily AVHRR data from NOAA polar-orbiting satellites are also used by many countries for similar purposes (Rango, 1985). In both cases, such data are usually supplemented by airborne gamma-radiation measurements and/or ground observations of snow depth and water equivalent. Cloud cover is always a problem in using visible and infrared satellite data, but passive microwave data can assist in solving this problem, albeit with a lower spatial resolution (Rango et al., 1989).

Past conditions

Last glacial maximum

For the last glacial cycle, global climates have been reconstructed by several groups using a variety of palaeoclimatic data extracted from cores of ocean sediments, lake and bog sediments, ice layers, tree rings, etc. (CLIMAP Project Members, 1976, 1981; COHMAP Members, 1988). Ocean temperatures estimated for the last glacial maximum, about 18,000 BP, based on planktonic foraminifera, show either modest cooling ($\leq 2°C$) or even warming of ocean surface waters in most low- and subtropical-latitude areas. In contrast, glacial-geomorphologic evidence of former glacier extent and downward displacement of vegetation zones in tropical mountain areas of the order of almost 1 km imply substantially greater cooling of 5–6°C. This pattern has, for example, been found for New Guinea (Bowler et al., 1976) and Hawaii (Porter, 1979).

Webster and Streten (1978), Rind and Peteet (1985), and LaFontaine (1988) discuss the evidence and possible explanations, if the interpretations of the palaeoclimatic records are assumed to be reliable. They suggest several physical causes: the atmospheric temperature lapse rate may have been steeper than now, or more frequent incursions of upper-level cold troughs from middle latitudes may have increased snowfall and lowered temperatures. Simulations of snowline lowering using the CLIMAP-reconstructed sea surface temperature data and Goddard Institute for Space Studies climate model (8° latitude × 10° longitude resolution) indicated lowerings only half (or less) of those observed, implying that the CLIMAP data may have a warm bias (Rind and Peteet, 1985). Subsequently, Rind (1988) has shown that model results must be used with great caution. By using a finer-grid resolution (4° × 5°) than in the previous study, he demonstrates that more atmospheric cooling and snowline lowering is simulated for the Last Glacial Maximum in the Hawaiian region, although little change occurred in the other tropical mountain areas (Colombia, New Guinea, and East Africa). The improved fit is attributable to the effects of the grid resolution on moist convection in the model. However, Rind emphasizes that better agreement between model and observations may not necessarily imply that the model results are more realistic!

Little Ice Age

Geomorphological studies show that there have been numerous advances and retreats of glacier margins since the large-scale retreat at the end of the last glacial maximum, dated around 18,000 years ago. The best known of these is the Little Ice Age of the sixteenth to nineteenth centuries in Europe (Grove, 1988), where there is supporting documentary and observational evidence (see above). However, similar fluctuations are reported in most

14

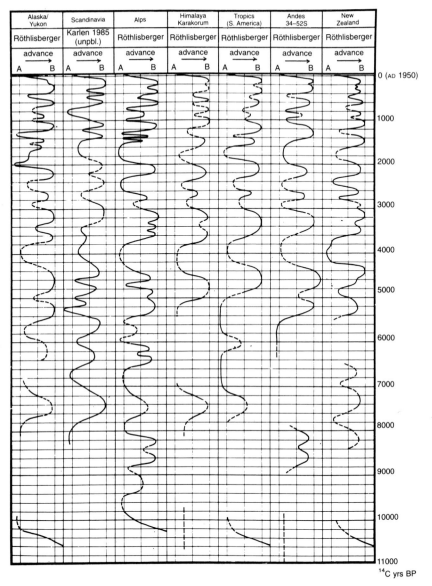

Alaska/ Yukon	Scandinavia	Alps	Himalaya Karakorum	Tropics (S. America)	Andes 34–52S	New Zealand	
Röthlisberger	Karlen 1985 (unpbl.)	Röthlisberger	Röthlisberger	Röthlisberger	Röthlisberger	Röthlisberger	
advance →	advance →	advance →	advance →	advance →	advance →	advance →	
A B	A B	A B	A B	A B	A B	A B	

Figure 1.2 Glacier fluctuations in the northern and southern hemispheres over the last 11,000 years. Each scale shows fluctuations relative to the 1980 minimum (left) and maximum neoglacial advances since 3700 BP (right) (from Röthlisberger, 1986).

mountain areas of the world, with slight differences of timing. For example, Figure 1.2 illustrates the glacier fluctuations identified in major mountain regions during the Holocene period (10,000 yr before present until now) by

Röthlisberger (1986). There are broad similarities, and the more detailed records of the last 1,000 years (dated mainly by lichenometry in New Zealand), indicate a general pattern resembling that of the Alps, but with a major fourteenth-century interval of advance (Gellatly *et al.*, 1988). In the central Asiatic mountains of the USSR, Kotlyakov *et al.* (1991) also find that there were important advances in the late fifteenth century to early sixteenth century. The glaciological expression of climate fluctuations depends strongly on the continentality of the climate. For example, the equilibrium line altitude was lowered only 50 m in the drier Tien Shan and Altai during the thirteenth century, but 150 m in the more humid Caucasus. Similar effects are evident in recent glacier fluctuations.

Apart from alpine glacier information, records of snow cover have been assembled for the Alps by Pfister (1985a and b). He reports that altitudes above 1,800–2,000 m in the vicinity of Mount Säntis became snowfree three to four weeks earlier in 1920–53 than in either 1886–1953 or 1954–80. The period 1920–53 experienced warm spring–early summer months. However, the later melt in 1954–80 was related primarily to 5 per cent more winter precipitation (falling as snow). In 1886–1919 there was an 8 per cent increase in spring precipitation. Surprisingly, 1821–51 was less extreme than either of these periods. Other records for Mount Pilatus, near Lucerne show a six-week difference between the 1930s and the end of the sixteenth century.

Pfister (1985b) shows that the climatic response to a given synoptic pattern may be quite different in the lowlands from that at higher altitudes. In winter, storm systems from the west bring rainfall and induce snowmelt in the lowland but snowfall in the Alps, whereas during anticyclonic situations inversions with fog may maintain lowland snow cover, but permit melt above the inversion on south-facing slopes. In spring, the increased amounts of precipitation fall as snow above 2,500 m, but only 50–60 per cent of the precipitation falls as snow at 1,200 m. As a result of these complexities, changes in snow cover at lower elevations are often poorly correlated with those on neighbouring mountains. Pfister finds little difference between the dates of snowmelt at 1,200 m for 1821–51 and 1913–32 on the north slope of Säntis and the Alps, respectively, in contrast to the situation at higher elevations (see Barry, 1990, figure 4).

The observational record

It is of interest to know what changes mountain regions have experienced. Mountain observations can provide an indication of changes in the free atmosphere for a much longer period than sounding data, but up to now such records have received limited attention. Measurements on the Zugspitze and Sonnblick in the Alps provide such long-term series, but even at these stations there are problems of data homogeneity due to changes of instrumentation or observing practice. Nevertheless, the unique 100-year

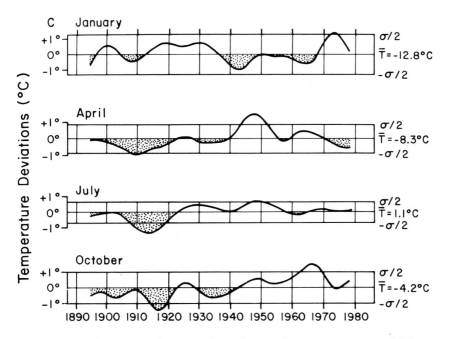

Figure 1.3 Temperature departures from the mean for 1887–1986 at Sonnblick (3,106 m) for January, April, July and October. Departures are smoothed by a 20-year Gaussian moving average. Monthly mean values (°C) and the standard deviation (s)/2 are indicated (after Auer *et al.*, 1990).

record on the Sonnblick, Austria (Auer *et al.*, 1990) shows the well-known early twentieth-century warming between 1900 and 1950 of about 2°C in spring and summer and 2.5°C between 1910 and 1960 in October (Figure 1.3). In winter, however, no clear trend is apparent, although the 1970s were warmer than the early part of the century. This contrast with the general northern hemisphere trend, where the warming is most pronounced in winter, may reflect the free-air lapse-rate conditions at the mountain summit. Records for five valley stations in the Canadian Rocky Mountains covering a common period of 1916–88 show important topoclimatic differences according to Luckman (1990). Only Banff, with almost a 100-year record, shows an overall positive trend in mean annual temperatures.

The records from mountain observatories also enable us to compare the degree of parallelism and relative amplitudes of trends in the mountains and on the adjacent lowlands. Analyses illustrating trends of sunshine and snowfall in lowland and mountain regions of Austria (Steinhauser, 1970, 1973) are among the most complete such records available. Annual sunshine totals at Sonnblick (3,106 m) and Villach (2,140 m) show fluctuations similar in timing and amplitudes to those at four lowland stations (see Barry, 1990,

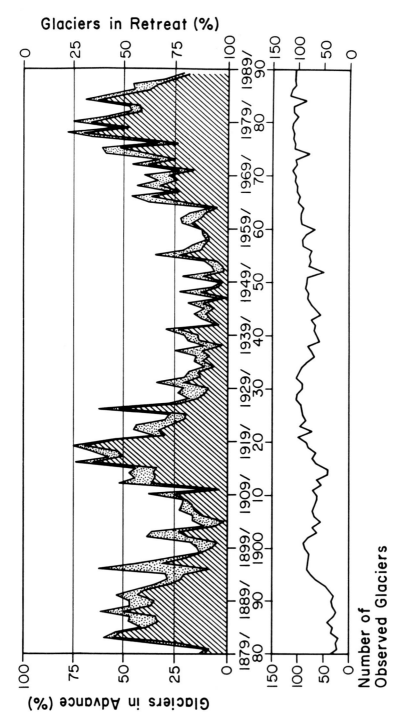

Figure 1.4 Percentage of glaciers advancing, retreating, or stationary in the Swiss Alps and number of glaciers observed (based on data supplied by the World Glacier Monitoring Service, ETH, Zürich).

figure 5). This agreement holds also on a seasonal basis, except in winter as a result of inversions associated with lowland fog and stratus. Variations in snow cover duration in Austria since 1900 show inter-regional differences, with stations in western Austria and the southern Alps displaying patterns that are different from those in the northern Alpine Foreland and north-eastern Austria (see Barry, 1990, figure 7). Temperature trends at Sonnblick for 1887–1979, compared with Hohenpeissenberg (994 m) and lowland stations in the surrounding region suggest that differences between them reflect spatial gradients rather than altitudinal contrasts (Lauscher, 1980).

For many alpine areas there are numerous observations of changes in ice masses. Retreat of glacier fronts and negative mass balances characterized most glaciers in the Alps (Figure 1.4) and other alpine areas from around 1900 until the 1960s–1970s with only brief interruptions (Wood, 1988; Hastenrath et al., 1989). Smaller glaciers in the European Alps underwent some advance in the 1980s, in response to increased winter accumulation and slightly cooler summers, but this interval appears to have ceased (Patzelt and Aellen, 1990). In contrast, large Alpine glaciers continued to retreat (Haeberli et al., 1989). Maritime glaciers in west coast areas of North America and Scandinavia showed similar advances associated with precipitation increases during the 1980s, whereas glaciers in continental interior locations mostly continued to retreat (Makarevich and Rototaeva, 1986).

The century-long meteorological records in the European Alps can be usefully compared with observations of changes in alpine glaciers. Chen (1990), for example, reconstructs the balance of eight glaciers using summer temperatures and annual precipitation records. For 1864–1987, he finds an area-weighted mean-annual mass balance of –287 mm. The annual mean air temperature (at 25 stations in or near the Alps) rose more than 0.6°C from the 1850s to 1988, culminating in the late 1940s with subsequent weak cooling. Chen and Funk (1990) demonstrate that the mass balance of the Rhône Glacier, which has been studied more or less continuously since 1874, responds primarily to summer temperature with sensible-heat flux and net radiation contributing almost equally to the energy input required for ablation.

Temporal changes of climate variables in the Alps show some interesting contrast with altitudinal effects. At Sonnblick (3,106 m) the average number of days with snow cover during May–September decreased from 82 days for 1910–25 to only 53 days for 1955–70 (Böhm, 1986). During the same interval, mean summer temperatures rose about 0.5°C. However, if we compare the difference in days with snow cover over an altitude difference of 100 m (corresponding to a mean lapse rate of 0.5°C/100 m), a decrease of 10–11 days, rather than 29 days, would be expected in the eastern Alps. Evidently, non-linearities associated with snowmelt effects and summer snowfalls must be involved.

There are quite long records of snow-depth data at mountain stations in

1. Jan.

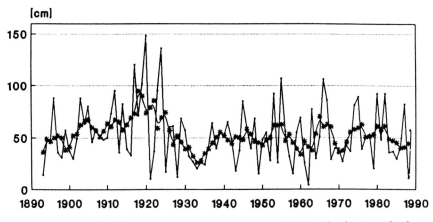

Figure 1.5 Snow depth on 1 January at Davos (1,540 m), Switzerland. Annual values for 1892–1989 and 5-year running means (Föhn, 1990).

the Alps and at snow courses in the North American Rocky Mountains and Cordillera. However, in only a few instances have these been digitized and archived. Consequently, these records have been under-utilized for climatic analyses of temporal change. Two recent studies using such information serve to illustrate the potential complexity of spatio-temporal changes.

Snow-depth data at Davos (1,540 m), Switzerland, are available since 1892. Föhn (1990) shows, using 5-year running means, that snow depth on 1 January increased from ≤ 50 cm in the 1890s to 80–90 cm around the 1920s, and declined to about 30 cm in the mid 1930s. Subsequently, depths increased again and have been more or less constant, around 50 cm, up to 1990 (Figure 1.5). Nevertheless, snowpack water equivalent on 15 April has substantially exceeded that on 1 January since about 1975, whereas the two values were similar from 1950 to 1965 (Föhn, 1990) indicating more snowfall in late winter. In the southern and western Alps, 1988 and 1990 were particularly poor snow seasons in the early winter but the century-long Davos record shows that such short-term anomalies are recurrent and need not indicate a warming trend.

Snow-course records on 1 April at 275 high-elevation sites in Idaho, Montana, Wyoming, Colorado and Utah for 1951–89 have been analysed by Changnon *et al.* (1991). They find that a shift in the north–south distribution of snowpack occurred in the mid-1970s. Winter seasons with heavy snowfall in the 'northern' Rockies and low snowfall in the 'southern' Rockies were confined to the period 1951–73, while seasons with the opposite pattern were likewise confined to the 1976–89 period. This appears to be a natural fluctuation associated with a change in the frequency of depression tracks.

20

Similar north–south contrasting patterns were identified in precipitation studies by Bradley (1976).

A different source of long-term information for the high Andes is provided by ice-cores collected from the Quelccaya Ice Cap in southern Peru (14°S, 71°W, 5,670 m above sea level). Annual layers for the last 1,500 years suggest that a recent decrease in precipitation may be associated with more frequent El Niño events in the western Pacific. The interval AD 1500–1720 was identified as the wettest part of the record (Thompson et al., 1984b; 1985).

FUTURE CLIMATE

Assessing future climate

Assessments of potential future climatic conditions can be made by following three approaches (see also Beniston, this volume).

Analogues

Until recently, the most commonly adopted method was an empirical study using analogues of previous events identified from observations, historical or geological records. For example, warm/cold years or wet/dry years have been grouped either from individual extreme years or longer intervals of one or other regime as a basis for scenarios of regional conditions (Jaeger and Kellogg, 1983; Lough et al., 1983). However, the observational record is considered too short to represent potential future greenhouse gas-induced changes adequately. The mid-Holocene and even Tertiary intervals have also been examined as possible analogues, of greenhouse warming, but Crowley (1990) concludes that none of these provide satisfactory analogues, especially for regional conditions.

General circulation models (GCMs)

GCMs can be used to simulate the effect on global climate of various changes in climatic forcing (for example in solar radiation, greenhouse gas concentration, surface boundary conditions due to deforestation and desertification). They can incorporate the major physical processes that interact to determine regional climates. Nevertheless, many of the processes involving ocean–atmosphere interactions and biogeochemical interactions are still treated rather crudely. Computing limitations also necessitate relatively coarse spatial resolution (equivalent to about 5° latitude typically) and therefore mountain terrain is often greatly smoothed. This is known to produce errors in the simulated wind fields and therefore in the distribution of clouds and precipitation in particular. Moreover, model climate outputs are conventionally presented for sea level and ignore conditions at the surface over elevated mountainous regions. To address this problem, local climatic conditions may be assessed empirically using statistical relationships for the observed present climate (Giorgi and Mearns, 1991). This assumes that similar relationships would still obtain under a different global regime. Brazel and Marcus (1991) illustrate an attempt to compare temperature obervations in a mountain area of Kashmir and Ladakh with standard GCM simulations of present climate.

Nested models

Recently, attention has turned to using a high-resolution limited area model nested within a GCM to examine regional climate change (Giorgi and Mearns, 1991). GCM output is used to provide the boundary conditions for a limited area model that can provide a resolution of 10–100 km over areas of 1–25 x 10^6 km^2. The results of such studies are illustrated by Dickinson *et al.* (1989) for the western United States and by Giorgi *et al.* (1990) for western Europe. Investigations of regional greenhouse-gas induced climate change scenarios are planned for the near future.

Climate scenarios and potential impacts

Up to now there have been a limited number of studies concerned with potential future climatic conditions in mountain regions. Attention has mainly focused on changes in glacier runoff. A glacier–climate model study of the Hintereisferner, Austria, suggests that over a 50-year period increasing greenhouse gas concentrations could increase runoff by 11 per cent, but the calculations assume the same summer radiation regime and neglect possible changes in cloudiness and precipitation (Grenell and Oerlemans, 1989). As illustrated in Figure 7 of their work, the possible hydroclimatic effects of greenhouse gases are likely to be complex.

Studies of changes in the hydrology of a medium-sized catchment in California, based on GCM simulations (Gleick, 1989; Lettenmaier and Gan, 1990) indicate some of the likely general characteristics that may be anticipated for a global warming. These include (1) decreased winter snowfall and reduced spring–summer runoff and soil moisture and (2) increased winter runoff, due to more winter rainfall and earlier snowmelt. The seasonal distribution of runoff in watersheds with seasonal snow cover is mainly sensitive to temperature, which accelerates snowmelt. However, total annual runoff is more sensitive to precipitation changes. Such investigations need to be carried for mountain watersheds in other parts of the world.

Potential significance of changes in climate for mountain environments

Mountain snowpacks are of major significance in a number of respects. First, they are a primary water resource for adjacent lowlands in terms of agricultural use (irrigation), industrial use, and drinking water, as in the western United States where about 70 per cent of the water required by agriculture originates from mountain snowpacks. A different water usage exists in Norway, for example, where snowmelt runoff is utilized for hydro-power generation. Second, adequate snow depths and duration of snow cover are the essential elements of winter sports developments, now widespread in alpine regions of both hemispheres. Third, the snowpack characteristics are important in providing habitat for small animals and in creating a mosaic of microenvironments for plant cover in mountainous terrain. In human terms, the snowpack properties and climatic elements also determine the potential for avalanche occurrence and for drifting and blowing snow. Climatic factors that may cause significant changes in mountain snow cover are now considered in the context of the above concerns.

Snow cover duration

The duration of continuous snow cover (D) typically varies linearly with altitude (h) since it depends mainly on mean temperature. The following relationships are cited by Slatyer *et al.* (1984):

$D = 0.173\,h - 204.3$ for the Australian Alps (36°S)
$D = 0.122\,h - 23.1$ for the Swiss Alps (47°N)
$D = 0.155\,h + 11$ for Britain (53°N)
where D is in days, h is in meters.

Slatyer *et al.* (1984) note that the area that has 90 days of snow cover in Switzerland (above 930 m) extends over nearly 22,900 km^2 while the corresponding area in Australia (above 1,700 m) covers only 40.5 km^2. A modest rise in mean snowline with a shorter snow cover duration would virtually eliminate the season for winter tourism in the Australian Alps (Galloway, 1988). Meteorological factors combined with the modest eleva-tion of the latter area also give rise to a large year-to-year variation in snow depths (Budin, 1985). Strong westerly regimes give snowy winters due to orographic uplift over the Australian Alps, whereas warm dry winds suppress snowfall when the subtropical ridge of high pressure is located over southeastern Australia, as during El Niño-Southern Oscillation (ENSO) warm phases.

The criteria used for assessing snow cover suitable for skiing in Canada are: the percentage probability of a day with at least 5 cm snow cover, a maximum temperature less than 4.5°C and no measurable liquid precipi-tation (including freezing precipitation). Conditions are considered unreliable if this percentile value is < 50, marginally reliable if it is 50–74, and reliable if it is ≥ 75 (Crowe *et al.*, 1977). Using climate simulations from two GCMs for CO_2-doubling, and climate data for two downhill skiing centres in Ontario, Harrison *et al.* (1986) conclude that the ski season at Lakehead on the northwest shore of Lake Superior could be reduced 31–44 per cent with economic losses approaching $2 million (Canadian) while in the Georgian Bay area further south, downhill skiing would be virtually eliminated. The Georgian Bay skiing resorts could lose $37 million annually and the service centre a further $13 million (Canadian). Such conditions occur there at present during mild winters.

Snow depth

For Switzerland, Witmer *et al.* (1986) provide a detailed analysis of snow depths in relation to altitude and aspect. They use data for 20 or 30 years for 209 stations between 200 m and 2,800 m, although only 5 stations are above 2,000 m. Particularly interesting for applied studies is their analysis of slope-aspect differences in snow depth. Compared with a horizontal surface, depths in late March are twice as great on a 20° north-facing slope, but only 30 per cent of horizontal on a 20° south-facing slope. Such aspect differences need to be considered in any assessment of potential future changes in snowfall amount and snow cover duration. Less readily detectable in standard climatological records would be the character of recent winter seasons in the European Alps where the snowfall shows little change in amount but the timing of significant accumulation is delayed beyond late December–early January (Föhn, 1990). Such shifts could have major effects on winter sports.

Snowmelt runoff

Since most precipitation at high elevations falls as snow, the timing of runoff is primarily determined by snowmelt and therefore spring temperatures and solar radiation. European runoff studies deal primarily with partly glacier-ized basins. Collins (1987, 1989), for example, shows that discharge from tributaries of the upper Rhône are inversely correlated with mean summer temperatures at valley stations such as Sion. A 1°C lowering in mean May–September temperature at Sion between 1941–50 and 1968–77 was associated with a 26 per cent reduction in mean summer discharge. Reductions in the glaciated area of a drainage basin such as the upper Rhône were investigated by Kasser (1973). A decrease in ice-cover of 165 km^2 (from 16.8 per cent of a 5,220 km^2 basin in 1916 to 13.6 per cent in 1968) caused an estimated reduction of 140 kg m^{-2} in April–September runoff at Porte du Scex compared with a mean value for 1916–55 of 887 kg m^{-2}. A further calculation for the same basin by Chen and Ohmura (1990a) which incorporates the influence of climatic changes, as well as the reduction in ice-covered area shows close agreement with Kasser's earlier results. The change in basin ice-cover is the dominant component of the runoff change. In a separate study, Chen and Ohmura (1990b) estimate the decreases in area and volume of ice-cover in the Alps over the last century. The area shrank from approximately 4,368 km^2 in the 1870s to 2,909 km^2 in the 1970s (a decrease of 15 km^2 yr^{-1}) while the volume decreased 57 ± 20 km^3 to 140 ± 10 km^3 in the 1970s (a rate of –0.574 km^3 yr^{-1}).

Direct anthropogenic impacts versus global climate changes

The relative importance of climatic factors versus human influences in causing environmental change is generally difficult to disentangle. This is illustrated by arguments over the nature and causes of desertification (Hare, 1983), for example. However, the removal of vegetation cover, particularly forests, by logging, and changes in species composition due to agriculture and grazing are more apparent. In a case study for the northern slope of the central Caucasus, Krenke et al. (1991) attempt to calculate the relative effects of anthropogenic versus natural changes on heat and water budgets. Between the second half of the nineteenth century and the 1980s, the forested area decreased from 32 per cent to 14 per cent in the Kabardin-Balkaria foothills between Mt Elbrus and Nalchik and the snow/ice-covered area shrank as the equilibrium line altitude rose. In the nival-glacial belt, a decrease in runoff has resulted mainly from an 11–12 per cent reduction in precipitation. In the formerly forested areas, now steppe or meadow, where precipitation amounts have increased by 5–10 per cent, evaporation has also decreased and runoff increased. At lower altitudes, anthropogenic impacts are more significant than the natural changes, whereas the converse is true at higher altitudes.

STATUS OF KNOWLEDGE AND RECOMMENDATIONS

Data

This survey indicates that there are significant gaps in the available data base on mountain climates. First, there is no convenient inventory of existing (or former) mountain climate stations, much less a catalogue of data collected by them. Assembling such information from national archives would be tedious but of modest cost relative to the value of a well-documented, accessible (computer-compatible) data base. It would need to be updated annually.

Second, there are many mountain areas, notably in the North and South American Cordillera, with very few permanent mountain stations. This situation will not be easily remedied, although automatic weather stations offer a possible solution, at least on a field 'campaign' basis.

Understanding of processes

Theoretical studies of mountain meteorology have focused on the synoptic-scale effects of mountain terrain (cyclone and precipitation modification on the windward slopes; and lee cyclogenesis downstream), or on mesoscale features such as local winds and lee waves. Studies within mountain regions have concentrated heavily on mountain-valley winds, the effects of topography on radiation, and weather modification. Many basic questions concerning the altitudinal variation of water balance components and energy budget components, for example, have been relatively neglected. Such questions are of great importance if climate model results are to be interpreted properly within areas of complex terrain.

Mesoscale numerical models suitable for studies of local circulation, diffusion and air quality, as well as of precipitation prediction, are well developed and have been applied to many regions of complex terrain, if not to high mountain areas (see also Pielke *et al.*, this volume). The problem of using GCMs for climate change scenarios may be capable of resolution via the nested model strategy, but such work is only just beginning. The availability of suitable observational data for the validation of such analyses is likely to be a serious deficiency.

Constraints

The limited progress in climatological studies in mountain regions is attributable primarily to insufficient funding. This has affected not only the maintenance of high-altitude stations, especially observatories, but also the training of scientific personnel. There has been a failure to recognize the need for climate studies in areas of complex terrain, where measurements may be difficult to interpret in a regional context and where operational costs may be high. Nevertheless, modern automatic weather stations and data loggers powered by solar panels are now available at modest cost and can operate reliably, unattended for long periods. Most climate research in mountain areas has been carried out in connection with the assessment of prediction of water resources, forestry management, and pollution monitoring. Biological and physiological studies of altitudinal and climatic effects on flora, fauna, and human populations have also been performed in several areas, although these have generally been short-term programmes.

Needed actions

Several recommendations can be made based on this review and assessment of the status of mountain climate research. The following tasks are essential:

- A thorough inventory of existing (or previously completed) climate observation programmes should be made. This survey should identify existing data, their period of record and variables measured, and the location and availability of existing data sets.

- These data should be compiled, quality controlled, and made widely available on convenient media (diskettes, tapes and/or CD-ROMs) preferably by a single centre working with various institutions that may hold some of the records.

- The importance of sustained measurement and monitoring programmes in mountain regions needs to be communicated to appropriate national and international agencies, especially those concerned with global change-related studies. These agencies include national programmes concerned with climate, water resources, and environmental quality, and international agencies such as the United Nations Environment Programme, the Man and Biosphere Programme and the Global Environmental Monitoring Programme of UNESCO, the World Climate Programme of WMO, and International Geosphere Biosphere Programme of ICSU.

- Building on such measurement and research programmes effectively will necessitate that efforts be made to develop the entire infrastructure of education and training programmes for mountain studies, including building and maintaining appropriate facilities in mountain locations. One or more Regional Research Centres for the IGBP could, for example, be devoted to problems of mountain environments.

REFERENCES

Allison, I. and Kruss, P. (1977) 'Estimation of recent climatic change in Irian Jaya by numerical modelling of its tropical glaciers', *Arctic and Alpine Research*, 9: 49–60.

Allison, K. and Peterson, J.A. (1976) 'Ice areas on Mt. Jaya: their extent and recent history', in G.S. Hope, J.S. Peterson, I. Allison and U. Radok (eds), *The Equatorial Glaciers of New Guinea*, A.A. Balkema, Rotterdam, pp. 27–38.

Auer, I., Böhm, R. and Mohnl, H. (1990) 'Die troposphärische Erwarmungsphase die 20.Jahrhunderts im Spiegel der 100-jährigen Messreihe des alpinen Gipfelobservatoriums auf dem Sonblick', *CIMA '88. Congresso 20° Meteorologica Alpina*, (Sestola, Italy), Italian Meteorological Service, Rome.

Aulitsky, H., Turner, H. and Mayer, H. (1982) 'Bioklimatische Grundlagen einer standortgemässen Bewirtschaftung des subalpinen Lärchen – Arvenwaldes. *Eidgenössische Anstalt für forstliches Versuchswesen Mitteilungen'*, 58 (4): 325–580.

Barry, R.G. (1978) 'H.B. de Saussure: the first mountain meteorologist', *Bulletin of the American Meteorological Society*, 59: 702–5.

—— (1980) 'Mountain climates of New Guinea', in P. van Royen (ed.), *The Alpine Flora of New Guinea, Vol. 1 General Part*, J. Cramer, Vaduz, Liechtenstein, pp. 75–109.

—— (1986) 'Mountain climate data for long-term ecological research', *Proceedings of International Symposium on the Qinghai-Xizang Plateau and Mountain Meteorology*, Science Press, Beijing, pp. 170–87.

—— (1990) 'Changes in mountain climate and glacio-hydrological responses', *Mountain Research and Development*, 10: 161–70.

—— (1992) *Mountain Weather and Climate*, 2nd edn, Routledge, London.

Baumgartner, A., Reichel, E. and Weber, G. (1983) *Der Wasserhaushalt der Alpen*, R. Oldenbourg, Munich.

Betancourt, J.L, van Devender, T.R. and Martin, P.S. (eds) (1990) *Packrat Middens: The Last 40,000 Years of Biotic Change*, University of Arizona Press, Tucson.

Böhm, R. (1986) *Der Sonnblick. Die 100 jährige Geschichte des Observatoriums und seiner Forschungstätigkeit*, Osterreichischer Bundesverlag, Vienna.

Bosch, J.M. and Hewlett, J.D. (1982) 'A review of catchment experiments to determine the effect of vegetation changes on water yield and evapotranspiration', *Journal of Hydrology*, 55: 2–23.

Bowler, J.M., Hope, G.S., Jennings, J.N., Singh, G. and Walker, D. (1976) 'Late Quaternary climates of Australia and New Zealand', *Quaternary Research*, 6: 359–99.

Bradley, R.S. (1976) *Precipitation History of the Rocky Mountain States*, Westview Press, Boulder.

Brazel, A.J. and Marcus, M.G. (1991) 'July temperatures in mountainous Kashmir and Ladakh, India', *Mountain Research and Development*, 11: 75–86.

Brookfield, H. and Allen, B. (1989) 'High-altitude occupation and environment', *Mountain Research and Development*, 9 (3): 201–9.

Browning, K.A. (1980) 'Structure, mechanism and prediction of orographically enhanced rain in Britain', in R. Hide and P.W. White (eds), *Orographic Effects in Planetary Flows. GARP Publ. Series*, No. 23, World Meteorological Organization, Geneva, pp. 85–114.

Budin, G.R. (1985) 'Interannual variability of Australian snowfall', *Australian Meteorological Magazine*, 33: 145–59.

Calder, I.R. (1990) *Evaporation in the Uplands*, J. Wiley & Sons, Chichester.

Cavelier, J. and Goldstein, G. (1989) 'Mist and fog interception in elfin cloud forests in Colombia and Venezuela', *Journal of Tropical Ecology*, 5: 309–22.

Changnon, D., McKee, T.B. and Doesken, R.J. (1991) 'Climate variability of mountain snowpacks in the central Rocky Mountains', *Proceedings 15th Annual Climate Diagnostics Workshop*, NOAA, US Department of Commerce, pp. 384–9.

Chen, J.Y. (1990) *Changes of Alpine Climate and Glacier Water Resources*, Doctoral dissertation no. 9243, Eidgenössische Technische Hochschule, Zürich.

Chen, J.Y. and Funk, M. (1990) 'Mass balance of Rhonegletscher during 1882/83–1986/87', *Journal of Glaciology*, 36: 199–209.

Chen, J.Y. and Ohmura, A. (1990a) 'On the influence of alpine glaciers on runoff', in H. Lang and A. Musy (eds), *Hydrology in Mountainous Regions. I. Hydrological Measurements. The Water Cycle*, International Association of Hydrological Science, Publ. no. 193, IAHS Press, Wallingford, pp. 117–26.

—— and —— (1990b) 'Estimation of Alpine glacier water resources and their change since the 1870s', in H. Lang and A. Musy (eds), *Hydrology in Mountainous Regions. I. Hydrological Measurements. The Water Cycle*, International Association of Hydrological Science, Publ. no. 193, IAHS Press, Wallingford, pp. 127–35.

CLIMAP Project Members (1976) 'The surface of the Ice Age earth', *Science*, 191: 1131–7.

—— (1981) 'Seasonal reconstruction of the earth's surface at the last glacial maximum', *Geological Society of America, Map Chart Series*, MC–36.

COHMAP Members (1988) 'Climatic changes of the last 18,000 years: Observations and model simulation', *Science*, 241: 1043–52.

Collins, D.N. (1987) 'Climatic fluctuations and run off from glacierized alpine basins', in S.I. Solomon *et al.* (eds), *The Influence of Climate Change and Climatic Variability on the Hydrologic Regime and Water Resources*, International Association of Hydrological Science, Publ. no. 168, IAHS Press, Wallingford, pp. 77–89.

—— (1989) 'Hydrometeorological conditions, mass balance and runoff from alpine glaciers', in J. Oerlemans (ed.), *Glacier Fluctuations and Climate Change*, Kluwer, Dordrecht, pp. 305–23.

Colton, D.E. (1976) 'Numerical simulation of the orographically induced precipitation distribution for use in hydrologic analysis', *Journal of Applied Meteorology*, 15: 1241–51.

Crowe, R.B., McKay, G.A. and Baker, W.M. (1977) *The Tourist and Outdoor Recreation Climate of Ontario*, vol. 3, Atmospheric Environment Service, REC-1–73, Downsview, Ontario.

Crowley, T.J. (1990) 'Are there any satisfactory geologic analogs for a future greenhouse warming?', *Journal of Climate*, 3: 1282–92.

Dickinson, R.E., Erico, R.M., Giorgi, F. and Bates, G.T. (1989) 'A regional climate model for the western United States', *Climatic Change*, 15: 383–422.

Fliri, F. (1982) *Tirol-Atlas. D. Klima*, Universitätsverlag Wagner, Innsbruck, 23 plates.

Föhn, P. (1990) 'Schnee und Lawinen', in D.D. Vishcher (ed.), *Schnee, Eis und Wasser der Alpen in einer wärmeren Atmosphäre*, Mitteilungen 108, Versuchsanstalt für Wasserbau, Hydrologie und Glaziologie, Eidgenössische Technische Hochschule, Zürich, pp. 33–48.

Forland, E.J. and Gjessing, Y.T. (1975) 'Snow contamination from washout/rainout and dry deposition', *Atmospheric Environment*, 9: 339–52.

Fritts, H.C. (1976) *Tree Rings and Climate*, Academic Press, New York.

Froehlich, W., Gil, E., Kasza, I. and Starkel, L. (1990) 'Thresholds in the transformation of slopes and river channels in the Darjeeling Himalaya, India', *Mountain Research and Development*, 10: 301–12.

Galloway, R.W. (1988) 'The potential impact of climate changes on Australian ski fields', in G.I. Pearman (ed.), *Greenhouse: Planning for Climate Change*, CSIRO, Aspendale, Australia, pp. 428–37.

Garnett, A. (1935) 'Insolation, topography and settlement of the Alps', *Geographical Review*, 25: 601–17.

Gellatly, A.F., Chinn, T.J.H. and Röthlisberger, F. (1988) 'Holocene glacier variations in New Zealand: a review', *Quaternary Science Review*, 7: 227–42.

Georgakos, K.P. and Bras, R.L. (1984) 'A hydrologically useful station precipitation model', *Water Resources Research*, 20: 1585–610.

Giorgi, F., Marinucci, M.R. and Visconti, G. (1990) 'Application of a limited area model nested in a general circulation model to regional climate simulation over Europe', *Journal of Geophysical Research*, 95: 18, 413–18, 431.

Giorgi, F. and Mearns, L.O. (1991) 'Approaches to the simulation of regional climate change: a review', *Reviews of Geophysics*, 29: 191–216.

Gleick, P.H. (1989) 'Climate change, hydrology and water resources', *Reviews of Geophysics*, 27: 329–44.

Grenell, W. and Oerlemans, J. (1989) 'Energy balance calculations on and near Hintereisferner (Austria) and estimate of the effect of greenhouse warming on ablation', in J. Oerlemans (ed.), *Glacier Fluctuations and Climate Change*, Kluwer, Dordrecht, pp. 305–32.

Grove, J.M. (1988) *The Little Ice Age*, Methuen, London, 498 pp.

Haeberli, W., Muller, P., Alean, J. and Bösch, H. (1989) 'Glacier changes following the Little Ice Age. A survey of the international data base and its perspectives', in

J. Oerlemans (ed.), *Glacier Fluctuations and Climate*, Reidel, Dordrecht, pp. 77–101.

Hanna, S.R. and Strimaitis, D.G. (1990) 'Rugged terrain effects on diffusion', in W. Blumen (ed.), *Atmospheric Processes over Complex Terrain*, Meteorological Monograph 23 (45), American Meteorological Society, Boston, MA, pp. 109–43.

Hare, F.K. (1983) *Climate and Desertification. A Revised Analysis*, WCP–44, World Climate Program, World Meteorological Organization/United Nations Environment Programme, Geneva.

Harrison, R., Kinnaird, V., McBoyle, G.R., Quinlan, C. and Wall, G. (1986) 'The resiliency and sensitivity of downhill skiing in Ontario to climatic change', *Proceedings 43rd Annual Meeting, Eastern Snow Conference*, pp. 94–105.

Hastenrath, S., Rostom, R. and Caukwell, R.A. (1989) 'Variations of Mount Kenya's glaciers, 1963–87', *Erdkunde*, 43: 202–10.

Hill, F.F. (1983) 'The use of average rainfall maps to derive estimates of orographic enhancement of frontal rain over England and Wales for different wind directions', *Journal of Climatology*, 3: 113–29.

Hope, G.S., Peterson, J.A., Radok, U. and Allison, I. (eds) (1976) *The Equatorial Glaciers of New Guinea*, A.A. Balkema, Rotterdam, 244 pp.

Hurni, H. and Stähli, P. (1982) 'Das Klima von Semien', in H. Hurni, *Hochgebirge von Semien – Athiopen Vol. II Kaltzeit bis zur Gegenwart*, Geographica Bernensia 13, University of Berne, Switzerland, pp. 50–82.

Jaeger, F. and Kellogg, W.W. (1983) 'Anomalies in temperature and rainfall during warm arctic seasons', *Climatic Change*, 5: 34–60.

Jarrett, R.D. (1990) 'Paleohydrologic techniques used to define the spatial occurrence of floods', *Geomorphology*, 3: 181–95.

Karlén, W. (1988) 'Scandinavian glacial and climatic fluctuations during the Holocene', *Quaternary Science Review*, 7: 199–209.

Kasser, P. (1973) 'Influence of changes in the glacierized area on summer run-off in the Porte du Scex drainage basin of the Rhône', *Symposium on the Hydrology of Glaciers*, International Association of Hydrological Science, Publ. no. 95, IAHS Press, Wallingford, pp. 221–5.

Kirchhofer, W. (ed. in chief) (1982) *Klimaatlas der Schweiz*, Schweizer Meteorologische Anstalt, Zürich.

Kotlyakov, V.M., Serebryanny, J.R. and Solomina, O.N. (1991) 'Climate change and glacier fluctuations in the southern mountains of the USSR', *Mountain Research and Development*, 11: 1–12.

Krenke, A.N., Nikolaeva, G.M. and Shamkin, A.B. (1991) 'The effects of natural and anthropogenic changes on heat and water budgets in the central Caucasus, USSR', *Mountain Research and Development*, 11: 173–82.

Kuhn, M. (1989) 'The effects of long term warming on alpine snow and ice', *Landscape Ecological Impact of Climate Change on Alpine Regions with Emphasis on the Alps*, Agricultural University of Wagenigen, Netherlands, pp. 10–20.

LaFontaine, C.V. (1988) *Comparison of the Simulated Climate and Geological Observations from Equatorial Land Regions for the Past 18,000 Years*, MS thesis (Meteorology), Center for Climatic Research, University of Wisconsin, Madison, 62 pp.

Lamb, H.H. (1977) *Climate Past, Present and Future Vol. 2. Climate History and the Future*, Methuen, London, 835 pp.

Lauscher, F. (1980) 'Die Schwankungen der Temperatur auf dem Sonnblick seit 1887 im Vergleich zu globalen Temperatureschwankungen', *16 International Tagung für Alpine Meteorologie* (Aix-les Bains), Société Météorologique de France, Boulogne-Billancourt, pp. 315–19.

Letréguilly, A. and Reynaud, L. (1989) 'Spatial patterns of mass-balance fluctuations of North American glaciers', *Journal of Glaciology*, 35 (120): 163–8.

Lettenmaier, E.P. and Gan, T.Y. (1990) 'Hydrologic sensitivities of the Sacramento–San Joaquin River basin, California, to global warming', *Water Resources Research*, 26: 69–86.

Lough, J.M., Wigley, T.M.L. and Palutikov, J.P. (1983) 'Climate and climate impact scenarios for Europe in a warmer world', *Journal of Climate and Applied Meteorology*, 22: 1673–84.

Luckman, B.H. (1990) 'Mountain areas and global change: A view from the Canadian Rockies', *Mountain Research and Development*, 10: 183–95.

Maddox, R.A., Hoxit, L.R., Chappell, C.F. and Caracena, F. (1978) 'Comparison of meteorological aspects of the Big Thompson and Rapid City floods', *Monthly Weather Review*, 106: 375–89.

Makarevich, K.G. and Rototaeva, O.V. (1986) 'Present-day fluctuations of mountain glaciers in the northern hemisphere', Soviet Geophysical Committee, Academy of Sciences of the USSR, Moscow, *Data of Glaciological Studies*, 57: 157–63.

Markgraf, V. (1980) 'Pollen dispersal in a mountain area', *Grana*, 19: 127–46.

Mielke, P.W., Jr, Brier, G.W., Grant, L.O., Mulvey, G.J. and Rosenzweig, P.N. (1981) 'A statistical reanalysis of the replicated CLIMAX I and II wintertime orographic cloud seeding experiments', *Journal of Applied Meteorology*, 20: 643–59.

Monasterio, M. (ed.) (1980) *Estudios Ecologicos en los Paramos Andinos*, Ediciones, Universidad de los Andes, Mérida, Venezuela, 312 pp.

Myers, V.A. and Morris, D.I. (1991) 'New techniques and data sources for PMP', in D.D. Darling, *Waterpower '91*, vol. 2, American Society of Civil Engineers, New York, pp. 1319–27.

Obrebska-Starkel, B. (1990) 'Recent studies on Carpathian meteorology and climatology', *International Journal of Climatology*, 10: 79–88.

Oerlemans, J. (ed.) (1989) *Glacier Fluctuations and Climate Change*, Kluwer, Dordrecht.

Oerlemans, J. and Hoogendorn, N.C. (1990) 'Mass-balance gradients and climatic changes', *Journal of Glaciology*, 35: 399–405.

Oki, T., Musiake, K. and Koike, T. (1991) 'Spatial rainfall distribution at a storm event in mountainous regions, estimated by orography and wind direction', *Water Resource Research*, 27: 359–69.

Parry, M.L. (1978) *Climatic Change, Agriculture and Settlement*, Dawson-Archon Books, Folkestone, 214 pp.

Patzelt, G. and Aellen, M. (1990) 'Gletscher', in D.D. Vischer (ed.), *Schnee, Eis und Wasser der Alpen in einer wärmeren Atmosphäre*, Mitteilungen 108, Versuchsanstalt für Wasserbau, Hydrologie und Glaziologie, ETH, Zürich, pp. 49–69.

Penman, H.L. (1963) *Vegetation and Hydrology*, Commonwealth Agricultural Bureaux (Technical Communication no. 53, Commonwealth Bureau of Soils), Farnham Royal.

Pfister, C. (1985a) *Klimageschiche der Schweiz, 1525–1860*, vol. 1, Academia Helvetica 6, P. Haupt, Berne.

—— (1985b) 'Snow cover snowlines and glaciers in central Europe since the 16th century', in M.J. Tooley and G.M. Sheail (eds), *The Climatic Scene*, G. Allen & Unwin, London, pp. 154–74.

Porter, S.C. (1979) 'Hawaiian glacial ages', *Quaternary Research*, 12: 161–87.

Rango, A. (1985) 'A survey of progress in remote sensing of snow and ice', in B.E. Goodison (ed.), *Hydrological Applications of Remote Sensing and Remote Data Transmission*, International Association of Hydrological Science, Publ. no. 145, IAHS Press, Wallingford, pp. 347–59.

Rango, A., Martinec, J., Chang. A.T.C., Foster, J.L. and van Katwijk, V.F. (1989) 'Average areal water equivalent of snow in a mountain basin using microwave and visible satellite data', *Institute of Electrical and Electronic Engineers. Transactions on Geoscience and Remote Sensing*, 27: 740–5.

Rhea, J.O. (1978) 'Orographic precipitation model for hydrometeorological use', *Atmospheric Sciences Paper*, no. 287, Colorado State University, Fort Collins.

Rind, D. (1988) 'Dependence of warm and cold climate depiction on climate model resolution', *Journal of Climate*, 1: 965–97.

Rind, D. and Peteet, D. (1985) 'Terrestrial conditions at the last glacial maximum and CLIMAP sea-surface temperature estimates: Are they consistent?', *Quaternary Research*, 24: 1–22.

Röthlisberger, F. (1986) *10,000 Jahre Gletschergeschichte der Erde*, Verlag Sauerlander, Aarau and Frankfurt.

Sarker, R.P. (1966) 'A dynamical model of orographic rain', *Monthly Weather Review*, 94: 555–72.

Schweingruber, F.H., Bräker, O.U. and Schär, E. (1979) 'Dendroclimatic studies on conifers from central Europe and Great Britain', *Boreas*, 8: 427–52.

Sellers, P.J. and Lockwood, J.G. (1981) 'A computer simulation of the effects of differing crop types on the water balance of small catchments over long time periods', *Quarterly Journal of the Royal Meteorological Society*, 107: 395–414.

Sissenwine, N. (1969) 'Standard and supplemental atmospheres', in D.F. Rex (ed.), *Climate of the Free Atmosphere*, World Survey of Climatology, vol. 4, Elsevier, Amsterdam, pp. 5–44.

Slatyer, R.O., Cochrane, P.M. and Galloway, R.W. (1984) 'Duration and extent of snow cover in the Snowy Mountains and a comparison with Switzerland', *Search*, 15: 327–31.

Slaymaker, O. (1990) 'Climate change and erosion processes in mountain regions of western Canada', *Mountain Research and Development*, 10: 171–82.

Smith, J.M.B. (1975) 'Mountain grasslands of New Guinea', *Journal of Biogeography*, 2, 27–44 and 87–102.

Smith, R.B. (1979) 'The influence of mountains on the atmosphere', *Advances in Geophysics*, 21: 87–230.

—— (1986) 'Current status of ALPEX research in the United States', *Bulletin of the American Meteorological Society*, 67: 310–18.

Solomon, A.M. and Silkworth, A.B. (1986) 'Spatial patterns of pollen transport in a mountain region', *Quaternary Research*, 25: 150–62.

Starkel, L. (1972) 'The role of catastrophic rainfall in the shaping of the relief of the lower Himalaya (Darjeeling Hills)', *Geographica Polonica*, 21: 103–60.

Steinhauser, F. (1970) 'Die säkularen Änderungen der Schneedeckenverhältnisse in Österreich', *66–67 Jahresbericht des Sonnblick-Vereines, 1970–1971*, Vienna, 1–19.

—— (1973) 'Die Änderungen der Sonnenscheindauer in Österreich in neurer Zeit', *68–69 Jahresbericht des Sonnblick-Vereines, 1972–1973*, Vienna, 41–53.

Steinhoff, H.W. and Ives, J.D. (1976) *Ecological Impacts of Snowpack Augmentaion in the San Juan Mountains Colorado*, College of Forestry and Natural Resources, Colorado State University Report CSU–FNR–7052–1, Fort Collins, Colorado, 489 pp.

Tabony, R.C. (1985) 'The variation of surface temperature with altitude', *Meteorological Magazine*, 114: 37–48.

Thompson, L.G., Mosley-Thompson, E. and Arnoa, B.M. (1984a) 'El Niño-Southern Oscillation events recorded in the stratigraphy of the tropical Quelccaya Ice Cap, Peru', *Science*, 226: 50–3.

Thompson, L.G., Mosley-Thompson, E., Grootes, P.M., Pourchet, M. and Hastenrath, S. (1984b) 'Tropical glaciers: Potential for ice-core paleoclimatic reconstructions', *Journal of Geophysical Research*, 89(D3): 4638–46.

Thompson, L.G., Mosley-Thompson, E., Bozon, J.F. and Koci, B.R. (1985) 'A 1500-year record of tropical precipitation in ice-cores from the Quelccaya Ice Cap, Peru', *Science*, 229: 971–3.

Wagenbach, D. (1989) 'Environmental records in alpine glaciers', in H. Oeschger and C.C. Langway, Jr (eds), *The Environmental Record in Glaciers and Ice Sheets*, J. Wiley & Sons, Chichester, pp. 69–83.

Webb, T., III and Clark, D.R. (1977) 'Calibrating micropaleontological data in climatic terms: a critical review', *Annals of the New York Academy of Sciences*, 288: 93–118.

Webster, P.J. and Streten, N.A. (1978) 'Late Quaternary ice age climates of tropical Australia: interpretations and reconstructions', *Quaternary Research*, 10: 279–309.

Wiesnet, D.R., Ropelewski, G.F., Kukla, J.G. and Robinson, D.A. (1987) 'A discussion of the accuracy of NOAA satellite – derived global seasonal snow cover measurements', in B.E. Goodison, R.G. Barry and J. Dozier (eds), *Large-scale Effects of Seasonal Snow Cover*, International Association of Hydrological Science Publ. no. 166, IAHS, Wallingford, pp. 291–304.

Witmer, U. (with Filliger, P., Kunz, S. and Kung, P.) (1986) *Erfassung, Bearbeitung und Kartierung von Schneedaten in der Schweiz*, Geographica Berneensia G25, University of Berne.

Wood, F.B. (1988) 'Global alpine glacier trends, 1960s to 1980s', *Arctic and Alpine Research*, 20: 404–13.

Zumbühl, H.J. (1988) 'Der Rhonegletscher in den historischen Quellen', *Die Alpen*, 64: 186–233.

2

THE ALPS UNDER LOCAL, REGIONAL AND GLOBAL PRESSURE

H. Grassl

INTRODUCTION

The Alps are the playing ground for Europe, the summer water reservoir for major European river systems such as the Danube, Rhine and Rhône, an important barrier for air-mass exchange, a unique unmanaged as well as managed ecosystem mosaic, and the largest European mountain range. As in the case of other mountainous areas, they display the strong influence of precipitation and temperature on vegetation type over short distances. Human settlements date back to the Stone Age and quite unique land-use patterns, for instance the semi-nomadic 'Almwirtschaft', have been relatively stable for hundreds of years, that is, they were sustainable.

During the last century, and especially during the last decades, the Alps have come under manifold new pressures. The local development of tourism and industry as well as the decay of traditional agriculture, the regional impact of air pollution, and the warming trend have strongly changed their appearance. Downhill skiing slopes replaced alpine forests, many glaciers disappeared (nearly all are shrinking), forest die-back has started especially on south-facing slopes, and many valleys have become densely populated.

If one compares forecast anthropogenic climate change (WMO/UNEP, 1992: an increase of 3 ± 1.5 K in 2100, if no measures are taken) with the 4 to 5 K shift from the ice age climax 18,000 years before the present (with fully ice-covered Alps) to Holocene climate, one might imagine what changes could occur soon. Thus, there is ample justification for a more thorough look into the synergy or amplification of local, regional and global pressures on the Alps in the coming decades.

Working Group II of the Intergovernmental Panel on Climate Change (IPCC), dealing with the impacts of global climate change, has as one of its main headings for the 1995 Second Assessment Report a specific chapter on mountain areas. The 1995 Report is primarily aimed at supplying scientific input to the International Negotiating Committee, whose task is the implementation of the United Nations Framework Convention on Climate

Change signed in Rio de Janeiro in June 1992, and which is now undergoing ratification. This is to say that the identification of mountain regions as a distinct entity confers a significant degree of importance to mountains as a system vulnerable to climate change.

WHAT IS A PRESSURE FOR AN ECOSYSTEM?

We may speak of pressure on an ecosystem if natural as well as anthropogenic changes surmount long-term mean natural variability, thus causing drastic changes in the ecosystems' biodiversity or species composition.

Global pressures

Examples of such an extreme variability, caused by a globally modified general circulation in the ocean and the atmosphere, driven naturally or anthropogenically are:

- The Younger Dryas period about 11,000 years ago, when nearly ice age climax climate conditions returned to northern and western Europe (from forests to tundra) for some 100, probably as a result of a nearly complete stop of deep water formation in the Northeast Atlantic. This is believed to have occurred because strong melt water inflow into the North Atlantic from the decaying Laurentian ice sheet sharply reduced salinity in the ocean.
- The anthropogenic greenhouse gas increase since the beginning of industrialization which amounts to the same increase as that from the ice age conditions to the present interglacial (Holocene). Carbon dioxide (CO_2) increased in 10,000 years from 190 to 280 ppmv and within the last 200 years to 357 ppmv (1993); for methane (CH_4) the respective numbers are: 0.35, 0.7, 1.7 ppmv (WMO/UNEP, 1990). The CO_2 increase has already certainly led to a rearrangement of species composition, and thus also has and will increase the numbers on the list of endangered or extinct species.
- The stratospheric ozone depletion caused by the catalytic destruction of ozone molecules by chlorine-containing decay products of chloro-fluoro-carbons increases UV-B radiation in the middle and high latitude troposphere of both hemispheres. This will mainly affect photo-chemistry in the troposphere as well as photosynthesis in high-altitude mountain areas. Here, UV-B radiation increases are not compensated by UV-B absorption as at lower elevations, where tropospheric ozone concentrations can be quite significant.

Regional pressures: acid rain and photochemical smog

The emission of many combustion products such as nitrogen oxides (NO_x), sulphur dioxide (SO_2), hydrocarbons and ammonia (NH_3) from intensive animal husbandry has caused continent-wide pollution in Europe. The consequence is acid precipitation after these emission products have either been chemically transformed into acids (mainly sulphuric acid particles and nitric acid, often deposited on to already existing aerosol particles) or have been dissolved in precipitation elements, thereby being partly transformed chemically also. What is of special interest is the possible enhancement of acid deposition in mountain areas because of a precipitation increase with altitude. An example taken from Smidt (1991) for two Austrian alpine sites in Figure 2.1 demonstrates that contamination of precipitation with ions mainly resulting from pollution decreases with altitude, but deposition does not, because of increased precipitation with height. The deposition of hydrogen ions (H^+) strongly increases with height. Since the concentration of basic anions and kations in precipitation is rather uniform over central Europe, the Alps have at least as heavy a burden as other areas because of increased precipitation, although they are not a major source area for pollutants.

Practically the same composition of pollutants emitted from power plants, domestic heating and traffic which leads to acid precipitation is also responsible for photochemical smog. Since NO_x emissions have not decreased appreciably despite environmental protection measures (such as the large incineration plants protocol in Germany), ozone (O_3) is the leading substance for this type of pollution, and as a consequence photochemical smog has become a major threat to humans, plants, and animals. O_3 concentrations in the lower troposphere often exhibit peaks in rather remote locations and at middle altitudes. These peaks are generally linked to pollution emanating from large industrial complexes in central Europe and northern Italy.

This trend is clearly reflected in the ozone soundings of Hohenpeissenberg and the measurements on Zugspitze (2,950 m above sea level) in the Bavarian Alps. There has been an upward trend in tropospheric ozone mixing ratios since measurements began in the late sixties. The trend peaks at slightly above 2 per cent per year in nearly the entire troposphere (see Figure 2.2). In the Alps this might be locally very different. While the upward trend is also present on Zugspitze and Wank (1,800 m), there is none observed in the polluted city of Garmisch nearby to the north but still within the Alps (Schneider, 1992), a phenomenon well known from measurements in other larger cities. With this ozone trend other dangerous oxidizing compounds like peroxacetyl-nitrate have also increased. Widespread damage reported from the Alpine forests is certainly due in part to the increasing burden of photochemical smog; and as always with environmental threats, the large

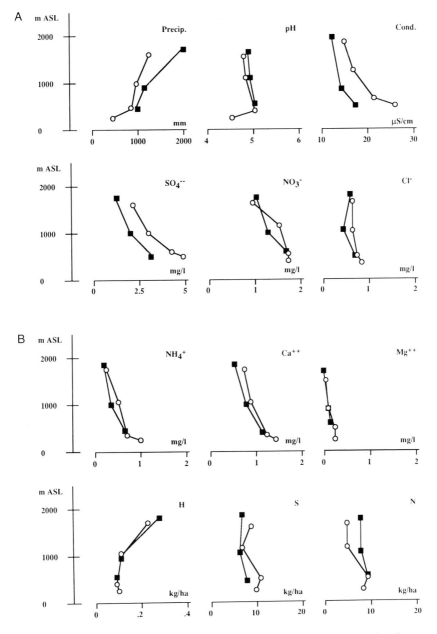

Figure 2.1 Contamination of precipitation at Austrian alpine sites: (A) height dependence of precipitation and pH-value, conductivity as well as sulphate, nitrate, and chlorine content of precipitation. (B) height dependence of ammonium, calcium and manganese anions in precipitation (upper portion) as well as deposition of hydrogen, sulphur and nitrogen with precipitation taken from Smidt, 1991.

37

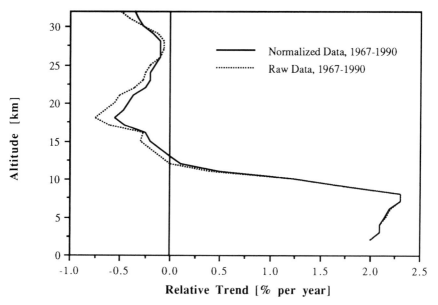

Figure 2.2 Ozone trend over Hohenpeissenberg for the period 1967–1990.
The trend is highly significant at all heights below 11 km and above 14 km;
from Vandersee *et al.* (1992).

time lag between load and full reaction when trees and soils are involved has
still not allowed us to fully grasp the entire impact.

Local pressures

It would be too lengthy to enumerate all the local pressures on the Alps and
their inhabitants. However, some dominant ones should be mentioned here,
and include:

- growing automobile and truck traffic crossing the Alps over an increasing number of traverses;
- construction of large skiing resorts, often without environmental protection measures such as adequate waste management;
- increasing number of misbehaving tourists and local people (heli-skiing,
  mountain-biking everywhere, cars on trails, etc.) injuring and sometimes
  destroying the habitats of many already endangered species;
- pollution of air and water by local industry and hotels, thus also
  reducing the attraction for tourists;
- construction of houses in endangered zones.

A typical scenario which combines direct and indirect anthropogenic
effects (positive feedback mechanism) can be illustrated by the following

example. A downhill ski run cut into a protecting forest increases the sensitivity of the trees at the edges of the ski run to direct air pollution, because they filter more than do trees in the interior of the forest. Thus, both acidification of the soil by higher deposition rates and oxidants' impact on pine-tree needles is increased. The weakened forest shows signs of forest die-back earlier, again loses part of its already reduced protection capacity against extreme precipitation events and avalanches, but is also increasingly endangered during less intense storm events. Since summertime heat episodes are likely to become more frequent as a result of global warming, extreme precipitation events normally occurring at the end of warm episodes will result in landslides and mud-flows, destroying human settlements and starting an intensified erosion thereafter for years. This erosion can only be halted by very costly protection measures.

Amplification of local, regional, and global pressures

There are many examples of positive and negative feedbacks in the climate system and in single ecosystems. It is often very attractive to mention only the positive feedbacks since they describe amplification processes and thus cause strong reactions to a very small stimulus. Not all human activities lead to amplification of stresses; some can lead to reductions in certain forcings. This is partly the case with the local UV-B radiation change. Ozone decrease in the stratosphere causes increased UV-B radiation entering the troposphere, where it is partly absorbed by the ozone increase in the troposphere, resulting in a smaller net UV-B radiation increase at the surface.

From the point of view of precautions it is, however, necessary to look especially carefully into amplification processes or positive feedbacks.

DO WE ALREADY SEE THE CONSEQUENCES?

The most convincing sign of a systematic change in the Alps is the shrinking and the loss of glaciers (see contribution by Haeberli in this volume), although we do not know whether this is partly or fully anthropogenic or has been damped by counteraction through natural processes. We clearly see also growing forest damage north and south of the crest of the Alps. Again we do not know the exact cause, though the burning of fossil fuels is probably the chief culprit. We also know that most of the planted forests (for example spruce) will suffer additionally from a summertime temperature increase, since they are adapted to colder climates. Thus, we should not only reduce fossil fuel burning but also reconsider certain forestry practices.

We cannot separate the different causes of a single landslide or a devastating avalanche. However, we can be sure that local felling of trees, regional weakening of a protection forest, and melting of permafrost soils in summertime heat episodes have all contributed, and many events would not have occurred if one contributing factor had been missing.

Should we act only if we see damage? The answer is definitely no, because soils, water bodies and trees are able to buffer stress impacts for some years or decades before they subsequently show signs of rapid degradation or disease.

HOW CAN THE THREATS OF ANTHROPOGENIC IMPACTS TO ECOSYSTEMS AND HUMAN SETTLEMENTS IN THE ALPS BE REDUCED?

Anthropogenic stresses on natural and socio-economic systems in the Alps could be mitigated by the following measures:

- Motivating all Alpine countries to take an active role in implementation of the UN Framework Convention on Climate Change and the UN Convention on Biodiversity.
- The same countries should be encouraged to implement European regulations for SO_2, NO_x, NH_3, and hydrocarbon emission reductions (preferably linked to the CO_2-reduction within the Climate Convention implementation); this is still in its infancy and is not being implemented rapidly enough on a continental scale.
- Halting the trend towards stronger per capita use of natural resources through tourism.

Which specific measures are needed for the Alps?

CO_2 emission reduction is especially needed for the hitherto exploding traffic sector. Thus, a shift of cargo from trucks to rail first of all within, but also outside, the Alps would be beneficial in a threefold manner: local, regional, and even global change impacts on the Alps would be reduced simultaneously. The attractions of the Alps, landscape, rare ecosystems, natural variability, possibility of personal satisfaction through individual activity as well as their Europe-wide significance have to be maintained. Some examples for local action are:

- In order that people learn to hear the call of nature once again, there should be a ban on heli-skiing, and roads in remote valleys have to be closed to private traffic.
- Tourist infrastructure needs to include adequate sewage treatment plants and clean-air facilities.
- Tourists should have easy access to high-quality information on the ecology of alpine areas.
- Energy from renewable sources, especially solar, is plentiful particularly in many higher elevations of the Alps and should be used for heating as well as for electricity generation in remote locations.

- The concentration on single activities in distinct resort areas should be curtailed.
- Production of artificial snow should be prohibited, even if snow is not abundant.

REFERENCES

Nodop, K. (1990) 'Weiträumige Verteilung und zeitliche Entwicklung säurebildender Spurenstoff in Europa, 1978 bis 1986', *Ber. d. Ins. f. Meteorologie und Geophysik*, Frankfurt, Nr., 81.

Schneider, U. (1992) 'Die Verteilung des troposphärischen Ozons in Bayrischen Nordalpenraum', Dissertation; University of Mainz.

Schönwiese, C.-D., Rapp, J., Fuchs, T. and Denkhard, M. (1993) *Klimatrend-Atlas Europa 1981–1990*, Bericht Nr. 20, Zentrum für Umweltforschung, ZUF-Verlag, Frankfurt.

Schweizerische Akademie der Naturwissenschaften (1991) *Die Alpen – ein sicherer Lebensraum, Ergebnisse der 171. Jahresversammlung*, Hrsg.: J.P. Müller, B. Gilgen, Desertina-Verlag, Disentis, Schweiz.

Smidt, S. (1991) *Messungen nasser Freilanddepositionen der Forstlichen Bundesver-suchsanstalt*, FBVA-Berichte, ISSN 1013–0713 50, Nasse Deposition, Austria.

Vandersee, W., Wege, K., Weigl, E. and Claude, H. (1992) *Untersuchungen zur Ozonkonzentration in orographisch gegliederten belände, zum täglichen Ozongang und zu möglichen Trends*, Final Report for the German Ministry for Research and Technology, Project 0744114, Bonn.

WMO/UNEP (1990) *Climate Change – the IPCC Scientific Assessment*, ed., J.T. Houghton, G.J. Jenkins, J.J. Ephraums, Cambridge University Press, Cambridge, 365 pp.

—— (1992) *1992 IPCC Supplement, Scientific Assessment of Climate Change*, ed., J.T. Houghton, B.A. Callander, S.K. Varney, Cambridge University Press, Cambridge.

3

USING MULTIPLE HIGH-RESOLUTION PROXY CLIMATE RECORDS TO RECONSTRUCT NATURAL CLIMATE VARIABILITY

An example from the Canadian Rockies

B.H. Luckman

INTRODUCTION

Mountain areas contain a variety of excellent natural archives that document former environmental changes with differing temporal and parameter resolution. These records involve both physical (e.g. glaciers, permafrost) and biological systems (e.g. trees) that respond to climate change and often occur in close juxtaposition, thereby providing complementary insights into former environmental changes. Since the time of Louis Agassiz, many key concepts about climate history have been developed from studies of the fluctuations of alpine glaciers (e.g. Penck and Bruckner, 1909; Matthes, 1939; Porter and Denton, 1967; Grove, 1987). Refining this history and supplementing it with new data, techniques and complementary sources (tree rings, lacustrine records, historical sources) allows us to document the magnitude and timing of past climate changes and benchmark natural climate variability at several timescales. Understanding and modelling this natural variability on decade-to-century timescales is a critical element in our ability to detect any anthropogenic signal in the relatively short instrumental climate record. It defines the background variability and trends upon which any future climate changes will be superimposed. Studies of historical system responses, particularly over the last century, also allow us to develop potential analogues to predict the future response of these systems.

In many mountain environments (Stone, 1992), human activity has modified natural systems (e.g. treelines in the Alps) making it difficult to discriminate between natural and anthropogenic controls of recent changes. In the Canadian Rockies large areas were set aside within National Parks prior to significant human impact and observed changes can be attributed solely to natural forcing. In addition, the absence of significant human disturbance has allowed the preservation of physical evidence of change (e.g. snags) that has been destroyed elsewhere. However, this absence of human activity also limits access and the period of instrumental or documentary records: contrary to the situation in Europe, there are scant documentary observations for the Canadian Rockies prior to 1900 and few areas have been studied in detail. Greater reliance must therefore be placed on the development of natural archives to document recent climate history. This paper comments on the two major sources of information about environmental changes over the late Holocene in the Canadian Rockies with particular emphasis on the last millennium.

THE GLACIAL RECORD

The geological record of glacier fluctuations (with limited palynological studies) has been the traditional source for Holocene climatic history in this area (Heusser, 1956; Osborn and Luckman, 1988). The Cavell Advance (Luckman and Osborn, 1979) is the regional Little Ice Age equivalent and is defined to include glacier advances of the twelfth and thirteenth centuries (Luckman and Osborn, 1979; Luckman, 1986). Therefore, in this review, the term Little Ice Age (LIA) is used to describe glacier advances between the twelfth and nineteenth centuries (see discussion in Bradley and Jones, 1992a, b). The glacial record has been reviewed in detail elsewhere (Luckman, 1992, 1993, 1994): the brief summary below is followed by a discussion of significant questions arising from this record.

In the Canadian Rockies the Little Ice Age was the most extensive Holocene glacial advance and evidence for events prior to the LIA maximum is fragmentary. Recent detailed work within glacier forefields has revealed palaeosols, detrital and in-situ wood remains that allow an initial reconstruction of the earlier glacier history. During the Hypsithermal glaciers were less extensive than at present: detrital wood recovered from adjacent glaciers at the Columbia Icefield indicates that mature forest grew upvalley of the present snouts of Athabasca and Dome Glaciers c. 7500–8300 and 6000–6500 yr BP, respectively (Luckman, 1988a; Luckman et al., 1993). The earliest evidence of Neoglacial glacier events is a palaeosol overridden at Boundary Glacier c. 4000 yr BP (Gardner and Jones, 1985). Subsequently treelines were higher than present c. 3000–3500 yr BP but trees were overridden and killed at four glaciers between c. 2800–3000 yr BP during the Peyto Advance (Luckman et al., 1993). At Peyto Glacier trees were overridden further downvalley c. 1550–1700 yr BP (ibid.). Evidence for the earliest LIA advance (c. 600–800 ^{14}C yr BP) has been recovered from overridden in-situ trees at three glaciers c. 0.5–1.0 km upvalley of the LIA maximum position. Calendar kill dates between 1246 and 1324 AD (Peyto) and 1142 to 1350 AD (Robson) have been determined from these trees at two sites (Luckman, 1986, 1993; Reynolds, 1992). Lichenometric and dendrodrochronological dating at over 30 glaciers indicates maximum LIA glacier extent and moraine building occurred either in the early 1700s or between *c.* 1825–1875 (Luckman and Osborn, 1979; Luckman, 1986, 1992). Subsequent, less extensive, glacier readvances took place in the late nineteenth and early twentieth centuries followed by rapid glacier retreat in the 1930s–1960s. During the 1970s and early 1980s glacier recession rates were reduced and some glaciers stabilized or advanced, particularly in the Premier Range, British Columbia (Luckman *et al.*, 1987). In the last few years glacier recession has resumed, but no detailed monitoring of glacier front positions is presently being carried out in this area.

Discussion

Most of the dating for LIA glacier fluctuations prior to 1900 AD comes from minimum lichenometric or tree-ring dates on moraine surfaces that may have error terms of 10–50 years. The development of long regional tree-ring chronologies (see below) will facilitate the dating of overridden snags and ice-damaged trees along glacier margins, thereby allowing more precise dating of glacier advances that extended below treeline. A greater sample depth of individual glacier histories is needed for this region and should be supplemented by studies that reconstruct ELA depression for LIA glaciers, using the techniques demonstrated by Maisch (1992) and Torsnes *et al.* (1993). Perhaps this will stimulate interest in contemporary glaciological studies which have been sadly neglected in this area since completion of the IHD (see Young and Ommanney, 1984).

Although pre-nineteenth-century moraines occur at the majority of glacier sites investigated to date (Osborn and Luckman, 1988), many of these older moraines are small, locally preserved, fragments situated on the margins or a short distance downvalley of nineteenth-century moraines. For example, although evidence of an eighteenth-century advance occurs at two localities at Athabasca Glacier (Luckman, 1988b), at least 90 per cent (by length) of the LIA maximum position was occupied by the glacier in the nineteenth century. In discussions of glacier chronology to date, too much emphasis has been placed on moraine chronology rather than estimates of the relative regional extent of ice-cover at these critical times. Although no specific calculations have been made, the maximum areal extent of ice-cover was probably during the early nineteenth century, even though older moraines are occasionally preserved. This is consistent with the hypothesis that the relative extent of known Holocene glacier events in the Canadian Rockies has increased throughout the Holocene and is related to the decreased input of summer solar radiation due to changing orbital parameters (see Luckman, 1993; Nesje and Johannesson, 1992).

Glacier recession in the last 100 years has been dramatic. A critical question that must be addressed is whether this rapid recession is exceptional or typical of previous Neoglacial glacial events. Unlike the Alps (see Zumbühl, 1986) there are no documentary sources available to indicate glacier positions prior to the LIA maximum in the Canadian Rockies. The few exposures of palaeosols or in-situ stumps reported from glacier forefields to date (see Luckman et al., 1993) indicate a decrease in age downvalley that is consistent with a history of progressively more extensive glacier expansion during the Neoglacial. The recent recovery of in-situ stumps, 2,900–3,200 years old, from freshly deglaciated sites at Peyto and Robson Glaciers (Luckman et al., 1993) suggests these sites may have been ice-covered for most of the intervening period. It is interesting to speculate that these recent finds in the Rockies may be due to the fact that glacier fronts have now receded to positions that they last occupied several thousand years ago, well upvalley of their pre-LIA position. The remarkable discovery of a Stone Age man, 5300 ^{14}C years old, adjacent to a small alpine glacier in the Oetztal (Haeberli, this volume) similarly indicates a reduction in snow and ice-cover to levels of about 5000 years ago in the Alps. Both of these observations suggest that the glacier recession over the last century may be rather more than simply a recovery from LIA conditions. Subsequent, more detailed work (and continued glacier recession) is needed to verify this hypothesis but, if correct, it suggests that glacier recession in the last century is unprecedented during the Neoglacial and, therefore, the last hundred years may reflect a significant turning point in the Neoglacial history of these glaciers. Whether or not subsequent research confirms this interpretation, historical reconstructions of changes in glacier extent, runoff and snow cover over the last century have significant implications for future effects of global warming in alpine environments.

THE TREE-RING RECORD

Although the glacial record has provided the framework that underpins studies of Holocene climate change in mountains (and promises considerable opportunities for further work), it does not supply a continuous, high-resolution record of past environmental changes. The enumeration of former periods of glacier advance provides only a partial (and biased) record lacking information about intervening (and often considerable) non-glacial intervals. Studies of natural climate variability on timescales of decades to centuries require long, annually resolved records to extend the meagre instrumental climate record. Although some limited work has been done on lacustrine cores (e.g. Leonard, 1986), tree-ring series offer the best potential to obtain such records in the Canadian Rockies.

Jacoby and co-workers have amply demonstrated that boreal species at the northern treeline retain a strong temperature signal in their ring-width records (Jacoby and Cook, 1981; Jacoby et al., 1988; Jacoby and D'Arrigo,

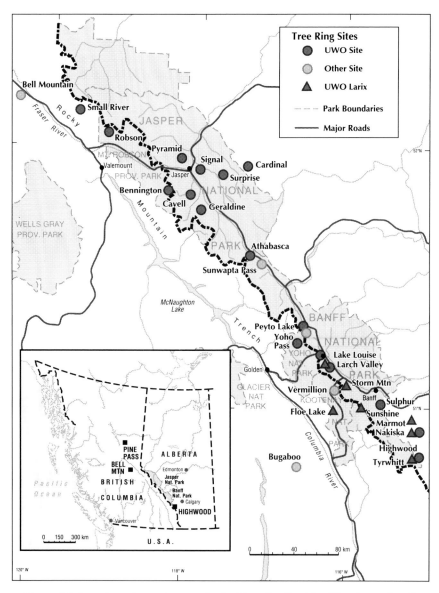

Figure 3.1 Location of tree-ring sites in the Canadian Rockies. The sites at Small River, Robson, Bennington, Cavell, Athabasca, Peyto Lake and Bugaboo are close to glaciers for which moraine chronologies have been developed (see Osborn and Luckman 1988; Luckman *et al.*, 1993). The location of the Pine Pass site is shown on the inset map.

46

1989; D'Arrigo and Jacoby, 1992). Similar preliminary results have been obtained from related species at alpine treeline sites in the Canadian Rockies (Luckman, 1990; Luckman and Colenutt, 1992; Colenutt and Luckman, 1991). Although some exploratory work on tree-ring–climate relationships has been carried out in this region, the emphasis to date has been on the development of long chronologies and a regional network of sites rather than palaeoclimatic reconstruction. The present distribution of sample sites is shown in Figure 3.1. Most of these sites are at treeline or adjacent to glaciers (or both). Sampling has focused on three species which attain maximal ages of over 700 years; *Pinus albicaulis* > 884 years; *Picea engelmannii* > 730 years; and *Larix lyallii* > 728 years. The following section presents some initial results from this work that indicate its potential for climatic reconstruction and give preliminary indications of climatic variability over the last millennium.

Picea and *Larix* chronology networks

Most chronologies in the Canadian Rockies utilize *Picea engelmannii* because it is the most widespread long-lived tree. This species has been neglected for dendrochronological work in alpine areas of western North America because of its generally low sensitivity (Fritts, 1976) and the high autocorrelation of tree-ring series in this species. Luckman and Colenutt (1992) report a mean sensitivity of 0.18 (range *c.* 0.13–0.23) for *picea* cores in the Canadian Rockies and autocorrelation values are usually in the range 0.7–0.9.

The present sampling network extends approximately 500 km along the Continental Divide (Figure 3.1). Treeline elevations range from *c.* 2,100–2,250 m in the south to 1,900–2,000 m in the north. Preliminary or completed *picea* chronologies are available for 23 sites (Table 3.1). During chronology development local master chronologies were developed using Cofecha (Holmes, 1992). Ring series are standardized using a cubic smoothing spline with a 50 per cent cutoff at 32 years. The persistence in these smoothed series is removed by autoregressive modelling. Although these series are usually used for crossdating they also provide a useful summary of the high frequency signal in these records. Inter-site correlation coefficients were calculated from Cofecha chronologies and are summarized in Figure 3.2. The larger matrix shows correlations between the 23 *picea* sites with living-tree chronologies listed in Table 3.1: the smaller matrix shows similar data for the six *larix* sites studied by Colenutt (1992). Figure 3.3 is a skeleton plot of narrow rings in the *picea* series, defined arbitrarily as those years > 1 standard deviation below the local mean of the series. In both diagrams the sites are positioned along an approximately north–south transect.

The data in Figure 3.2 show that, generally, correlation decreases with increasing inter-site distance and that, probably because of their geographical

Table 3.1 Network of *picea* chronology sites in the middle Canadian Rockies

Site	Lat.	Long.	Elev.	Record	Sampled by
Athabasca	52 12	117 15	2000	1072–1987	UWO
Bell Mt	53 20	120 40	1530	1649–1983	FHS
Bennington	52 42	118 21	1900	1104–1991	UWO
Bugaboo	50 54	117 47	1600?	1580–1976	FCC
Cardinal Pass	52 53	117 15	2050	1528–1990	UWO
Cavell Lake	52 41	118 03	1800	1540–1990	UWO
Geraldine	52 35	117 56	2000	1510–1990	UWO
Highwood	50 36	114 59	2250	1613–1987	UWO
Lake Louise	51 25	116 10	1650	1690–1982	UWO
Larch Valley	51 18	116 11	2200	1567–1987	UWO
Nakiska	50 50	115 15	2160	1619–1987	UWO
Peyto Glacier	51 42	116 31	1850	760–1990	UWO
Peyto Lake	51 43	116 30	2050	1634–1983	FHS
Pine Pass	55 30	122 40	780	1700–1983	FHS
Pyramid	52 58	118 10	2000	1608–1990	UWO
Robson	53 09	119 07	1690	1567–1983	UWO
Robson Snag	53 09	119 07	1690	865–1350	UWO
Signal	52 52	117 58	2050	1489–1990	UWO
Small River	53 10	119 29	1900	1569–1989	UWO
Sulphur Mtn	51 08	115 34	2250	1644–1990	UWO
Sunwapta Pass	52 15	117 00	2050	1608–1983	FHS
Surprise	52 48	117 38	1800	1552–1990	UWO
Vermillion	51 10	116 10	1500	1686–1983	FHS
Yoho Pass	51 29	116 29	1950	1536–1990	UWO

Notes: All chronologies utilize *Picea engelmannii* except Pine Pass (*Picea glauca*) and Vermillion Pass which includes both *Picea engelmannii* and *Pinus contorta*. Chronology collection: UWO = University of Western Ontario; FHS = F.H. Schweingruber (see Schweingruber, 1988); FCC = Forintek Canada Corporation (M.L. Parker and L.A. Jozsa). The elevation of the Bugaboo sampling site is estimated from a photograph of the site provided by L. Jozsa. The Peyto Glacier and Bennington Chronologies are from Reynolds (1992) and McLennan (1993) respectively.

  The Schweingruber sample site network (1988) includes a *Pseudotsuga menziessii* chronology from Bear Lake (54 30 N, 122 30 W, 650 m) between Bell Mountain and Pine Pass. This chronology has a mean correlation (1780–1980) of 0.14 with the *picea* sites listed above, 0.20 with Bell Mountain and −0.04 with Pine Pass. It was therefore excluded from the analyses used in this paper.

position in the centre of the range of sites, the highest mean correlations are found in the sites between Banff and Jasper (Sulphur to Pyramid). There are, however, some obvious anomalies. The Pine Pass site is clearly quite different from the others (both in correlation and marker ring sequences) and the sites at Vermillion, Lake Louise and Robson Glaciers have mean correlations with all other sites of 0.50 or less. These four sites are all at least 200–300 m below local treeline (Table 3.1) and indicate that sites within the subalpine forest do not yield as strong a common signal as the treeline sites. This effect is also seen in the two Peyto chronologies: the Peyto Lake chronology (from an

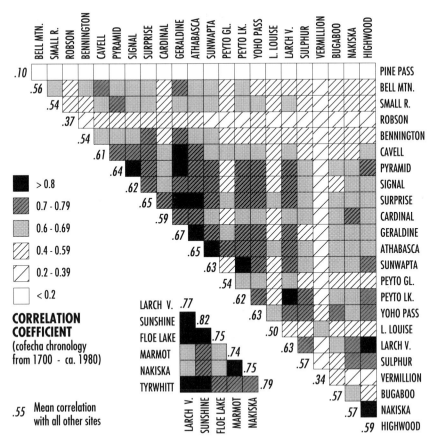

Figure 3.2 Cross-correlation matrix for Cofecha ring-width chronologies. The larger matrix is for the 23 *picea* chronologies listed in Table 3.1. The smaller matrix is for the *larix* sites studied by Colenutt (1992). For the spruce sites correlations were based on the chronologies between 1700–1980 for all pairs of sites except those involving Bugaboo (1700–1975). The numeric values on the left of each row are the mean correlation of individual chronologies with all other chronologies for the common period of record (1700–1975). For the larch chronologies the common period 1700–1985 was used for both the paired comparisons and mean inter-site correlations. The sites are arranged in an approximately north–south sequence.

interfluve, treeline site) correlates more strongly with other treeline sites (i.e. the regional signal) than it does with the nearby Peyto Glacier site on the valley floor near the LIA terminus of the glacier. The only exception to this altitudinal generalization is the Surprise site which, though 200–300 m below local treeline, correlates well with other sites. However, at Surprise, the sampled trees are growing in an open stand on the crest of a blocky Holocene rockslide deposit rather than in a closed forest situation. These results

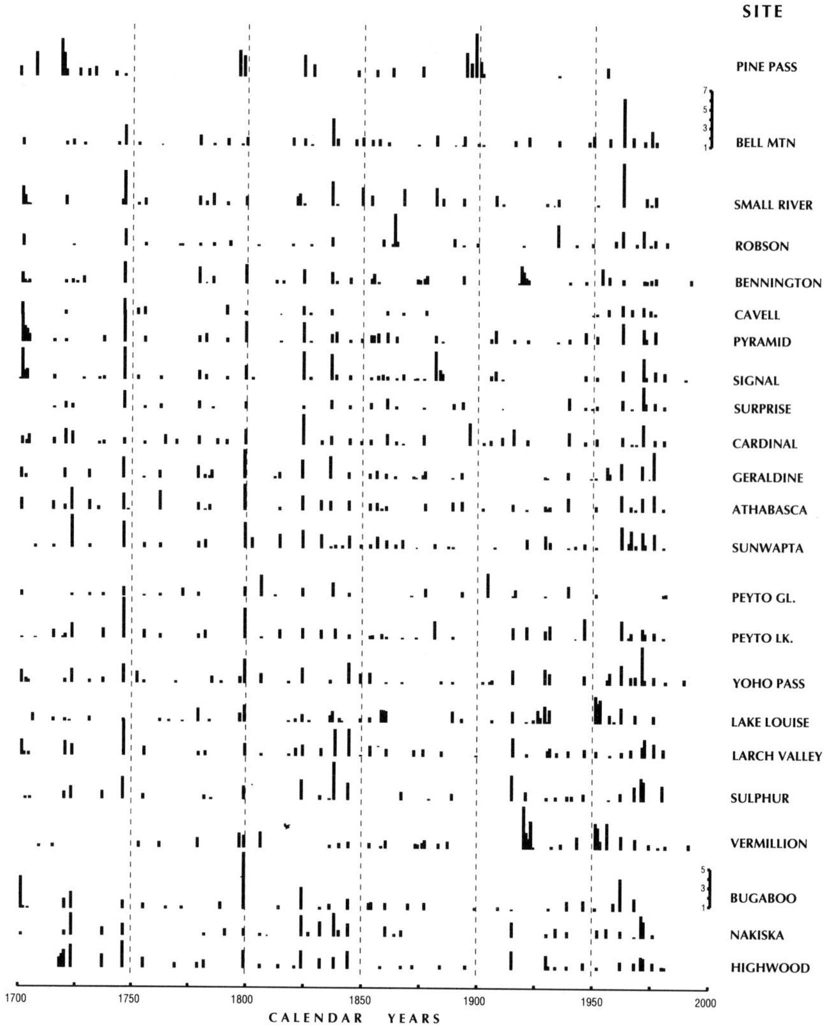

Figure 3.3 Narrow rings from *picea* chronologies in the Canadian Rockies, 1700–1990. Only rings > 1 standard deviation from the local mean of the series are shown. The height of the bar is scaled in standard deviation units from the Cofecha chronology for the site. The sites are arranged in an approximately north–south sequence.

indicate that careful site selection is critical to maximize the climatic signal in these records.

One of the principal problems in palaeoenvironmental studies is separation of the local and regional components of the record at a site. Tree-ring chronologies of 300–400 years can be obtained relatively easily for many sites in the Rockies and the data shown in Figures 3.2 and 3.3 allow (i) the identification of sites at which a strong regional signal is present and (ii) the assessment of the characteristics of such sites, thereby allowing better sample site selection in future network development. The limited number of longer chronologies makes it more difficult to apply this methodology to the assessment of older records (i.e. prior to 1600). However, it seems reasonable to assume that sites which show a strong regional signal over the 1700–1980 period will also be representative in the earlier part of that record.

Although the ring-width series in these chronologies have low sensitivity (*sensu* Fritts, 1976) and high autocorrelation, there are nevertheless very strong common patterns of year-to-year variation in ring-width between sites. Correlations between Cofecha chronologies of the treeline sites usually average 0.6 to 0.7 and correlations exceeding 0.8 may be found for sites over 100 km apart. The Highwood and Pyramid sites are over 350 km apart but have a mean correlation of 0.71 for the 1700–1980 period. The smaller larch chronology network has a mean inter-site correlation of 0.78 over a maximum distance (Larch Valley–Tyrwhitt) of 100 km. Mean correlation among a comparably close grouping of six spruce sites near Jasper (Cardinal, Surprise, Geraldine, Cavell, Pyramid and Signal; see Figure 3.1) is 0.71.

Figure 3.3 shows the marker rings of the *picea* series. Using the best correlated 20 sample sites (r > 0.49, Figure 3.2) as a data base, 3 narrow rings occur at all sites (1746, 1799 and 1824). Another nine years meet the arbitrary 1 SD threshold at 75 per cent of the sites (1701, 1720, 1755, 1779, 1836, 1854, 1853, 1962 and 1972, Figure 3.3). Significant patterns in the spatial distribution of marker rings also indicate variations on a sub-regional scale, both for individual years (e.g. 1723, 1844 and 1915 are much narrower south of Athabasca whereas 1779 and 1836 are narrower north of the Icefields) and for longer periods of time (compare the regional distribution of marker rings during the intervals, 1799–1850, 1850–1900 and 1900–1950).

Pinus albicaulis chronologies

Only two *Pinus albicaulis* chronologies have been developed so far: the common period of record is shown in Figure 3.4. The trees at both sites grow on blocky east (Peyto) or south-facing (Bennington) talus slopes overlooking Little Ice Age lateral moraines. The Bennington record is developed mainly from living trees with some snag material in the early part of the record. The Peyto chronology is derived entirely from crossdated snag material lying at the base of the talus slope, adjacent to the moraine. These preliminary chronologies were developed using ARSTAN (Holmes 1992) using only negative exponential or straight-line indexing without secondary modelling to preserve the low-frequency record.

These two sites are 170 km apart on opposite sides of the Continental Divide. The mean correlation of the two chronologies between

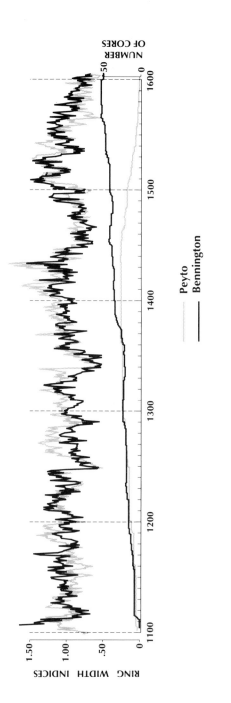

Figure 3.4 Comparison of indexed ring-width chronologies for *Pinus albicaulis* sites near Bennington and Peyto Glaciers. The upper graph shows indexed ARSTAN ring-width chronologies and the lower graph shows sample depth for the chronologies.

1200–1500 AD is 0.68 (0.63 for the Cofecha chronologies). The coincidence of both the long-term trends of these records and some of the fine detail (e.g. the abrupt ring-width reductions between 1246 and 1248 or 1443 to 1445) are particularly striking. This indicates a strong regional component to both chronologies with some local and unique elements in each record. When the chronologies differ it is difficult to tell which site (if either) carries the regional signal without comparison to other long records such as the Athabasca chronology (see below).

These two chronologies cover much of the earlier part of the last millennium beyond the range of routine 300–400 year chronologies from living trees and indicate the potential for snag material to provide critical information about this timeframe. Studies of these snag populations also demonstrate the timing and duration of changes in treeline position over this period; for example treelines at Athabasca Glacier appear to have been above present levels during the fourteenth to the seventeenth centuries and there is limited evidence of higher treelines *c.* 1000 yr BP (see Luckman, 1994).

Long tree-ring chronologies

Figure 3.5 summarizes the present development of long chronologies and indicates their potential to provide a detailed palaeoenvironmental record in the Canadian Rockies. The chronologies shown are ARSTAN ring-width chronologies (similar to and including those from Figure 3.4) that have been smoothed with a 25-year running mean to emphasize long period change over the last 1,000 years. Two of the three *picea* chronologies (Peyto and Robson) are based mainly on snag material recovered from glacier forefields and overridden by LIA glaciers. The two *pinus* chronologies are repeated from Figure 3.4 and the *larix* chronology shown combines the records from the six sites investigated by Colenutt (1992) to increase replication in the earlier part of the record. The moraine chronologies and dated periods of glacier activity are from Osborn and Luckman (1988) and Luckman (1993).

The clear synchronicity in the variations in ring-width patterns in three different species at sites spread over several hundred kilometres indicates a strong regional signal that reflects climatic forcing. More extensive climate calibration studies, possibly using regional climate data bases, are needed to define the exact nature of this signal, but the treeline location of these sites and previous work (e.g. Jacoby and Cook, 1981; Colenutt and Luckman, 1991) suggest that it is basically a temperature record. The obvious coincidence of narrower ring-widths with periods of glacier advance in the last two centuries is strong confirmation of this interpretation. However, although considerable further work is needed before quantitative proxy climate records are derived from these data, they do permit some general observations about climate over the last millennium. The tree-ring record in Figure 3.5 clearly indicates that there have been no extended periods (> 100

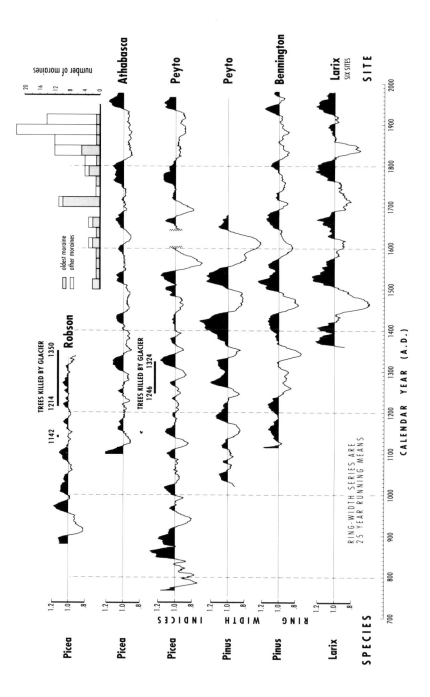

Figure 3.5 (opposite) Long tree-ring chronologies and glacier fluctuations over the last millennium in the Canadian Rockies. The six indexed ring-width chronologies plotted have the same vertical scale and are smoothed with a 25-year mean. Periods when tree-ring indices exceed the mean for each record are shaded. Two Peyto *picea* chronologies are shown: 1630–1970 from Schweingruber's Peyto Lake site and *c.* 800–1600 from Reynolds' (1992) Peyto Glacier chronology, the latter being mainly snag material. The *larix* chronology is the mean of six *larix* chronologies (Colenutt, 1992). Dates of Little Ice Age moraines are based on Luckman (1986) plus more recent data for Peyto and Robson Glaciers. The lightly shaded columns refer to data for the outermost moraine at each site; the open histograms include all dated moraines (diagram revised from Luckman, 1993).

years) of suppressed or favourable conditions for ring-width growth over the last millennium. The pattern of variation seen in the last 200–300 years (and mirrored in the glacial record) extends back to at least 1200 AD. Prior to that time the signal is less clear, possibly due to the limited sample base of the records presently available prior to 1100 AD. Inspection of Figure 3.5 indicates periods of suppressed growth in several of these records, possibly in the mid-900s, but particularly *c.* 1140–1160; 1230–1250s; 1330–1350s; 1440–1500; *c.* 1580–1620; *c.* 1690–1750 and for most of the nineteenth century. Low replication in the *larix* chronology pre-1550 and both old Peyto chronologies after 1550 (see Figure 3.4) may exaggerate the amplitude of response in these records. Nevertheless, the suppression events of the late 1300s, 1400s and 1500s are well marked, synchronous, and clearly comparable to the better replicated eighteenth- and nineteenth-century events. These ring-width records indicate that there may have been two or three periods of glacier advance between 1300 and 1700 for which well-dated evidence has not yet been recovered. Work in progress by Smith and McCarthy (1991) suggests moraine evidence of this age may be preserved in front of small glaciers in the Kananaskis area (between Sunshine and Highwood, Figure 3.1).

CONCLUDING REMARKS

The preliminary results presented above demonstrate the high potential of tree-ring series to provide precisely dated, annual proxy climate data that span the last millennium in the Canadian Rockies. Ecotonal sites containing long-lived species are optimal; in addition to the treeline record, studies at the lower forest margin using *Pseudotsuga menziessii* (e.g. Robertson and Jozsa, 1988; Drew, 1975) or *Pinus flexilis* (G. MacDonald, personal communication, 1992) may yield complementary records more sensitive to precipitation. Although the records developed to date only utilize ring-width, future chronologies may involve a greater range of data such as densitometry and isotope dendroclimatology once the chronologies have been established and appropriate calibration studies carried out. The episodic glacier record provides an independent verification of the interpretation of the ring-width records. These preliminary data suggest that the level and pattern of climate variability during the eighteenth and nineteenth centuries extends back over the last millennium and that the 'Little Ice Age' probably included several periods of glacier advance for which evidence is not preserved in the morphological record. Parallel high-resolution palaeoenvironmental records could be developed from sedimentological or palaeolimnological studies that complement or further extend these records (e.g. Leonard, 1986; Cumming and Smol, 1993).

Prediction of future climate trends must be based on a detailed under-standing of the controls and variability of natural climate systems. This case study demonstrates the potential of alpine areas to provide the detailed palaeoenvironmental records needed for the study of climate variability: these historical studies benchmark the natural climate system and produce graphic evidence of the response of biological and physical systems to past changes. They also supply a key to the understanding and modelling of the response of these systems to future, probably anthropogenically driven climate changes. The vertical zonation of life zones and geomorphic systems in mountains allows the development of complementary proxy climate records from a variety of environments in close juxtaposition. This provides opportunities for the independent verification of palaeoenvironmental reconstruction that are rarely available in more homogeneous environments where perhaps only one type of high resolution record is available. Utilization of the proxy data records from several natural archives may also reveal information about different components (e.g. seasons) of the past climate record allowing a more complete reconstruction of these former environments. Alpine areas are therefore ideally suited for the development of diverse, sensitive, high-resolution, palaeoenvironmental records with appropriate timeframes for global change studies that should be fully exploited and targeted by palaeoclimate studies within IGBP.

ACKNOWLEDGEMENTS

This research programme in the Canadian Rockies has been supported by grants from the Natural Sciences and Engineering Research Council of Canada and by the University of Western Ontario. The assembly of tree-ring data bases is labour intensive and I would particularly like to thank the following for measurement of tree-ring data; I.M. Besch, D. Boyes, M.E. Colenutt, S. Daniels, G.W. Frazer; Geography 312, 1990–2; J.P. Hamilton, R. Heipel, T. Kavanagh, P.E. Kelly, D.P. McCarthy, W. Quinton, J.R. Reynolds, B. Schaus, J. Seaquist, E. Seed, D. Smith, C. Somr, A. Tarrusov, V. Wyatt and R. Young. Several of the preceding, plus F.F. Dalley, C.J. Rowley and D.C. Luckman, also assisted in the field. M.E. Colenutt, L.A. Jozsa (Forintek Canada Corporation), J.R. Reynolds and F.H. Schweingruber provided tree-ring chronologies. I thank Parks Canada and the BC and Alberta Provincial Parks for support and the granting of research permits: Gordon Shields, Geography, UWO, for the cartography, and M. Beniston for the invitation to contribute to this volume.

REFERENCES

Bradley, R.S. and Jones, P.D. (1992a) 'When was the "Little Ice Age"?', in T. Mikami (ed.), *Proceedings of the International Symposium on the Little Ice Age Climate*, Department of Geography, Tokyo Metropolitan University, Japan.

—— and —— (eds) (1992b) *Climate since A.D. 1500*, New York, Routledge.

Colenutt, M.E. (1992) 'An investigation into the dendrochronological potential of alpine larch', unpublished M.Sc. thesis, University of Western Ontario.

Colenutt, M.E. and Luckman, B.H. (1991) 'Dendrochronological studies of *Larix lyallii* at Larch Valley, Alberta', *Canadian Journal of Forest Research*, 21: 1222–33.

Cumming, B.F. and Smol, J.P. (1993) 'Development of diatom-based salinity models for paleoclimatic research from lakes in British Columbia, Canada', *Hydrobiologica*, 269/270: 176–96.

D'Arrigo, R. and Jacoby, G.C. (1992) 'Dendroclimatic evidence from northern North America', in R.S. Bradley and P.D. Jones (eds), *Climate since A.D. 1500*, Routledge, New York.

Drew, L.G. (ed.) (1975) 'Tree-ring chronologies of Western America, VI. Western Canada and Mexico', Laboratory of Tree-ring Research, Chronology Series 1, University of Arizona, Tucson.

Fritts, H.C. (1976) *Tree Rings and Climate*, Academic, London.

Gardner, J.S. and Jones, N.K. (1985) 'Evidence for a neoglacial advance of the Boundary Glacier, Banff National Park', *Canadian Journal of Earth Sciences*, 22: 1753–5.

Grove, J. (1987) *The Little Ice Age*, Cambridge University Press, Cambridge.

Haeberli W. (1993) 'Accelerated glacier and permafrost change in the Alps' (this volume).

Heusser, C.J. (1956) 'Postglacial environments in the Canadian Rocky Mountains', *Ecological Monographs*, 26: 253–302.

Holmes R.W. (1992) *Dendrochronology Program Library*, Installation and Program Manual (January 1992 update), unpublished manuscript, Tree-Ring Laboratory, University of Arizona, Tucson, 35 pp.

Jacoby, G.C. and Cook, E.R. (1981) 'Past temperature variations inferred from a 400-year tree-ring chronology from Yukon Territories, Canada', *Arctic and Alpine Research*, 13: 409–18.

Jacoby, G.C. and D'Arrigo, R. (1989) 'Reconstructed northern hemisphere annual temperature since 1671 based on high-latitude tree-ring data from North America', *Climatic Change*, 14: 39–49.

Jacoby, G.C., Ivanciu, I.S. and Ulan, L.D. (1988) 'A 263-year record of summer temperature for northern Quebec reconstructed from tree-ring data and evidence of a major climatic shift in the early 1800s', *Palaeogeography, Palaeoclimatology, Palaeoecology*, 64: 69–78.

Leonard, E.M. (1986) 'Use of lacustrine sedimentary sequences as indicators of Holocene glacial history, Banff National Park, Alberta, Canada', *Quaternary Research*, 26: 218–31.

Luckman, B.H. (1986) 'Reconstruction of Little Ice Age events in the Canadian Rockies', *Geographie physique et Quaternaire*, 40: 17–28.

—— (1988a) '8000-year-old wood from the Athabasca Glacier', *Canadian Journal of Earth Sciences*, 25: 148–51.

—— (1988b) 'Dating the moraines and recession of Athabasca and Dome Glaciers, Alberta, Canada', *Arctic and Alpine Research*, 20: 40–54.

—— (1990) 'Mountain areas and global change – a view from the Canadian Rockies', *Mountain Research and Development*, 10: 183–95.

—— (1992) 'Glacier and dendrochronological records for the Little Ice Age in the Canadian Rocky Mountains', in T. Mikami (ed.), *Proceedings of the International Conference on the Little Ice Age Climate*, Tokyo Metropolitan University.

—— (1993) 'Glacier fluctuation and tree-ring records for the last millennium in the Canadian Rockies', *Quaternary Science Review*, 12.

—— (1994) 'Evidence for climatic conditions between ca. 900-1300 A.D. in the southern Canadian Rockies', *Climate Change*, 30: 1–12.

Luckman, B.H. and Colenutt, M.E. (1992) 'Developing tree-ring series for the last millennium in the Canadian Rocky Mountains', in T.S. Bartolin, B.E. Berglund, D. Eckstein, and F.H. Sweingruber (eds), *Tree Rings and Environment, Proceedings of the International Dendroecological Symposium*, Ystad, South Sweden, 3–9 September 1990. Lundqua Report 34, Department of Quaternary Geology, Lund University.

Luckman, B.H., Harding, K.A. and Hamilton, J.P. (1987) 'Recent glacier advances in the Premier Range, British Columbia', *Canadian Journal of Earth Sciences*, 24: 1149–61.

Luckman, B.H., Holdsworth, G. and Osborn, G.D. (1993) 'Neoglacial glacier fluctuations in the Canadian Rockies', *Quaternary Research*, 39: 144–53.

Luckman, B.H., and Osborn, G.D. (1979) 'Holocene glacier fluctuations in the middle Canadian Rocky Mountains', *Quaternary Research*, 11: 52–77.

McLennan, J. (1993) 'Development of a new spruce ring-width chronology for the Bennington Valley, British Columbia', unpublished BA Thesis, University of Western Ontario.

Maisch, M. (1992) 'How fast will alpine glaciers disappear? Estimates and speculations on the glacier retreat in the next century', *Swiss Climate Abstracts*, special issue (Berne, September 1992) for the International Conference on Mountain Environments in Changing Climates, Davos, Switzerland, 11–16 October, 1992.

Matthes, F.E. (1939) 'Report of the Committee on Glaciers, April 1939', *Transactions of the American Geophysical Union*, 20: 518–23.

Nesje, A. and Johannesson, T. (1992) 'What were the primary forcing mechanisms of high frequency Holocene glacier and climate fluctuations?', *Holocene*, 2: 79–84.

Osborn, G.D. and Luckman, B.H. (1988) 'Holocene glacier fluctuations in the Canadian Cordillera (Alberta and British Columbia)', *Quaternary Science Reviews*, 7: 115–28.

Penck, A. and Bruckner, E. (1909) *Die Alpen im Eiszeitaler*, Tauchnitz, Leipzig.

Porter, S.C. and Denton, G. (1967) 'Chronology of neoglaciation in the North American Cordillera', *American Journal of Science*, 265: 177–210.

Reynolds, J.R. (1992) 'Dendrochronology and glacier fluctuations at Peyto Glacier, Alberta', unpublished B.Sc. thesis, University of Western Ontario.

Robertson, E.O. and Josza, L.A. (1988) 'Climate reconstruction from tree rings at Banff', *Canadian Journal of Forest Research*, 18: 888–900.

Schweingruber, F.H. (1988) 'A new dendroclimatic network for western North America', *Dendrochronologia*, 6: 171–80.

Smith, D.J. and McCarthy, D.P. (1991) 'Little Ice Age glacial history of Peter Lougheed Provincial Park, Alberta', *Program with Abstracts*, Canadian Association of Geographers Annual Meeting, Queen's University, Kingston, Ontario.

Stone, P.B. (ed.) (1992) *The State of the World's Mountains. A Global Report.* Zed Books, London.

Torsnes, I., Rye, N. and Nesje, A. (1993) 'Modern and Little Ice Age equilibrium-line altitudes on Outlet Valley Glaciers from Jostedalsbreen, western Norway: an evaluation of different approaches to their calculation', *Arctic and Alpine Research*, 25: 106–16.

Young, G.J. and Ommanney, C.S.L. (1984) 'Canadian glacier hydrology and mass balance studies; a history of accomplishments and recommendations for future work', *Geografiska Annaler*, 66A: 169–82.

Zumbühl, H.L. (1980) *Die Schwankungen des Grindelwaldgletscher in den historischen Bild-und Schriftguellen des 12 bis 19 Jahrhunderts*, Birkhauser Verlag, Basel.

4

CLIMATE IN EUROPE DURING THE LATE MAUNDER MINIMUM PERIOD (1675–1715)

C. Pfister

INTRODUCTION

Palaeoclimatic studies help us to gain valuable perspectives and insights into the nature and possible origins of present-day climatic variations that are beyond the reach of instrumental weather observations and allow us to view the relatively short instrumental record in the context of longer-timescale variability. In view of the growing concern over a CO_2 enhanced greenhouse effect, there is an urgent need to verify this anthropogenic signal projected by climate models as early as possible, by means of observational statistics (Schönwiese, 1990). This implies a better understanding of natural climatic variability and forcing on two fundamental timescales: decadal and secular. On the decadal timescale, there are natural climatic fluctuations that centrally involve air–sea–ice interactions, solar and volcanic forcing, as well as man-induced changes to the atmosphere and to the land surface. On the century timescale substantial global warming is likely to occur due to the cumulative effects of anthropogenic forcing (Stocker and Mysak, 1992). It is well known that the primary impacts of climate on society result from extreme events on short timescales from hours to years, such as storms, floods, killing frosts, heat and cold waves, and droughts, rather than from changes in the mean climate, and that the sensitivity is relatively greater the more extreme the event (Kates *et al.*, 1985). Therefore, palaeoclimatic studies should also examine how changes in the frequency and severity of extreme events are related to changes in average conditions (Mitchell *et al.*, 1990).

The ultimate tool for unravelling the patterns and causes of decade to century-scale climate variability and for assessing their impacts upon contemporary societies would be a global array of accurately dated palaeoclimatic time series (Rind and Overpeck, 1993). Such a data base should at least:

- include a substantial part of the Holozene
- cover the entire globe

- be strongly controlled by climatic variables
- have a seasonal or monthly time resolution
- cover all seasons of the year
- disentangle temperature and precipitation
- adequately describe outstanding anomalies and natural hazards.

Needless to say, only a small number of the Holocene time series will have the accuracy needed to study decadal-scale variability, at least on a global level. However, by limiting the investigation to Europe and by focusing upon the period from the High Middle Ages to the beginning of meteorological network observations, there is a good chance that a sufficient amount and density of high-resolution proxy data might be available for a detailed reconstruction of climatic changes in time and space. In a first step efforts should be directed towards integrating all kinds of pertinent data into a large multi proxy data base. This involves data from two different bodies of evidence:

1 Continuous long chronologies of high resolution proxy data such as tree-rings, varves, or ice-cores that are drawn from natural archives are needed for documenting long-term changes for the spring–summer period.
2 Descriptive information and various kinds of proxy data that are drawn from historical archives are needed to document the frequency and nature of anomalies and long-term changes in winter climate.

In order to synthesize and integrate these two kinds of evidence their strengths and weaknesses should be properly assessed. Proxy data from natural archives allow formal calibration with instrumental series and the derivation of mathematical transfer functions (e.g. Briffa, in Frenzel et al., 1994). Many series go back in time beyond the present millennium, some even beyond the ice age. However, most of them are related to parts of the year – mostly subperiods of the growing season – and they are bound to the ecological parameters of their specific site.

Documentary data have an absolute dating control, a high (seasonal, monthly, or daily) time resolution and they allow disentangling the effects of temperature and precipitation. Moreover observations are made all year round; the data are not bound to a specific season. The evidence for the winter half-year is of particular importance, because the maximum temperature increase due to global warming is believed to occur in the mid-latitude and polar winter (Schlesinger, 1990) and because documentary data are often the only and certainly the most accurate proxies for assessing conditions in the time from October to April. Finally, as documentary data reflect the reaction of society to climatic events, they are highly sensitive to anomalies and natural hazards and often include short-lived events such as violent hailstorms or severe killing frosts. The more extreme an event, the more often and the more fully it is described.

On the other hand, documentary data are known to have a number of weaknesses. First, many of them are discontinuous, heterogeneous, and may be biased by the selective perception of the observer. Second, the available compilations of documentary data are quite often made up of an incoherent mix of reliable and unreliable data and the entries are not fully documented; this makes an analysis costly and time consuming. Third, the task of integrating material from all over the continent into a coherent body of evidence involves understanding the full range of languages and regional dialects which are spoken in Europe, and this not only for the present, but also for the historic past. Finally, according to the nature of the data the mathematical techniques of elaboration are relatively simple and robust (Banzon *et al.*, 1992), because the data, except those that are available in the form of continuous time series (e.g. vine harvest dates), are not suited for sophisticated statistical analyses.

This paper outlines the broad lines of a pilot project called Euro-Climhist which aims at realizing some of the objectives mentioned earlier. It is part of the Program 'European Palaeoclimate and Man since the Last Glaciation' set up by the European Science Foundation (ESF). It involves a number of geographers, historians, meteorologists and dendrochronologists aiming at reconstructing past climate variability over Europe. Some of the background and further details of this work may be found in Frenzel *et al.* (1992), Pfister (1992) and Frenzel *et al.* (1994).

The Maunder Minimum period was selected as an appropriate time window for testing the feasibility of such an approach following a recommendation of the IGBP PAGES Program (Eddy, 1992).

THE LATE MAUNDER MINIMUM

One of the most outstanding fluctuations in our millennium on a decadal scale is known to have occurred at the end of the seventeenth century. The seven decades from 1645 to 1715 AD are known for their extra low numbers of sunspots. Eddy (1977) proposed that this period of lowered sunspot activity, which was called the Maunder Minimum, correlated with the period of generally lowered temperature and in this way he included it in the Little Ice Age concept of Lamb.

However, the correlation between sunspot minimum and the proposed period of climatic cooling is not convincing and straightforward (Stuiver and Braziunas, 1992; Mörner, 1992) and as a result of recent findings from many parts of the world, it is doubtful whether this was really a global cooling event (and, if so, its duration in time) or whether there was a compensating redistribution of heat between different parts of the globe (Bradley and Jones, 1992; Mikami, 1992). It is in fact not even certain whether the period was thoroughly cold in all parts of Europe, that is:

- did the cooling affect the whole of the European continent from the Iberian Peninsula to the Urals or was there a compensating redistribution of heat between different parts of the continent?
- did the cooling mainly affect particular seasons of the year or was it uniformly distributed throughout the year?
- was the cooling associated with wet or with dry conditions?
- how are these patterns related to the global circulation and to natural forcing factors (solar, volcanic, air–sea interactions) in the climate system?
- how exceptional was this period in the wider context of the last centuries?
- how (if at all) did the frequency and severity of outstanding anomalies and natural hazards change?

Given the limitations in time and funding it was decided not to investigate the entire period of the Maunder Minimum, but to attempt a multi-season reconstruction for the latter part of this period, which, in western Europe, is known to include a number of unusually cold years and severe anomalies at the end of the seventeenth century and a subsequent warming period at the beginning of the eighteenth century.

A SURVEY OF THE EVIDENCE

In the following the main sources are briefly described:

In the marginal environment of Iceland, variations of the weather played a significant role in people's lives who mostly depended on animal husbandry and thus on the annual hay crop. This interest in weather, plus a high level of literacy, and a tradition of writing, has resulted in a wide variety of documents in which the nature of the weather over a given year, season, month, or day is recounted in some detail (Ogilvie, 1992).

In Denmark in 1670, King Christian V ordered entering of logbooks several times a day on the naval ships. The enterings should among other things consist of wind and weather observations. Many of these logbooks have survived until today. The bulk of data included in the data base is from the Oresund area (Frydendahl et al., 1992).

In England, several compilations of weather data are known to exist (Jefferey, 1933; Jones et al., 1984). If the known manuscript sources are included, daily observations of wind and weather are available for England since 1669. Changes in monthly temperatures are known from the long series from Central England (Manley, 1974). The history of weather and rainfall patterns (Siegenthaler, in Frenzel et al., 1994) draws mainly from printed sources prior to 1697. Between 1697 and 1715, time series of rainfall measured at Kew near London (Wales-Smith, 1971) were used.

A verified compilation of documentary evidence for the winter months was undertaken by Buisman (1984) for the Netherlands. A long series of winter temperatures from 1634 includes administrative proxy data concerning the freezing of Dutch canals (1634–1734) and the well known series of temperature measurements carried out at De Bilt (van den Dool *et al.*, 1978).

A unique instrumental diary was recently discovered in France. Louis Morin (1635–1715), who was a physician in Paris, was perhaps the most outstanding among the pioneers of early instrumental meteorology. His meteorological journal covers 48 years (1665–1713) and contains three readings of air temperature and pressure a day, as well as – among others – daily observations of the direction and speed of clouds and a record of the duration and intensity of precipitation (Pfister, Bareiss, in Frenzel *et al.*, 1994).

The abundant information for Switzerland is contained in the work of Pfister (1984). It includes several series of proxy data (e.g. beginning of vine flowering, volume of vine harvests and regular observations of snow-cover) as well as observations carried out in the Alps. The full documentation is listed in Pfister (1985).

In Germany, historical climatology has a long tradition, and a new effort was recently launched by a research group at the Geographical Institute at Würzburg University (Glaser and Hagedorn, 1991). The main sources for the period 1675–1715 include two weather diaries from Tübingen and Ulm (Lenke, 1961) as well as daily observations and measurements carried out from 1677 to 1774 by the Kirch family in Leipzig, Guben and Berlin (Brumme, 1981).

The climate history research group in Brno is collecting evidence from the area of Bohemia and Moravia (Brazdil and Dobrovolny, 1992). Climate history in the Carpathian Basin mainly draws from a compilation of source material which was verified by Racz (1992).

An early instrumental diary was kept in the Silesian town of Breslau (today Wrocław in Poland) by the physician David Grebner from 1692 to 1721 (Grebner, 1722). Besides the usual observations, he focused upon wind direction, that was noted down up to three or four times a day. His readings of temperature from 1701 have been homogenized by Mlostek (1988).

Forest limit pines in northern Fennoscandia constitute a valuable source of palaeoclimatic data, as their ring-width is indicative of summer temperatures. Year-by-year values of mean summer (April–August) temperature have been reconstructed by a collaborative tree-ring project, involving a number of European dendrochronologists (Briffa, in Frenzel *et al.*, 1994; Eronen, Lindholm, Zetterberg, in ibid.). In Finland and the Baltic, one has to draw upon proxy information (harvest fluctuations, ice-break-up dates of ports and rivers, and direct and indirect phenological indicators) concerning the growth and harvest of rye, which was by far the most important crop (Vesajoki, in Frenzel *et al.*, 1994; Tarand, in ibid.).

Russian chronicles are the most important sources for the history of natural events in the Russian plain until the late seventeenth century (Borisenkov, 1992). They include the compilation of more ancient sources which makes it difficult to distinguish between contemporary and non-contemporary observations.

The documentation of weather patterns in the Mediterranean is of paramount significance for reconstructing a history of weather patterns in Europe, but this region is not so well covered with data as it would deserve. Records from Spain are often connected to the occurrence of rainfall anomalies. In times of continuous droughts or rainy periods, it was a tradition to go on pilgrimages to local patron saints, and when the desired change in the weather had taken place, to sing a Te Deum (Tarrats, 1971, 1976, 1977; Font Tullot, 1988).

Evidence for Italy was provided from several sources. The well known climate history databank in Padua contains chronological information on weather and climate related environmental events (Camuffo, Enzi, in Frenzel *et al.*, 1994). Weather information for the central and eastern Mediterranean was mainly extracted from records found in the Venetian Archives and in the State Archive in Palermo (Grove and Conterio, in Frenzel *et al.*, 1994). In spite of Greece's long history and extensive monastery records, very little work has been done to archive extreme events for this country (Repapis *et al.*, 1989; Chrysos, in Frenzel *et al.*, 1994).

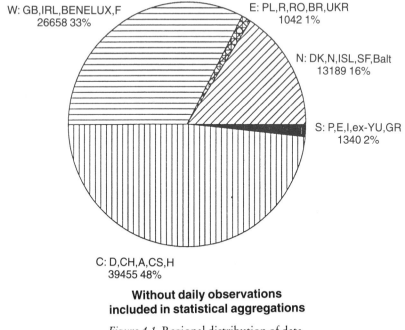

W: GB,IRL,BENELUX,F
26658 33%

E: PL,R,RO,BR,UKR
1042 1%

N: DK,N,ISL,SF,Balt
13189 16%

S: P,E,I,ex-YU,GR
1340 2%

C: D,CH,A,CS,H
39455 48%

**Without daily observations
included in statistical aggregations**

Figure 4.1 Regional distribution of data.

Results for western Europe and the Mediterranean originating from dendroclimatic research carried out at Marseille under the guidance of the late Françoise Serre-Bachet (*et al.*, 1992; Guiot , 1992) include among others reconstructions for annual temperatures and for the April to September temperature for a couple of gridpoints in the Jones *et al.* (1985) data network.

As a whole this body of evidence yielded more than 220,000 single data entities ('observations'). They are regionally concentrated in central, western and northern Europe. Data from the Mediterranean and from eastern Europe are not adequately represented (Figure 4.1) for assessing the location of the main depressions and anticyclones in every month.

The Euro-Climhist approach, which is described in more detail in Schüle and Pfister (1992), includes the following steps:

1 Collecting and verifying manuscript and printed documentary information.
2 Standardizing this evidence by means of a coding system.
3 Providing the following operations by computer:
    (a) translation of the content without a loss of information;
    (b) full documentation for every single observation: time, place (including the altitude asl), name of observer, type of event, source, style (Julian, Gregorian, other), reliability, owner of the source;
    (c) transformation of non-Gregorian to Gregorian styles;
    (d) frequency analysis;
    (e) inclusion of instrumental observations and of climate estimates obtained from the analysis of natural proxy-data.
4 Summarizing the available information on a regional level in terms of numbers (indices) (Pfister, 1992). They should allow a comparison over time and between regions as well as the inclusion of data into palaeoclimatological, palaeoecological and historical models.
5 Mapping the most significant part of the information on different scales and levels of analysis.

An example of the provisional layout of the data listing is displayed in Figure 4.2.

The main part of the data was transformed back from the numerical code into standard expressions (Schüle and Pfister, 1992). At present decoding routines have been prepared in twelve languages – English, German, French, Italian, Danish, Norwegian, Icelandic, Finnish, Estonian, Czech, Russian, and Hungarian. Thus, every research partner can study the information in his or her own language. Key quotations are left in the original language as footnotes and explained by an English summary. For the ESF meeting of September 1992 in Berne a full set of 480 monthly and 160 seasonal weather charts was produced, plus some examples for charts at the level of five-day and ten-day periods.

1696 Winter

HU:N: mild (;[g]) R:733(Pozsony (—) T1P3) (S:RA-7411)
HU:P0: T-index ;normal {0} R: (entire Hungary T1P3) (S:RA-ind)
HU:P0: P-index ;normal {0} R: (entire Hungary T1P3) (S:RA-ind)
HU:TRNS: mild (;[g]) R:209(Erdély T1P3) (S:RA-7134)
HU:TRNS: little snow (;[g]) R:209(Erdély T1P3) (S:RA-7134)
I:P0: T-index ;normal {0} R: (entire country T1P0) (S:CE-IDX)
CH:P0: little snow R:10(Stans (452m) T1P1) (S:PF-3120)
CH:P0: T-index ;above normal {+1} R:0(several regions T1P1) (S:PF-3993)
CH:P0: P-index ;dry {-2} R:0(several regions T1P1) (S:PF-3993)
CF:M: hot (;[j,+9]) R:SM(Bystřice nad Pernštejnem (—) T1P3) (S:DO-0043)
PL:P0: warm (;[j,+9]) R:9(not sepcified T1P3) (S:SM-1000)
D:S: warm (;[j,+9]) R:DR-13-01(Franken (Landschaft) T1P3) (S:GM-16N1)[2]
D:S: no snow (;[j,+9]) R:DR-13-01(Franken (Landschaft) T1P3) (S:GM-16N1) [2]
RU:P12: see note (;[j,+9]) R:R122(рег. Онежское озеро TAPA) (S:BO-0011)[3]
RU:SW: very cold (;[j,+9]) R: (juго-запад T0P0) (S:BO-0002)[5]
RU:SW: no snow (;[j,+9]) R: (jugo-zapad Russia T0P0) (S:BO-0002) [5]
UCR:P0: no snow (;[j,+9]) R: (Ukraine T0P0) (S:BO-0011) [3] R: (Ukraine T0P0) (S:BO-0012)
      [4]
IS:P0: T-index ;cold {-2} R: (entire country T-P-) (S:TJ-IDX)
IS:W: wind.11.1.NNA;strong (;[j,+9]) R:40(Eyri (—) T-P-) (S:TJ-1501)[6]
IS:W: mainly frost.some snowfall and rain (;[j,+9]) R:40(Eyri (—) T-P-) (S:TJ-1501) [6]
IS:W: famine (;[j,+9]) R:30(Hestur (—) T-P-) (S:TJ-1604)[7]
EST:P0: early filedwork (;[j,+9]) R:K(Eestimaa (hist.) T1P6) R:L(Liivimaa (hist.) T1P6)
      (S:TK-0135)[8]
EST:P0: grain.sowing (=D_345;[j,+9]) R:K(Eestimaa (hist.) T1P6) R:L(Liivimaa (hist.) T1P6)
      (S:TK-0135) [8]

Figure 4.2 Example of Euro-Climhist printout.

REGIONAL TRENDS AND ANOMALIES

In the subsequent section regional trends and spatial anomalies are discussed. The backbone of the reconstruction relies on three regions – England, Switzerland, and Hungary – for which continuous information on both temperature and precipitation is available on a monthly level. This yields a west-to-east profile from the London area over a distance of 2,200 km to the Carpathians. The rather sporadic observations from the adjacent zones of northern Europe and the Mediterranean are needed to reconstruct the spatial pattern of anomalies.

The monthly information for England, Switzerland, and Hungary is set up in terms of Graduated Temperature Indices $GTI_{[+3,-3]}$ and Graduated Precipitation Indices $GPI_{[+3,-3]}$ (Pfister, 1992). On a seasonal level GTI and GPI are defined as the average of the monthly indices, which yields gradations of 0.3 between –3 and +3. The term 'remarkable' was applied to those seasons which had a GTI or a GPI of $> = +1$ or $< = -1$.

The average number of 'remarkable' months displays some notable features between the three countries (Table 4.1).

The following points need to be mentioned:

Table 4.1 Frequency of remarkably warm, cold, wet, and dry seasons in Iceland, England, Switzerland, and Hungary, 1675–1714

	Iceland				England				Switzerland				Hungary			
	Warm	Cold	Total	%	Warm	Cold	Total	%	Warm	Cold	Total	%	Warm	Cold	Total	%
Temp.	68	33	101	63	2	53	55	34	15	50	65	41	15	29	44	28
	Wet	Dry	Total	%	Wet	Dry	Total	%	Wet	Dry	Total	%	Wet	Dry	Total	%
Precip.	21	17	38	24	30	4	34	21	17	30	47	29	44	11	55	34

%: Percentage of remarkable seasons (all seasons = 100%)
Data: Pfister in Frenzel, Pfister, Glaeser (eds) (1994)

- The percentage of cold months decreased from west to east, in particular from Switzerland to Hungary.
- The percentage of warm months increased from England to Switzerland.
- In Hungary and in England the period was mainly wet, in Switzerland (as in France) it was prevailingly dry.

In order to highlight changes over time, the number of remarkable seasons was arranged into five-year periods for each season and the entire year (Figures 4.3 to 4.7).

Winter

From the available data (Figure 4.3), we make the following observations and draw the following conclusions:

The period of severe winters displays a shift over time from west to east. In England, it started rather abruptly after 1675 and ended the same way around 1700. No severe winter was observed over the following fifteen years (except that of 1708–9). In Switzerland the increase and decrease in the frequency of cold winters was more gradual, beginning after 1680 and ending before 1705. A seasonal examination of the evidence for Hungary and Bohemia/Moravia reveals similarities as well as contrasts to the picture obtained for western Europe. In Hungary winters were cold throughout the 40-year period; a first run of cold seasons stands out around 1685, a second, more pronounced one in the 1700s runs against the warming trend observed in western Europe. The number of severe winters increased to a maximum in 1710–14. Winter precipitation was more abundant than in France and Switzerland. In the Baltic area winter temperatures were not higher than in the previous period after 1700 as can be deduced from observations of the break-up of the ice-cover in the port of Tallinn (Tarand, 1992).

There seems to be another contrast between the situation in the mid-latitudes and that in the high latitudes. In Iceland warm winters were more common than cold ones (except from 1695 to 1699). A preliminary analysis of synoptic situations revealed that Iceland was repeatedly situated in the track of warm airflows from the southwest connected to a strong anticyclone centred at about 60°N somewhere between the Norwegian Sea and the Baltic. In this situation central Europe was dominated by cold airflows from the northeast which would explain the high number of long winter droughts.

In the Mediterranean extreme cold or long periods of snow are reported from 16 out of 25 winters between 1675 and 1700 (information is lacking from Spain after this date) (Table 4.2). The winter is listed according to the year in which January fell.

In eight years (1679, 1681, 1689, 1691, 1693, 1694, 1695, 1698) the cold air-mass reached down from central Europe to the Iberian Peninsula. In some

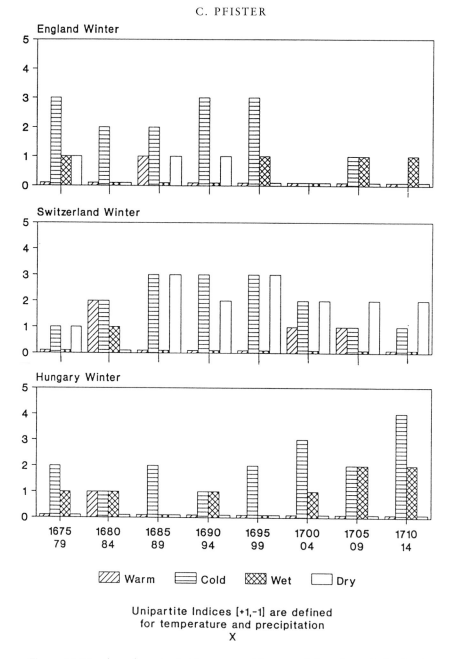

Figure 4.3 Number of unmistakably warm, cold, wet, and dry winters in England, Switzerland, and Hungary 1675–1715.

Table 4.2 Cold winters in the Mediterranean 1675–1700

Year	Iberian Peninsula	Italy	Greece
1679	C,S		R
1680	D	S (N)	**C**
1681	C,S	D (N) C (S)	
1682	**D**	**D**	C
1683	C,D	C	C
1684	**C** (N)	S	
1685	**C** (N,S)		
1686		R	
1687		**C**	
1688			
1689	C	R	
1690			
1691	C,D	C,D	
1692	S (S)		
1693	C	C (N,M)	R
1694	C	C (N,M)	
1695	C,S	R (N,M)	
1696			
1697	C	S	C
1698	C,S	R	R

C = Cold; D = Drought; R = Rainy; S = Snow; bold letters indicate enhanced intensity
Italy: N = North, M = Centre, S = South

of those winters (1681, 1691, 1693, 1694) it included Italy, in others the cold in Spain coincided with long periods of rain in Italy (1689, 1695) or in the eastern Mediterranean (1679, 1693, 1698). In three years (1684, 1685, 1692) temperatures were lowest in Italy, in 1680, 1682 and 1687 rivers froze over in Greece. In 1683 and 1697 the entire Mediterranean Basin was dominated by a cold air-mass. A systematic analysis of the synoptic situation in these outstanding winters would be both possible and rewarding.

In Castilia winters and springs were extremely dry in the late 1670s and early 1680s. In 1680 it did not rain until the beginning of May, many cattle died from hunger, and in the Cuenca area people had to obtain drinking water from snow fields in the mountains. In 1682, not a single drop of rain fell during the spring in Madrid. In 1683 the drought prevailed until the beginning of May.

In the eastern Mediterranean an extreme drought occurred in winter and spring 1713–14. The drought which caused a widespread failure of the harvests very likely continued during the summer months in the normal way (Grove and Conterio, in Frenzel *et al.*, 1994). In Switzerland the same winter 1713–14 is among the ten driest of the last five hundred years. At the end of this long dry spell rivers and lakes fell to extremely low levels (Pfister, 1985).

The 1680s were the period of highest variability in western and central

Winter 1696

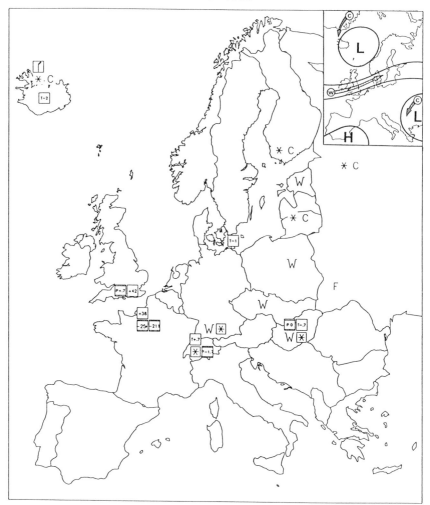

Figure 4.4 Seasonal weather chart and reconstruction of mean surface pressure pattern for winter 1695–6.

Europe which seems to be connected to a frequent redistribution of heat on the continent. The winter 1680, to take an example, was extremely dry in Castilia, rather warm in large parts of central Europe including Poland, but very cold and snowy in Russia and rainy in Greece. In 1684, to take another example, the Thames was covered with ice and fairs were held on the frozen river. However, the winter 1685–6, just two years after the Great Frost of 1683–4 is the fourth warmest from 1659 to 1973 (Manley, 1974).

Moreover, the period included multisecular cold extremes. Considering the freezing of lakes in the Swiss Middleland, the winter of 1694–5 (together with that of 1572–3) was the severest in the last 475 years in central Europe. 1684–5 was colder than any subsequent winter since the beginning of thermometrical measurement, and the severity of 1683–4 may be compared to 1829–30. On the other hand, even this period of extreme winter cold includes isolated mild and moist 'westerly' winters (Figure 4.4).

In France the famous winter of 1708–9 seems to have had the lowest minimum temperatures. The progression of the arctic air across the country from north to south from 5 January to 7 January was reconstructed with a time resolution of three hours (Lachiver, 1991). In northern Italy this season was the most severe one during the last 500 years with exceptionally heavy snowfall of 1.5 m in some areas. Large rivers and the Venice Lagoon froze over and were crossed by carriages and artillery; wells, wine, and oil froze; people, animals and trees died (Salmelli, 1986; Camuffo and Enzi, in Frenzel et al., 1994). The cold air-mass extended from Finland and Russia across the continent beyond the Pyrenees, but it did not invade the eastern Mediterranean, because no evidence has been found of any winters at the turn of the seventeenth century which were hard enough to cause widespread damage to olive trees and failure of the oil crop (Grove and Conterio, in Frenzel et al., 1994).

Spring

In England spring temperatures plunged after 1686 and stayed on a low level till 1701. Between 1685 and 1703 no spring exceeded the 1901–1960 average temperature. The situation is similar in Switzerland (Figure 4.5).

After 1705, the spring seasons warmed up in England, whereas in central Europe this season continued to be considerably colder than today. Likewise there is no indication for a warming of springs in Finland after 1700, if we consider the ice break-up of the Tornionjoki river (Vesajoki in Frenzel et al., 1994). In Iceland cold springs were somewhat more frequent than warm ones, except in the five-year period 1690–4.

The prevalence of cold springs, typical for western and northwestern Europe, was less pronounced in Hungary.

Wet springs were more frequent than dry ones in England (Siegenthaler in Frenzel et al., 1994) and in Italy (Camuffo and Enzi in ibid.) and in Hungary. In northern France and in Switzerland this season was rather dry.

The raininess of spring 1693 was outstanding by modern standards. The anomaly is well documented in England, France, Southern Germany, Switzerland, Silesia, Hungary, and the rainy spell included even northern Italy down to Bologna. The spring of 1698 which in western Europe was one of the coldest on record, was associated with a very long spell of rainfall in central Italy in April and May of that year. This situation may be related to

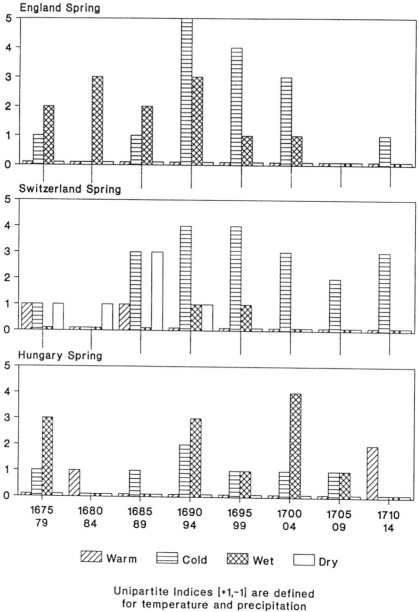

Figure 4.5 Number of unmistakably warm, cold, wet, and dry springs in England, Switzerland, and Hungary 1675–1715.

a stable cyclone over the central Mediterranean associated with a blocking northwesterly situation in Europe.

In the 1680s Spain suffered from droughts in five out of ten springs, in 1683 the weather in Catalonia remained cold and dry until mid-May, in 1684 long spells of rain affected the harvests. The lack of records indicating prayers for rain in the early 1690s suggests that rainfall was more abundant, or that temperatures were somewhat lower. In 1697 the pattern of frequent droughts was resumed. Cold anomalies were restricted to the eastern Mediterranean. In 1687 spring was extremely chilly in Greece. Little lakes in the northern part of the country remained frozen until the end of April. Large amounts of snow fell in March and April, 1700, and the harvest of barley was extremely delayed.

Summer

In England summers became slightly cooler from the second half of the 1670s until the mid 1690s. Wet seasons prevailed from 1680 to 1696, just two summers in that period (1676, 1684) being below the 1901–60 average (Figure 4.6).

In Switzerland and in France summer temperatures and rainfall did not deviate from the trends of the present century, and the number of anomalies was rather small.

Summers in Hungary prior to 1680 and after 1700 (e.g. 1701, 1709) were often warm, often rainy (e.g. 1708, 1710), and sometimes very dry (1684, 1710).

It needs to be pointed out that no warm and dry summer occurred in the entire region from 1684 to 1705. This suggests that stable summer anticyclonic situations did not develop over central and western Europe and that the Azores anticyclone was very weak during that time. As a second characteristic of summer climate at that time, the relatively rare occurrence of cold extremes needs to be stressed. This could be associated with a rather low level of blocking. The most pronounced low temperature anomaly occurred in 1675. In that summer, western and central Europe were flooded by unseasonably frigid air masses. In the Alps it snowed down to 900 m in July and even in August. In Corfu the salterns were damaged by 'copious rain' in June, and in July the Venetian Governor complained that 'a very sharp wind which takes off the burning heat from the sun' had kept the temperature down so that 'the heat of the summer has not been felt yet' (Grove and Conterio, in Frenzel et al., 1994). In western and central Europe the vine was harvested at the end of October or in early November (Le Roy Ladurie, 1972). This is equivalent to 1628 or 1816 which were the coldest summers of the last 500 years.

In Iceland, summers were prevailingly warm and rather dry during the 1690s, it can be concluded from the coded data in Euro-Climhist which points to the frequent manifestation of an anticyclone not far from the island.

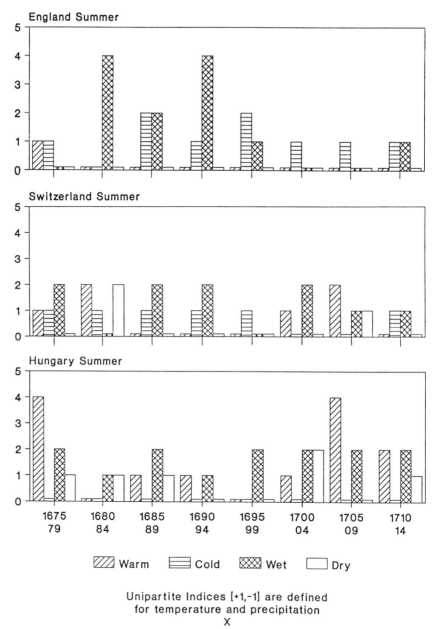

Figure 4.6 Number of unmistakably warm, cold, wet, and dry summers in England, Switzerland, and Hungary 1675–1715.

From the analysis of melt-water features in the South Greenland ice, Kameda *et al.* (1992) have concluded that summer temperatures around 1680–1710 were about 0.5°C below the average summers of the reference period (AD 1550 to 1988). In Northern Fennoscandia the entire period 1675–1715 was marked by low summer temperatures when compared to a modern reference period (Briffa in Frenzel *et al.*, 1994), but the 1690s had positive anomalies (Briffa, 1992). From the tree-ring record that goes back to 1204 it was concluded that the weather in Finland since the Middle Ages has never been as unfavourable as it was in 1696. Some trees suffered so much during those severe years that many years elapsed before they regained their earlier growth rate (Eronen *et al.*, in Frenzel *et al.*, 1994).

Over the entire period dry warm summers were more frequent than rainy ones in Italy (Camuffo and Enzi, in Frenzel *et al.*, 1994). Climatic anomalies in the Mediterranean included heavy rainfall in summer. During a sequence of years in the 1690s even harvests in Sicily suffered from heavy summer rain. In 1684 the walls of the fortress of Suda in northern Crete were damaged from the continuous rain (Grove and Conterio, in Frenzel *et al.*, 1994). In 1693 the season on the Spanish Meseta was cloudy and cool until mid-July, so that prayers for clear weather were held. Likewise in July 1695, prayers were held in Galicia for good weather because of prevailing cloudiness.

Autumn

In England autumns were prevailingly cold until 1700, and the tendency towards colder conditions increased in the 1690s. Over the following decade temperatures reached the level of the twentieth century. Autumns in Switzerland and Hungary did not reveal any clear tendencies, except that this season was somewhat rainy over most of the period in Hungary (with an extreme peak in 1709). Iceland had a net excess of warm months from 1690 to 1715.

Spain suffered from drought in 1676, 1683, and 1685. In central Europe the autumn of 1680 was extremely warm. In October and November pears and apples were again flowering and fruits of the size of a walnut developed which points to a warm anomaly of 4°C or more. Drought was extreme. It hardly rained from September to early December (Pfister, in Frenzel *et al.*, 1994). During that time Castilia and Catalonia were suffering from rain.

Summary

Figure 4.7 displays the annual frequencies of remarkable months. A first wave of cold seasons occurred in 1675–9, a second one, more severe and longer lasting, began in 1685–9 and reached a maximum in the mid-1690s. For those living at the time, 1695 was the most memorable year, to judge from the severity of anomalies in many parts of Europe (Lindgren and

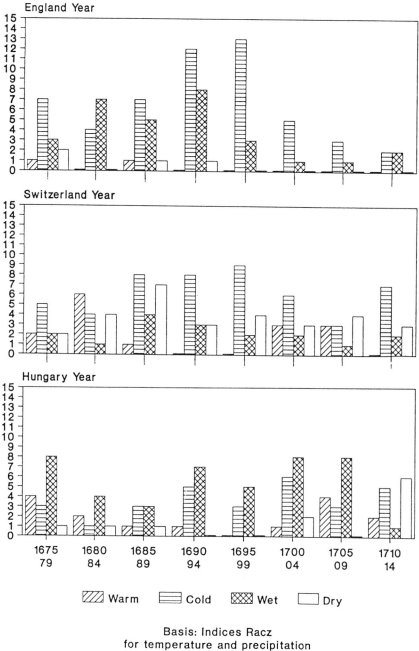

England Year

Switzerland Year

Hungary Year

⬚ Warm　☰ Cold　▨ Wet　☐ Dry

Basis: Indices Racz
for temperature and precipitation
X

Figure 4.7 Number of unmistakably warm, cold, wet, and dry years in England,
Switzerland, and Hungary 1675–1715.

Neumann, 1981). The reversal of the trend after 1700 was first felt in England, where in 1700–4 the number of cold months had declined to the level of the early 1680s. In Switzerland warming was still moderate at that time, and the number of cold months even increased in Hungary. In that region the overall thermal picture of the seasons was not balanced before 1705–9.

In some European mountain areas the cooling was associated with a lowering of snowlines. We have travellers' reports of permanent snow on the tops of the Cairngorms in the Scottish Highlands. These observations seem to require temperatures 1.5–2.0°C below twentieth-century values averaged over the year, a lowering twice to three times as great as that which has been substantiated in central England from actual thermometer readings (Lamb, 1982). A lowering of snowlines was inferred from proxy data for the period 1688–1703 in the Alps. In 1675, 1692, 1695 and 1698, when the remaining part of the summer was also cloudy and cool, summer crops in the mountains were buried under snow cover before they could be harvested, and grapes in the lowlands did not ripen fully. In the 1690s the average duration of the growing season in Einsiedeln (882 m asl) was reduced to only half of its twentieth-century average (Pfister in Frenzel *et al.*, 1994). At the turn of the seventeenth century, snow lay longer and in greater quantity on the mountains of Crete than during the present century (Grove and Conterio, in Frenzel *et al.*, 1994).

During the cold phase the record displays a considerable increase in variability over the period 1685–1700. Low-temperature events occurred more frequently and some of them rate among the most severe ones within this millennium (e.g. the winter 1683–4 in England, that of 1684–5 and 1694–5 in Switzerland, that of 1708–9 in most parts of Europe, the cold and rainy summers of 1695 and 1696 in Finland, in the Baltic, and in northwestern Russia). It has been shown that extreme events are more sensitive to the variability of climate than to its average, and that this sensitivity is greater the more extreme the event (Katz and Brown, 1992). It is well known that the primary impacts of climate on society result from extreme events, and considering the relatively low level of economic development at the time, we may imagine the load of human suffering.

Scotland was probably the worst affected: in the upland parishes the harvests (largely oats) failed in seven out of the eight years between 1693 and 1700 which caused a mass emigration (Lamb, 1982). Likewise the famines in western and northern Europe are related to repeated harvest failures. After the very cold summer of 1692 had caused a first harvest failure in western Europe, the continuous rain in spring of 1693 was one of the chief reasons for the second one. This cumulation of climatic stress lead to a subsistence crisis in 1693–4 which was one of the severest in the centuries up till then (Mattmüller, 1987; Lachiver, 1991; Pfister, 1992). In Finland the springs of 1695 and 1696 were so cold that sowing could not be completed before

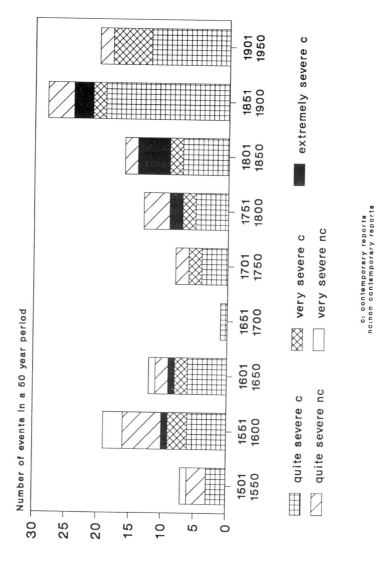

Figure 4.8 Floods in the Swiss Alps 1500–1950.

midsummer, the crops were destroyed by early frosts in late August or early September before they could be harvested. This resulted in a severe famine that killed about a third of the population in Finland (Vesajoki and Thornberg, in Frenzel *et al.*, 1994).

Another indication that points to the unique nature of the situation at the end of the seventeenth century is the frequency of floods over the last 475 years in the Cantons of Uri, Valais, Ticino, and Grisons, situated around the Gotthard Pass (Figure 4.8). After the one-in-a-century event in August 1987, some 1,900 reports on floods contained in documentary sources were systematically collected and classified according to the spatial range and to the magnitude of the damage (Pfister and Hächler, 1991).

In Figure 4.8 the number of floods is given according to three grades of severity, distinguishing between contemporary and non-contemporary reports. It is striking that in the second half of the seventeenth century just one important flood impact was observed in the area. This cannot be explained from the long-term rise in the frequency of reporting, because the number of observations was considerably higher in the preceding century. Rather this seems to be connected to the specific climatic situation which involved smaller amounts of melt-water in spring and a lower frequency of wet summers.

DISCUSSION

We have used a large number of documentary high-resolution proxy data to describe the patterns of climate change in Europe at the end of the seventeenth century and in the early eighteenth century, distinguishing between regional and seasonal aspects as well as between temperature and precipitation. To a certain extent the results have modified the picture of this period, as it is known from the work of Manley and Lamb, in the following three aspects:

- Cooling was at a maximum in extreme western Europe, probably in Scotland, decreasing inland to the Carpathian Basin. In particular, summers and autumns were considerably less cold in Switzerland and Hungary than in England. The 1680s were the period of highest variability. No cooling was observed in the western high latitudes (Iceland), whereas in the Mediterranean Basin outbreaks of cold continental air-masses became more common in winter. This implies that the cooling at the end of the seventeenth century was primarily a North Atlantic phenomenon.
- Precipitation in spring and summer was higher in England than in the twentieth century. It decreased inland to western central Europe, where all seasons were drier than today except summers, whereas in the Carpathian basin, spring and winter were somewhat wetter. In the

Mediterranean some prolonged periods of cloudy and cool weather and untimely rainfall were reported in summer. Spain often suffered from severe droughts in winter and spring in the 1680s and after 1697. These changes are related to shifts in the atmospheric circulation to be discussed below.

- The cooling began in the mid-1670s in the western part of the continent and spread farther east in time. This is demonstrated most clearly from the time of maximum winter severity which was in the 1680s in England, in the 1690s in western central Europe and in the 1700s in the Carpathian Basin. In a similar way the trend towards warmer conditions was first observed in England from 1697, around 1703 it became manifest in Switzerland and after 1705 in Hungary. A shift from west to east is even observed in the wave of severe famines in the 1690s. Western central Europe got the worst part in 1692–4, the Baltic, Finland, and Russia in 1695–7.

The observed shift in the prevailing winds in western Europe at the end of the seventeenth century is consistent with the variations in both temperature and moisture balance. Lamb (1972) has reconstructed a series of southwesterly surface wind frequencies for 'England' to the present time. Since southwesterly surface winds in England are closely related to the occurrence of general westerly situations over the British Isles, it is possible to obtain from this record a general idea of the circulation over the region. During the 40-year period 1675–1715 the SW wind frequencies have some notable features: a maximum in the 1670s followed by a sharp decrease to a minimum around 1690, after which the frequencies increase again to reach another peak of zonal circulation in the 1730s (Kington, in Frenzel et al., 1994). This record agrees well with Morin's observations of cloud directions in Paris (Pfister, Bareiss, in Frenzel et al., 1994).

Lamb (1977) has attempted to reconstruct the atmospheric circulation changes at that time from the evidence available to him. On his putative weather charts for the Januarys of the 1690s he obtained a centre of high pressure over Scandinavia. This implies northeasterly flows and a dominance of cold continental air-masses in western and central Europe with frequent outbreaks into the Mediterranean. The months of July in the 1690s exhibited a southward displacement of the Icelandic low and a truncation of the Azores anticyclone; this implies the prevalence of northwesterly isobars over central Europe. This result is qualitatively consistent with results obtained from the Euro-Climhist approach. According to the reconstruction of the wind-system in Denmark from ship's logs (Frich and Frydendahl, in Frenzel et al., 1994), the proportion of winds from the NW and N in summer was almost 200 per cent over the whole period 1675–1715 compared to 1951–80, mainly at the expense of winds from the NE and E. Moreover, as noted above, no warm and dry summer was recorded in western and central Europe over the two decades 1685–1704. More effort is needed to understand the changes in atmospheric circulation. In particular, classifying the monthly evidence available in Euro-Climhist according to the scheme of European Weather Types (Grosswetterlagen) would allow a meaningful and substantial quantification of this data.

The causal factors which underly these variations in both temperature and moisture balance are not yet understood. There are at least five hypothesized causes of decadal to century-scale climate variability. These include:

- the inherent ('random') variability in the atmosphere;
- the inherent or forced variability in the atmosphere–ocean system;
- solar variability;
- variability in volcanic aerosol loading of the atmosphere;
- atmosphere trace gas variability (e.g. CO_2).

Modelling experiments conducted for each of these potential mechanisms show that they have different signatures in time and space which may allow for discrimination in the climate record (Rind and Overpeck, 1993). In this context the issues of solar variability and inherent and forced variability in the atmosphere–ocean system will be discussed only because these two hypotheses seem to have the highest explanatory value for this particular period.

Much effort dealing with the period of the Maunder Minimum was put into the issue of solar forcing. The usual tests involved the statistical analysis of palaeoclimatic series with indicators of solar activity. For northern Fennoscandia, a series for 'summer' (April–August) temperature has been reconstructed from tree-ring data extending from AD 500. Comparing the reconstructed temperatures during the other five sunspot minima thought to have occurred since AD 1000, reveals that it was only during the Maunder Minimum (1645–1715) that mean summer (April to August) temperatures were significantly lower relative to a reference period of 1951–80. Conditions were, however, notably cooler in northern Fennoscandia during the 50 years preceding 1645 (Briffa, in Frenzel et al., 1994). Based upon tree-ring data from western and southwestern Europe Serre-Bachet (in Frenzel et al., 1994) has concluded that the cooling of the subperiod 1675–1715 – as far as annual and summer temperatures are concerned – was not exceptional within the context of the 351-year interval 1500–1850. In Switzerland, the cooling of this subperiod is related to lower temperatures in winter and spring in the two decades from 1685–1704 which, as far as these two seasons are concerned, are the coldest and the snowiest of the last 450 years. Annual temperatures. however, were somewhat lower from 1805–24, from 1842–61 and from 1878–97, because in those years the cooling in summer and autumn was more pronounced (Pfister, in Frenzel et al., 1994).

From the results of GCM studies mentioned above, it has been concluded that the characteristic signature related to solar forcing is global. Significant cooling occurs in the tropics without an obvious high latitude amplification and the surface temperature reduction is largest in central Eurasia and becomes smaller towards the Atlantic Ocean (Rind and Overpeck, 1993). This is just the opposite of the effects that are observed for the late seventeenth century in Europe. However, this does not rule out solar forcing altogether as a viable cause of cooling. It should be stressed that sunspot numbers only represent surface processes that have little or no significance for variations in luminosity and terrestrial climate. On the other hand a very good time correlation between the period of most severe climate in western Europe (about 1680–1700) and decreased deflection of incoming cosmic-ray (about 1675–98) is observed (Beer *et al.*, 1988; Mörner, in Frenzel *et al.*, 1994).

Changes in ocean heat transport, associated with changes in wind-driven circulation or large-scale thermohaline circulation, represent another obvious method of inducing climate variability (Rind and Chandler, 1991).

There is a good temporal agreement between the changes in weather patterns over Europe and ice-sheet growth in the North Atlantic consistent with a heat transport reduction. During the 1680s the sources recount four occurrences in sea ice around Iceland, the worst years being 1684 and 1685 (Ogilvie, 1992). In February 1684, the appearance of pack-ice in the Channel is described by Richard Freebody of Lydd (Kent) in a letter to his nephew. It seems that ice, broken away from the main pack, had been brought by the cold East Greenland Current into the North Sea and then on into the Channel (Lamb, 1982). The growth of sea ice around Iceland continued until 1695, then it declined (Ogilvie, 1992).

Lamb (1977) connects the cold period at the end of the seventeenth century to sea-surface temperature cooling in the North Atlantic, arguing from the historical record on cod fisheries. The cod, which thrives best in rather cold waters at between 4° and 7°C, seems to be a valuable indicator in this connection, because its kidneys fail at temperatures below 2°C and it therefore cannot venture into colder seas. In the Faroe Islands, there were no cod thereabouts for thirty years between 1675 and 1704. In 1695 cod became scarce also in Shetland waters and disappeared off the entire coast of Norway. Lamb (1982) has inferred that at that time the water from the cold East Greenland current had spread across the surface of the whole Norwegian Sea.

Lower sea-surface temperatures at this latitude would also be consistent with higher windspeeds as a consequence of enhanced temperature gradients. Lamb and Frydendahl (1991: 23f.) concluded that storms between 1694 and the early eighteenth century were at the same time more frequent and more severe. Indeed between 1688 and 1702 the British Isles were ravaged by at least eight violent storms. It is true that only a few heavy storms were observed between 1675 and 1687, and after the Great Storm of December 1703, the frequency and severity of these events lessened. The analysis of

Danish ships' logs for summer supports this argument. It was demonstrated that in the Copenhagen area winds of Beaufort 7 to 8 were more frequent from 1675–1715 than in the modern period. Moreover some storms of Beaufort 9 to 11 were observed in the earlier period, but have not been known since 1951 (Frich *et.al.*, in Frenzel *et al.*, 1994).

According to GCM simulations the characteristic climatic signature associated with the colder sea-surface temperature is a maximum cooling in extreme western Europe decreasing inland, a higher pressure over the Atlantic, and a lower precipitation over Europe (Rind and Overpeck, 1993) which agrees well with the spatial reconstructions based on the Euro-Climhist Multi Proxy Mapping experiment for the latter part of the Maunder Minimum period. Recent results of ocean and climate models indicate that decadal and century timescale cycles need not be due to external (solar) forcing but may well be the manifestation of natural variability within the coupled atmosphere–ocean-ice system (Stocker and Mysak, 1992).

CONCLUSIONS AND IMPLICATIONS FOR THE FUTURE

From the early 1970s our understanding of climatic change in Europe in the Little Ice Age was moulded by the influential work of English scientists such as Gordon Manley (1974) and Hubert Lamb (1977), who mainly draw from the example of the British Isles. It was not clear, however, how far these geographically restricted results could be taken to extrapolate to other regions of Europe or to the globe. The Euro-Climhist project of the European Science Foundation allowed broadening this data base for a time-window within the late Maunder Minimum (1675–1715) to include most of the European continent, and producing a set of monthly and seasonal weather charts (Multi Proxy Mapping). Besides some tree-ring series, the evidence consisted of verified documentary data mostly originating from central, western and northern Europe. Regarding the value of this approach the following observations and conclusions are made:

- A Multi Proxy Mapping approach involving data from historical and natural archives may yield comprehensive spatial reconstructions on a monthly or on a seasonal level. Chronologies of natural proxy data are needed to assess changes in the spring–summer period. Documentary data are in most cases the only proxy for assessing conditions in the winter half-year from October to April. The approach implies a certain level of data density and quality. In particular, care must be taken to include the bordering regions of the continent, in particular Iceland and the Mediterranean.
- The spatial images obtained from this data may be interpreted in terms of weather situations. Trends of temperature and precipitation patterns

are obtained by using simple statistical techniques such as temperature and precipitation indices.

- The approach is particularly well suited for the analysis of anomalies, because those were more fully and more often described than ordinary weather spells.
- Multi Proxy Mapping turned out to be a useful approach to test hypothesized causes of decadal-scale climate variability. Modelling experiments conducted for the potential forcing mechanisms (solar, ocean-atmosphere, etc.) show that each of them has its own characteristic spatial fingerprint (Rind and Overpeck 1993). Thus, a comparison of results with high-resolution reconstructions in time and space allows for discrimination of each in the climate record. In the case of the late seventeenth century the results of GCM experiments for a cooling of the sea-surface temperature in the North Atlantic were in good agreement with the changes obtained from Multi Proxy Mapping of documentary and tree-ring data.

Understanding the climate system and its mechanisms may greatly benefit from further co-operative efforts of Multi Proxy Mapping and modelling in the future.

ACKNOWLEDGEMENTS

Euro-Climhist is supported by Swiss National Science Foundation, NFP-31 Program on Climatic Change and Natural Hazards. Heinz Wanner, Berne, Rudolf Brazdil, Brno and John Kington, Norwich, have critically read earlier versions of this paper and made helpful suggestions. In addition to the partners of Euro-Climhist, the following persons have contributed data to the program: Professor Emmanuela Guidoboni, Bologna; Professor Dr Maciej Sadowski, Warsaw. Valuable hints to compilations from Spain were provided by Dr Ramon Guardans, Madrid.

REFERENCES

Banzon, V., de Franceschi, G. and Gregori, G. (1992) 'The mathemathical handling and analysis of non homogeneous and incomplete multivariate historical data series', in B. Frenzel, C. Pfister and B. Glaeser (eds) *European Climate Reconstructed from Documentary Data: Methods and Results*, Gustav Fischer, Stuttgart, pp. 137–51.

Beer, J., Siegenthaler, U., Bonani, G., Finkel, R.C., Oeschger, H., Suter, M. and Wölfli, W. (1988) 'Information on past solar activity and geomagnetism from 10Be in the Camp Century ice-core', in *Nature*, 332: 675–9.

Borisenkov, E.P. (1992) 'Documentary evidence from the U.S.S.R.' in R.S. Bradley and P.D. Jones (eds), *Climate since AD 1500*, Routledge, London, pp. 171–83.

Bradley, R.S. and Jones, P.D. (1992) 'Climatic variations over the last 500 years', in

R.S. Bradley and P.D. Jones (eds), *Climate since AD 1500*, Routledge, London, pp. 649–66.

Brazdil, R. and Dobrovolny, P. (1992) 'Possibilities of the reconstruction of the climate of Bohemia during the last millennium on the basis of written sources', in *Proc. of the Symposium: Use of Direct and Indirect Data for Reconstruction of the Climate in the Last Two Thousand Years*, Brno, pp. 142–65.

Briffa, K.R. (1992) 'Dendroclimatic reconstructions in Northern Fennoscandia', in T. Mikami (ed.), *Proc. of the International Symposium on the Little Ice Age Climate*, Metropolitan University, Tokyo, pp. 5-10.

Brumme, B. (1981) 'Methoden zur Bearbeitung historischer Mess- und Beobachtungsdaten (Berlin und Mitteldeutschland 1683–1770)', in *Archiv für Meteorologie, Geophysik und Bioklimatologie*, Serie B. Bd. 29., Springer, Heidelberg, pp. 191-210.

Buisman, J. (1984) *Bar en boos. Zeven eeuwen winterweer in de Lage Landen.*, Bosch and Keuning NV, Baarn.

Dool, H.M. van den, Schuurmans, C.J.E. and Krijnen, H.J. (1978) 'Average winter temperatures at De Bilt (The Netherlands), 1634–1977', in *Climatic Change* 1: pp. 319–30.

Eddy, J.A. (1977) 'Climate and the changing sun', in *Climatic Change*, 1: pp. 173–90.

—— (ed.) (1992) 'The Pages Project: Proposed implementation plans for research activities', Report No. 19, *An Expanded Explanation of the PAGES Project Plan Described in IGBP Report No.12*, IGBP, Stockholm.

Font Tullot, I. (1988) *Historia del Clima de Espana. Cambios climaticos y sus causas*, Instituto Nacional de Meteorologia, Madrid.

Frenzel, B., Pfister, C. and Glaeser, B. (eds) (1992) *European Climate Reconstructed from Documentary Data: Methods and Results*, Gustav Fischer, Stuttgart.

——, —— and —— (eds) (1994) *Climate in Europe 1675–1715, Special Issue*, ESF Project European Palaeoclimate and Man, Gustav Fischer, Stuttgart.

Frydendahl, K., Frich, P. and Hansen, C. (1992) 'Danish weather observations 1675–1715', in *Danish Met. Inst. Technical Report*, 92–9, Danish Meteorological Institute, Copenhagen.

Geurts, H.A.M. and van Engelen, A.F.V. (1992) 'Beschrijving antieke meetreksen', in *Historische Weerkundige Waarnemingen*, Deel V, Koninklijk Nederlands Meteorologisch Institut, publication 165–V, De Bilt.

Glaser, R. and Hagedorn, H. (1991) 'The climate of Lower Franconia since 1500', in *Theoretical and Applied Climatology* 43: 101–4.

Glogger, B. (1992) *Die Schweiz im Treibhaus. Regionale Auswirkungen der globalen Klimabedrohung*, Verlag NZZ, Zürich.

Grebner, D. (1722) 'Davidis von Grebner Caesar. Medic. Ephemerides Meteorologicae Vratislaviensis ab anno MDCXCII ad annum huius seculi secondum and vigesimum, Atque notationes barometricae et thermometricae anno MDCCX inceptae ad annum praesentem continuatae', *Haeredes Neumann*, Vratislaviae.

Guiot, J. (1992) 'The combination of historical documents and biological data in the reconstruction of climate variations in space and time', in B. Frenzel, C. Pfister, and B. Glaeser (eds), *European Climate Reconstructed from Documentary Data: Methods and Results*, Fischer, Stuttgart, pp. 93–104.

Holzhauser, H. (1984) 'Rekonstruktion von Gletscherschwankungen mit Hilfe fossiler Hölzer', *Geographica Helvetica* 39: 3–15.

Jefferey, R. (1933) 'Was it wet or fine? An account of English weather from chronicles, diaries and registers', Unpublished manuscript in the Library of the Meteorological Office, Bracknell.

Jones, P.D., Ogilvie, A.E.J. and Wigley, T.M.L. (1984) 'Riverflow data for the United

Kingdom: reconstructed data back to 1844 and historical data back to 1556', Climatic Research Unit, University of East Anglia, Norwich.

Jones, P.D., Santner, B.D., Cherry, B.S.G., Goodess, C., Bradley, R.S., Diaz, H.F., Kelly, P.M. and Wigley, T.M.L. (1985) 'A grid point surface air temperature data set for the Northern Hemisphere, 1851–1984', *DOE Tech. Rep. TRO22*, US Dept of Energy, Washington DC.

Kameda, T., Narita, H., Shoji, H. and Nishio, F. (1992) '450-year summer temperature record from melt feature profile in South Greenland ice core (Abstract)', in T. Mikami (ed.), *Proc. of the International Symposium on the Little Ice Age Climate*, Tokyo, p. 101.

Kates, R.W., Ausubel, J.H. and Berberian, M. (eds) (1985) *Climate Impact Assessment. Studies of the Interaction of Climate and Society*, John Wiley & Sons, Chichester.

Katz, R.W. and Brown, B.G. (1992) 'Extreme events in a changing climate: variability is more important than averages', in *Climatic Change* 21: 289–302.

Lachiver, M. (1991) *Les années de misère*, Paris: Fayarad.

Lamb, H.H. (1972) *Climate: Present, Past and Future. Vol. 1*, Methuen, London.

—— (1977) *Fundamentals and Climate Now. Vol. 2, Climatic History and the Future*, Methuen, London.

—— (1982) *Climate, History and the Modern World*, London and New York, Methuen.

Lamb, H.H. and Frydendahl, K. (1991) *Historic Storms of the North Sea, British Isles and Northwest Europe*, Cambridge University Press, Cambridge.

Lauer, W. and Frankenberg, P. (1986) 'Zur Rekonstruktion des Klimas im Bereich der Rheinpfalz seit Mitte des 16.Jh mit Hilfe von Zeitreihen der Weinquantität und Weinqualität', Gustav Fischer, Stuttgart.

Le Roy Ladurie, E. (1972) *Times of Feast, Times of Famine: A History of Climate since the Year 1000*, translated by Barbara Bray, Allen & Unwin, London.

Legrand, J.P. and Le Goff, M. (1992) 'Les observations météorologiques de Louis Morin', in *MonoFiguresie* 6, 2 vols, Direction de la Météorologie Nationale, Météo France, Trappes.

Lenke, W. (1961) 'Bestimmung der alten Temperaturwerte von Tübingen und Ulm mit Hilfe von Häufigkeitsverteilungen. Anhang: Die Beobachtungen von Camerarius, Tübingen (1691–1694) und Algöwer, Ulm (1710–1714)', in *Berichte des Deutschen Wetterdienstes*, Nr. 75, Bd.10, Offenbach a. M.

Lindgren, S. and Neumann, J. (1981) 'The cold and wet year 1695. A contemporary German account', in *Climatic Change* 3: 173–87.

Manley, G. (1974) 'Central England temperatures: monthly means 1659 to 1973', in *Quarterly Journal of the Royal Meteorological Society*, 100 (425), July: 389–405.

Mattmüller, M. (1987) *Bevölkerungsgeschichte der Schweiz. Teil I: Die frühe Neuzeit, 1500–1700, Bd. 2: Wissenschaftlicher Anhang*, Helbing and Lichtenhahn AG, Basel/Frankfurt a. M.

Mikami, T. (ed.) (1992) *Proc. of the International Symposium on the Little Ice Age Climate*, Tokyo Metropolitan University, Tokyo.

Mitchell, J.F.B., Manabe, S., Meleshko, V. and Tokoika, T. (1990) 'Equilibrium climate change and its implications for the future', in J. R. Houghton, G.J. Jenkins and J.J. Ephraums (eds), *Climate Change: The IPCC Scientific Assessment*, Cambridge University Press, Cambridge, pp. 131–72.

Mlostek, E. (1988) *Warunki Termiczne Wroclawia i Okolicy u XVIII u . Thermal Conditions in Wrocław and neighbourhood in the 18th century*, Ph.D. Thesis, University of Wrocław.

Mörner, N. A. (1992) 'Global change: the last millennia', *Palaeogeogr. Palaeoclim. Palaeoecol. Global Planetary Change*, in press.

Mörner, N.A. and Karlén, W. (eds) (1984) 'Climatic changes on a yearly to millennial basis: geological, historical and instrumental records', *Proc. 2nd Nordic Symposium on Climatic Changes*, Stockholm, 16–20 May, 1983, D. Reidel, Dordrecht, Boston, Lancaster.

Ogilvie, A.E.J. (1984) 'The past climate and sea-ice record from Iceland. Part 1: Data to AD 1780', in *Climatic Change* 6: 131–52.

—— (1992) 'Documentary evidence for changes in the climate of Iceland, AD 1500 to 1800', in R.S. Bradley and P.D. Jones (eds), *Climate since AD 1500*, Routledge, London, pp. 92–117.

Pfister, C. (1979) 'Getreide-Erntebeginn und Frühsommertemperaturen im schweizerischen Mittelland seit dem 17. Jahrhundert', in *Geographica Helvetica* 34: 23–35.

—— (1984) *Das Klima der Schweiz von 1525–1860 und seine Bedeutung in der Geschichte von Bevölkerung und Landwirtschaft. Band 2: Bevölkerung, Klima und Agrarmodernisierung*, Haupt, Berne.

—— (1985) *Witterungsdatei CLIMHIST. Bd. 1 Schweiz 1525.1863*, METEOTEST, Fabrikstr. 29 a, 3012 Berne.

—— (1992) 'Monthly temperature and precipitation in central Europe from 1525–1979: quantifying documentary evidence on weather and its effects', in R.S. Bradley and P.D. Jones (eds), *Climate since AD 1500*, Routledge, London, pp. 118–42.

Pfister, C. and Hächler, S. (1991) 'Überschwemmungskatastrophen im Schweizer Alpenraum seit dem Spätmittelalter. Raum-zeitliche Rekonstruktion von Schadenmustern auf der Basis historischer Quellen', in R. Glaser and R. Walsh (eds), *Historical Climatology in Different Climatic Zones*, Würzburger Geograph, Arbeiten H. 80, Würzburg, pp. 127–48.

Racz, L. (1992) 'Variations of climate in Hungary (1540–1779)', in B. Frenzel, C. Pfister and B. Glaeser (eds), *European Climate Reconstructed from Documentary Data: Methods and Results*, Fischer, Stuttgart, pp. 125–36.

Repapis, C.C., Schuurmans, C.J.E., Zerefos, C.S. and Ziomas, J. (1989) 'A note on the frequency of occurrence of severe winters as evidenced in monastery and historical records from Greece during the period 1200–1900 AD', in *Theoretical and Applied Climatology* 39: 213–17.

Rind, D. and Chandler, M. (1991) 'Increased ocean heat transports and warmer climate', in *Journal of Geophysical Research* 96, No. D4, 7437–61.

Rind, D. and Overpeck, J. (1993) 'Hypothesized causes of decade-to-century-scale climate variability: climate model results', in *Quaternary Science Reviews* (submitted).

Sahsamanoglu, H.S., Makrogiannis, T.J. and Kallimopoulos, P.P. (1991) 'Some aspects of the basic characteristics of the Siberian anticyclone', in *International Journal of Climatology* 11: 827–39.

Salmelli, D. (1986) 'L'alluvione e il freddo: il 1705 e il 1709', in Roberto Finzi (ed.), *Le Meteore e il Frumento. Clima Agricoltura, Meteorologia a Bologna nel '700*, Mulino, Bologna, pp. 17–98.

Sammlung (1718–26) Sammlung von Natur-und Medicin- wie auch hierzu gehöriger Kunst- und Literatur Geschichten. So sich Anno', in *Schlesien und anderen Ländern begeben, ans Licht gestellt von Einigen Academ. Naturae Curios. in Breslau.*

Schlesinger, M. (ed.) 1990 *Proc. of the US Dept of Energy Workshop on Greenhouse Induced Climatic Change*, US Department of Energy, Washington DC.

Schönwiese, C.D. (1990) 'Detecting the anthropogenic greenhouse effect by means of observational data', in R. Brazdil (ed.), *Climatic Change in the Historical and Instrumental Period*, Brno, pp. 64–70.

Schönwiese, C.D. and Runge, K. (1988) 'Der anthropogene Klimagaseinfluss auf das globale Klima. Erweiterte statistische Abschätzungen im Vergleich mit Klimamodell-Experimenten', *Berichte des Instituts für Meteorologie und Geophysik der Universität Frankfurt / Main*, Nr. 76, internal institute report, Frankfurt.

Schove, J. (1962) 'The reduction of annual winds in north western Europe, AD 1635–1960', in *Geografiska Annaler* 44: 303–27.

Schüle, H. and Pfister, C. (1992) 'EURO-CLIMHIST – outlines of a Multi Proxy Data Base for investigating the climate of Europe over the last centuries', in B. Frenzel, C. Pfister and B. Glaeser (eds), *European Climate Reconstructed from Documentary Data: Methods and Results*, Fischer, Stuttgart, pp. 211–28.

Schweingruber, F.H., Bartholin, T., Briffa, K. and Schär, E. (1988) 'Radiodensitometric-dendroclimatological conifer chronologies from Lapland (Scandinavia) and the Alps (Switzerland)', in *Boreas*, 17: 559–66.

Serre-Bachet, F., Guiot, J. and Tessier, L. (1992) 'Dendroclimatic evidence from southwestern Europe and northwestern Africa', in R.S. Bradley and P.D. Jones (eds), *Climate since AD 1500*, Routledge, London, pp. 349–65.

Stocker, T.F. and Mysak, L.A. (1992) 'Climatic fluctuations on the century time scale: a review of high resolution proxy data and possible mechanisms', in *Climatic Change* 20: 227–50.

Stuiver, M. and Braziunas, T.F. (1992) 'Evidence of solar activity variations' in R.S. Bradley and P.D. Jones (eds), *Climate since AD 1500*, Routledge, London, pp. 593–605.

Tarand, A. (1992) 'Ice cover in the Baltic region and the air temperature of the Little Ice Age', in T. Mikami (ed.), *Proc. of the International Symposium on the Little Ice Age Climate*, Tokyo, pp. 94–100.

Tarrats, J.M.F. (1971) *Entre el Cardo y la Rosa. Historia del Clima en las Mesetas*, Typoskript, Madrid.

—— (1976) *Historia del Clima en Cataluña. Noticias. Antiquos, medievales y en especial de los siglos XV, XVI, y XVII*, Typoskript, Madrid.

—— (1977) *Historia del Clima del Finis-Terrae Gallego [Galicia]*, Typoskript, Madrid.

Wales-Smith, B.G. (1971) 'Monthly and annual totals of rainfall representative of Kew, Surrey, from 1697 to 1970', in *Meteorological Magazine* 102: 345–62.

Zumbühl, H.J. (1980) 'Die Schwankungen der Grindelwaldgletscher in den historischen Bild- und Schriftquellen des 12- bis 19. Jahrhunderts. Ein Beitrag zur Gletschergeschichte und Erforschung des Alpenraums', *Denkschr d. Schweiz. Natf. Ges.* 92, Birkhäuser, Basel.

Zumbühl, H.J. and Holzhauser, H. (1988) 'Alpengletscher in der Kleinen Eiszeit', in *Die Alpen*, 64.Jg., H.3. *Sonderheft zum 125 jährigen Jubiläum des Schweizer Alpen-Clubs*, Stämpfli, Berne.

5

ACCELERATED GLACIER AND PERMAFROST CHANGES IN THE ALPS

W. Haeberli

INTRODUCTION

Changes in mass and temperature of glacier and permafrost ice are results of changes in the mass and energy balance at the earth's surface. Rates and ranges of such glacier and permafrost changes can be determined quantitatively over various time intervals and expressed as corresponding energy fluxes. This permits direct comparison with estimated effects of anthropogenic greenhouse forcing. Glacier and permafrost changes are thus linked to changing atmospheric conditions via important filter, memory and enhancement functions. As a consequence, they are among the clearest signals evident in nature of ongoing warming trends and potential acceleration tendencies related to the enhanced greenhouse effect (Haeberli, 1990; WGMS, 1993; Wood, 1990).

Documentation of useful quality and completeness is available for glaciers in the Alps (Haeberli, 1991). Average rates of mass change can now be estimated for the vanishing of Ice Age glaciers (20,000 to 10,000 ka BP), for recent secular warming (1880–1980), and for the last decade of accelerated melting (1980–1990). Systematic collection of information on alpine permafrost only began as recently as 1970 (Haeberli *et al.*, 1993), but first estimates of warming and degradation rates can be made with respect to approximately the same time intervals as for glaciers.

An attempt is made in the following to outline briefly the basic concepts, the available data basis, and the first results of quantitatively assessing the apparently accelerating changes in alpine ice occurrences.

PROCESSES AND CONCEPTS

Glacier reactions to climatic change involve a complex chain of processes (energy balance – mass balance/englacial temperature – geometry/flow – length change). In areas of cold firn predominating at polar latitudes, in regions of continental climate and at very high altitudes, atmospheric

91

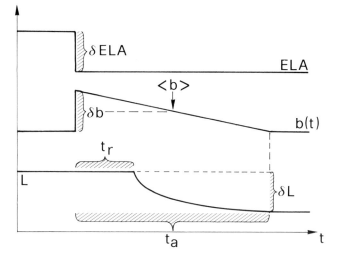

Figure 5.1a Scheme of the glacier reaction to a step change in equilibrium line altitude (ELA) and mass balance (b) as a function of time (t): glacier length (L) reaches a new equilibrium position (L + δL) with b = 0 after an initial reaction time (t_r) and the response time (t_a). Average mass balance during t_a is close to 1/2 × δb.

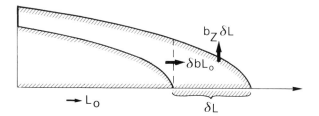

Figure 5.1b Schematic long profile of the glacier terminal area before and after full response to a step change in ELA and (b): the original mass excess δb over the entire glacier length L_0 must be compensated for by the ablation at the terminus (b_t) over the distance of length change δL.

warming does not directly lead to mass loss through melting/runoff but to firn warming and thereby produces corresponding signals in firn temperature profiles with depth (Robin, 1983). In areas of temperate firn which predominate at lower latitudes/altitudes, on the other hand, atmospheric warming mainly causes changes in mass and geometry of the glaciers (Figure 5.1). An assumed step change (δ) in equilibrium line altitude (ELA) causes an immediate step change in specific mass balance (b = total mass change divided by glacier area). The specific mass balance is thus the *direct, undelayed reaction of a glacier to climatic forcing*. A change in specific mass balance (δb) is calculated from the shift in equilibrium line altitude (δELA),

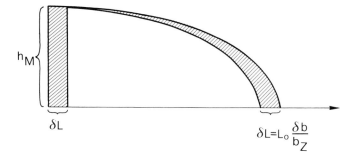

Figure 5.1c Scheme for deriving typical time scales for the glacier response time (t_a) from a simple geometric consideration: shifting the ablation area into the new equilibrium position opens a hole with the dimensions $h_m \times \delta L$ with h_m = maximum glacier thickness (at the equilibrium line) and $\delta L = L_0 \times \delta b/b_t$. This hole must be filled with $\delta b \times L_0$ so that $t_a \times \delta b \times L_0 = h_m \times L_0 \times \delta b/b_t$, or $t_a = h_m/b_t$. After Johannesson *et al.* (1989).

the gradient of mass balance with altitude (db/dH), and the distribution of glacier surface area with altitude (hypsography). The hypsography represents the local or individual topographic part of the glacier sensitivity, whereas the mass balance gradient mainly reflects the regional or climatic part (Kuhn, 1990). As the mass balance gradient tends to increase with increasing humidity (Kuhn, 1981), the sensitivity of glacier mass balance is generally much higher in areas with humid/maritime than with dry/continental climatic conditions. Cumulative mass changes lead to ice thickness changes which, in turn, exert a positive feedback on mass balance and at the same time influence the dynamic redistribution of mass by glacier flow.

The complex chain of dynamic processes linking glacier mass balance and length changes can, at best, be understood and numerically simulated for a few individual glaciers only, which have been studied in great detail (cf., for instance, Kruss, 1983; Oerlemans, 1988; Oerlemans and Fortuin, 1992; Greuell, 1992). The complications, however, disappear if the time intervals analysed are sufficiently long. After a certain reaction time (t_r) following a change in mass balance, the length of a glacier (L_0) will start changing and finally reach a new equilibrium ($L_0 + \delta L$) after the response time (t_a). The length change is thus the *indirect, delayed reaction of a glacier to climatic forcing*. After full response, continuity requires that (Nye, 1960)

$$\delta L = L_0 \times \delta b/b_t$$

with b_t = (annual) ablation at the glacier terminus. This means that, for a given change in mass balance, the length change is a function of the original length of a glacier, and that the change in mass balance of a glacier can be quantitatively inferred from the easily observed length change and from estimates of ELA and db/dH. The response time, t_a, of a glacier is related to the ratio between its maximum thickness (h_{max}) and its annual ablation at the terminus (Johannesson et al., 1989)

$$t_a = h_{max}/b_t$$

Corresponding values for alpine glaciers are typically several decades to slightly less than a century. Aletsch Glacier as the largest alpine glacier, for instance, with h_{max} close to 900 m and b_t around 12 m/a has an estimated response time of some 70 to 80 years. During the response time, the mass balance b will adjust to zero again so that the average mass balance ‹b› is close to $1/2 \times \delta b$. The secular mass change of glaciers estimated in this way, ‹b›, can be compared directly with the few measured long-term mass balance series existing in the Alps.

Climatic changes cause *changes in mountain permafrost* at various scales of time and space (Haeberli et al., 1993). Along vertical profiles with depth at individual points, reactions of permafrost to climatic changes are supposed to take place in three main forms (Figure 5.2):

1   changes in active layer thickness and thaw settlement/frost heave in ice-supersaturated material at the permafrost table as an immediate response (time scale: year(s));
2   disturbance of temperature profiles within the permafrost, i.e., between the permafrost table and the permafrost base, as an intermediate response (time scale: years to decades); and
3   vertical displacements of the permafrost base ($h_0 \rightarrow h_1$ in Figure 5.2) as a final response (time scale: decades to centuries or even millennia).

With regard to the 3-dimensional dynamics of complex landscapes, two more types of reactions at highly variable time scales can be envisaged:

4   modification of permafrost distribution patterns, involving
5   adjustment of geomorphic, hydrological and nivo-glaciological pro-cesses such as permafrost creep, frost heave/thaw settlement, thermo-karst, erosion and slope instability on thaw-destabilized slopes, runoff variations in time, drainage characteristics, snow-cover evolution and metamorphism, and avalanche formation.

Hence, the most important parameters to be observed in relation to climate change effects are surface heave/subsidence in permafrost areas and the depth to the permafrost table beneath the surface, the vertical temperature profile within the permafrost, the local/regional distribution pattern of permafrost, the horizontal and vertical deformation of ice-bearing ground, geomorphic forms, and runoff characteristics in the periglacial belt. Evidently, a combina-tion of methods must be applied (cf. King et al., 1992) to document appropriately this entire set of phenomena and associated changes.

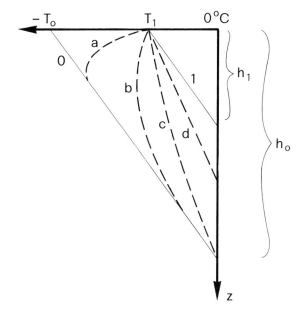

Figure 5.2 Scheme of the evolution of a temperature profile (0 → a → b → c → d → 1) within permafrost as a reaction to a step change in mean surface temperature from T_0 to T_1. Original permafrost thickness (h_0) and new permafrost thickness (h_1) correspond to steady states.

DATA BASIS

Mass balances since the late nineteenth century have been measured for six glaciers in the European Alps by repeated precision mapping (Table 5.1). The average annual mass loss over the entire period varies between about 0.2 and 0.6 m water equivalent. The overall reduction in ice depth since the Little Ice Age is thus measured in tens of metres. The representativity of the small sample of direct measurements can be confirmed by quantitative inter-pretation of glacier length changes. Table 5.2 illustrates the agreement between directly measured secular mass changes and estimates based on quantitative interpretation of length changes for the Rhône glacier and the Hintereisferner. Figure 5.3 is a compilation of cumulative length changes of 15 glaciers in the Swiss Alps: glacier length change is indeed a function of glacier length and again furnishes values for average secular mass losses ranging between –0.25 and –0.5 m w.e./a. Furthermore, the unique quality of this length change signal becomes obvious: the high-frequency (interannual) 'noise' is filtered out by the effect of latent heat, the 'memory' of the perennial ice bodies enables cumulation of effects for decades and the thickness change of a few decameters is 'amplified' into a length change

Table 5.1 Secular mass balances of alpine glaciers

Glacier	Observation period	Coordinates	Alt (m)	Area (km²)	w. e. loss
Rhone	1882–1987	4637/0824	2940	17.38	−0.25
Vernagt	1889–1979	4653/1049	3228	09.55	−0.19
Guslar	1889–1979	4651/1048	3143	03.01	−0.26
N Schnee	1892–1979	4725/1059	2690	00.39	−0.35
S Schnee	1892–1979	4724/1058	2604	00.18	−0.57
Hintereis	1894–1979	4648/1046	3050	09.70	−0.41

Sources: Chen and Funk (1990), Finsterwalder and Rentsch (1980), IAHS (ICSI)/UNEP/UNESCO (1988)

Table 5.2 Comparison of measured and estimated secular glacier mass changes

Glacier	Rhône Glacier	Hintereisferner
Measured average b (m w.e./a)	−0.25	−0.41
Total length today (km)	10	7.7
Length change (km, c. 1890–1980)	1	1.2
Ablation at snout (m w.e./a)	5.5	5
Inferred balance change (m w.e./a)	0.55	0.78
Inferred average b (m w.e./a)	−0.25 to −0.3	−0.4

Sources: Funk (1985), Haeberli (1991), IAHS (ICSI)/UNEP/UNESCO (1988), IAHS (ICSI)/UNESCO (1985), WGMS (1991, 1993)

measured in hundreds of metres if not a few kilometres. Application of similar model calculations to a sample of 65 glaciers in the Valais Alps (Scheuer, 1992) further confirms the representativity of the measured average rates of secular glacier mass loss in the Alps. Assuming a 30–50 per cent loss in glacierized surface area since about 1850 (Patzelt and Aellen, 1990), a present-day glacierized surface area of about 2,900 km² (IAHS (ICSI)/UNEP/UNESCO 1989) and an average present-day glacier thickness of 50–70 m (Haeberli, 1991; Müller et al., 1976), total secular mass loss and total remaining ice mass of Alpine glaciers can roughly be estimated at some 50 per cent and 150–200 km³, respectively. The spectacular loss of about half the original mass within just one century is most likely to be a result of an estimated 0.5°C rise in mean annual air temperature and an upward shift in equilibrium line altitude of some 100 m (Kuhn, 1990; Maisch, 1992).

Alpine glacier mass balances – with strongly negative values – for the decade 1980–90 are available from IAHS (ICSI)/UNESCO (1985), IAHS (ICSI)/UNEP/UNESCO (1988) and WGMS (1991, 1993). The average decennial mass balance of eight regularly observed Alpine glaciers was close to −0.6 m w.e. per year (Table 5.3). Quantitative information on rates of glacier mass loss for the Late Pleistocene warming period have been compiled and compared with twentieth-century data by Haeberli (1991). As a main

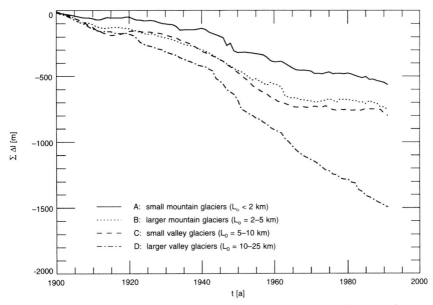

Figure 5.3 Cumulative length changes of glaciers in the Swiss Alps since the beginning of the twentieth century. Three-year running means are averaged for the size categories A to D in order to eliminate short-term and local noise in the data and to reflect the influence of glacier length on glacier length change as a reaction to mass balance changes. The individual glaciers of the four size categories are: A = Pizol, Sardona, Punteglias, Basodino, B = Grand Désert, Lavaz, Kehlen, Valsorey, C = Trient, Saleina, Morteratsch, Zinal, D = Rhone, Gorner, Aletsch. Data basis: Kasser *et al.* (1986) and M. Aellen (personal communication).

conclusion from this comparison, modern glacier shrinkage appears to be considerably faster than on average for the Late Pleistocene.

Systematic collection of information on Alpine permafrost only began recently (Haeberli *et al.*, 1993) but first estimates of warming and degradation rates can be made with respect to approximately the same time intervals as for glaciers. High-precision photogrammetry of creeping rock glacier permafrost started at Gruben in 1970. Repeated analyses in 1975 and 1979 (Haeberli and Schmid, 1988) showed that growth and degradation of permafrost can take place simultaneously at different places within the same rock glacier, but that an overall thinning tendency at a rate of a few centimetres per year occurred in the purely periglacial part of the rock glacier. New analyses were carried out from aerial photographs taken in 1985 and 1991. The number of grid points was thereby increased by reducing the mesh width to half its original value (50 to 25 m) in order to improve the signal-to-noise ratio relating to the investigated changes in surface altitude. In comparison with the 1970s, average annual rates of surface subsidence have accelerated by a factor of 2 to 3 in the warm 1980s to early 1990s (Figure

Table 5.3 Alpine glacier mass balances during 1980–1990 (in cm)

Years	Careser	Silvretta	Gries	Hintereis-ferner	Sonn-blickees	Kessel-wandf.	Sarennes	Saint Sorlin
1980/81	−840	350	−350	−173	414	161	40	310
1981/82	−1680	−290	−910	−1240	−1282	−620	−100	−1020
1982/83	−790	−530	−580	−580	−535	−182	−70	−340
1983/84	−590	360	−70	32	338	178	−40	−120
1984/85	−760	510	−1210	−574	−281	−8	−1210	−120
1985/86	−1140	−270	−690	−732	−1432	−494	−1790	−1730
1986/87	−1640	−210	−940	−717	−525	−243	−920	−810
1987/88	−1010	−580	−1100	−945	−711	−265	−690	310
1988/89	−820	−250	−1040	−637	252	−151	−2590	−2490
1989/90	−1580	−530	−1890	−995	−561	−242	−2140	−1400
mean	-1085	−144	−878	−656	−432	−187	−951	−741
maximum								
	−590	510	−70	32	414	178	40	310
minimum								
	−1680	−580	−1890	−1240	−1432	−620	-2590	−2490

5.4). This is probably caused by increased melting of ice at the permafrost table and may therefore be superimposed on to an assumed long-term trend of slow melting at the permafrost base. Results from similar surveys since 1981 are available from Muragl rock glacier in the Grisons (Vonder Mühll and Schmid, 1993) and confirm the generally small rates of geometric change in alpine permafrost. Additional results from repeated photogrammetry can soon be expected from the rock glaciers Réchy, Furggen/Gemmi and Gufer/Aletsch in the Valais and from Murtèl/Corvatsch and Ursina/Pontresina in the Grisons.

During the 1987 drilling through the permafrost within the active rock glacier Murtèl/Corvatsch, the first borehole was equipped for long-term observations in the Swiss Alps (Haeberli et al., 1988; Vonder Mühll and Haeberli, 1990). Until 1992, permafrost temperatures at 10 m depth have increased at a rate of 0.5 to 1°C per decade (Figure 5.5). This fast warming of near-surface permafrost is a result of the especially warm late 1980s and early 1990s. Two more boreholes were installed in 1990 at the nearby site of Ursina/Pontresina (see Vonder Mühll and Holub, 1992). Temperature profiles measured after dissipation of thermal disturbance from drilling within all three boreholes give mean annual permafrost temperatures at 10 m depth of −0.5, −1.3 (Ursina) and −2°C (Murtèl) with a corresponding permafrost thickness of 40, 70 (Ursina) and possibly more than 100 m (Murtèl). The last value is a rough estimate based on a discussion of groundwater influence in a talik between 52–56 m depth, which can also be assumed to be the cause of the strongly elevated value for the overall temperature gradient and vertical heat flow in the Murtèl borehole (Vonder

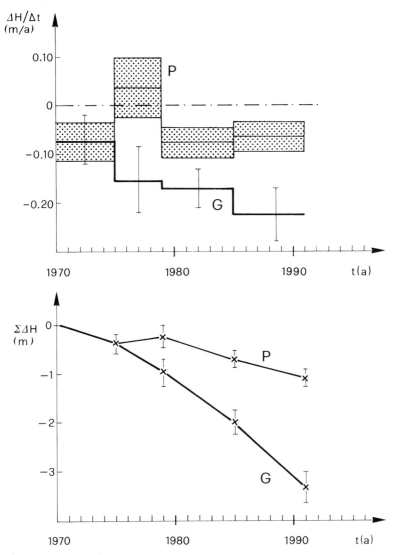

Figure 5.4 Changes in surface altitude at Gruben rock glacier from aerial photogrammetry since 1970: annual velocities (ΔH/Δt, top) and cumulative effect (ΣΔH, bottom). P = periglacial part, G = glacier-affected part with dead ice remains. Surface lowering (thaw settlement) is significant in both parts and considerably faster in the warm 1980s than in the 1970s.

Mühll, 1992). The less disturbed temperature gradients at depths below about 20 m in the two Ursina boreholes indicate vertical heat flow values close to steady state conditions and, hence, point to relatively stable surface temperatures between about 1950 and the early 1980s (see also Vonder Mühll,

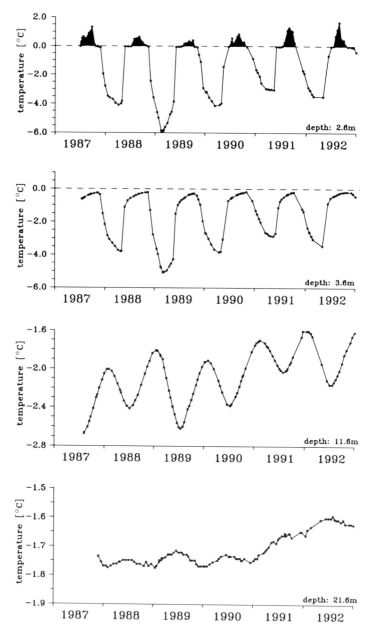

Figure 5.5 Borehole temperatures in the permafrost of the active rock glacier Murtèl/Corvatsch (Grisons) at various depths. The permafrost table is between 2.6 and 3.6 m depth. Recent warming trends from the exceptionally warm 1980s and early 1990s are most clearly visible at 10.6 and 21.6 m depth. The accuracy of the measurements is about ±0.05°C.

1993). This is in general agreement with an earlier (first) attempt to statistically reconstruct the secular evolution of mean permafrost temperature in the Alps (Haeberli, 1985), which indicates permafrost warming by about 1°C from 1880 to 1950 and more or less stable permafrost temperature from 1950 to 1980. Late glacial ground warming in the periglacial part of Switzerland and central Europe is assumed to be 15°C and typical thicknesses of Pleistocene permafrost in periglacial areas are probably around 100 to 200 metres (see Frenzel *et al.*, 1992).

RATES OF CHANGE AND RANGES OF PRE-INDUSTRIAL VARIABILITY

Table 5.4 summarizes the measured and reconstructed/estimated changes of alpine glaciers and permafrost. *Rates of change* are expressed as the energy fluxes required to melt glacier and permafrost ice and to warm previously frozen ground (see also Ohmura, 1988). The heat capacity of rocks was taken to be the same as the one for ice, an assumption which is quite well justified for crystalline rocks (see Vonder Mühll and Haeberli, 1990). The energy required for warming cold parts of the glaciers considered was in all cases assumed to remain within the uncertainty of estimates for latent heat of glacier melting. The term 'degradation' includes ground warming as well as melting of permafrost ice. The values given are best estimates for means of several glaciers/boreholes etc. The uncertainty of these estimates, however, is difficult to define because of the representativity problems involved. Estimated anthropogenic greenhouse forcing is from IPCC (1992).

Average rates of glacier mass loss in the Alps during twentieth-century warming were about one order of magnitude higher than in late glacial times. They further increased by about 50 per cent during 1980–1990 with respect to the secular average. Present-day melt rates represent a most effective energy sink, considerably in excess of the estimated anthropogenic greenhouse forcing. Most of the excess, however, can probably be attributed to the mass balance/altitude feedback (around 50 per cent of measured average balances). Warming of alpine permafrost since the late 1980s as observed in the borehole Murtèl appears to indicate an acceleration by a factor of about 5 to 10 as compared with reconstructed secular permafrost warming. Moreover, melting of ground ice as inferred from surface subsidence in permafrost areas determined by high-precision rock glacier photogrammetry seems to have accelerated markedly as well, in 1980–1990 as compared to 1970–1980. Today, warming and melting of alpine permafrost together reflect a much less effective energy sink than glacier melt and clearly remain below assumed anthropogenic effects, probably due to the long response times involved with heat penetration into frozen ground. A considerable amount of heat may, in fact, now be stored in near-surface ground layers and thereby have the potential of leading to future warming and corresponding melting of frozen layers at greater depths.

101

Table 5.4 Thermal characteristics, energy fluxes, and total amounts of energy
involved in alpine glacier and permafrost changes

Latent heat of fusion	3.338×10^5 J/Kg
Heat capacity of ice/rocks	2.100×10^3 J/(kg; °K)
Typical long-term geothermal heat flow	0.1 W/m²
Estimated anthropogenic greenhouse forcing	2.5 W/m²
Ice Age glacier melting	0.3 W/m²
Twentieth-century glacier melting	4.2 W/m²
1980–90 glacier melting	6.9 W/m²
Ice Age permafrost degradation	0.2 W/m²
Twentieth-century permafrost degradation	0.3 W/m²
1980–90 permafrost warming	0.1 W/m²
1980–90 permafrost melting	0.5 W/m²
1980–90 permafrost degradation	0.6 W/m²
Total energy in glacier melting since 1900	10^{10} J/m²
Total energy in permafrost decay since 1900	5×10^8 J/m²

The higher rates of recent secular glacier mass losses as compared to late glacial averages may be attributed first of all to (a) different lengths of considered time intervals, (b) faster modern warming rates, and (c) higher sensitivity of more humid modern glaciers in the Alps (higher mass balance gradients). Most of the recent secular change took place during the second half of the nineteenth century and the first half of the twentieth century, that is in times of weak anthropogenic forcing. These short intervals of fast warming until about 1980 may therefore have been predominantly natural but could have included anthropogenic effects as well. With due consideration to the feedback mechanisms affecting glacier mass balance and the delayed response related to permafrost warming and degradation, the overall secular trend of alpine glacier and permafrost changes could well be more or less in accordance with estimated rates of anthropogenic forcing.

Ranges of preindustrial and prehistoric variability are best documented for glaciers. Reconstruction of past glacier length changes from direct measurements, old paintings, written sources, moraines, pollen analysis, tree-ring investigation, etc. indicates that even in earlier times glacier extent had been reduced as much as today and that the rates of change observed during the twentieth century were probably not uncommon for the Holocene (Zumbühl and Holzhauser, 1988). On the other hand, alpine glacier extent has varied over the past millennia within a range defined by the extremes of the Little Ice Age maximum extent and today's reduced stage (Gamper and Suter, 1982, Figure 6). This means that the situation seems to be evolving towards or even beyond the 'warm' limit of natural Holocene variability. The fundamental importance of glacier length change again becomes evident: as one of the key indicators for fluctuations in

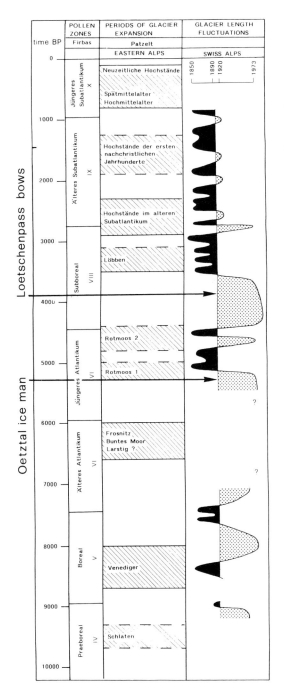

Figure 5.6 Holocene history of glacier length changes in the Alps and its relation to recent archaeological findings from melting ice in saddle configurations (Oetztal ice man/Hauslabjoch, three bows/Lötschenpass). Modified from Gamper and Suter (1982).

the mass and energy balance at the earth surface, this phenomenon allows for quantitative assessments of base levels and variabilities extending far back in time and, thus, greatly helps in disentangling the strong and rapidly increasing human impact on the climate system from the simultaneously occurring externally or internally forced fluctuations.

Most recently, extraordinarily important evidence also emerged from sites other than glacier snouts, that is, from the top of glacier accumulation areas. Even at low altitudes, wind-exposed ice crests and firn/ice divides are not temperate but slightly cold and frozen to the underlying (permafrost) bedrock (Haeberli and Funk, 1991). Such glaciological conditions (reduced heat flow through winter snow, no meltwater, no basal sliding, low to zero basal shear stress at firn/ice divide) explain the perfect conservation of the 'Oetztal ice man', whose body had been buried by snow/ice in a small topographic bedrock depression on such a crest/saddle at Hauslabjoch (Austrian Alps; 3,200 m a.s.l.) more than 5,000 years ago and thereafter remained in place until it melted free in 1991 (VAW, 1993). At an even lower altitude (2,700 m a.s.l.) but at a comparable site (Lötschenpass, Swiss Alps) three well-preserved wooden bows and a number of other archaeological objects were discovered as early as 1934 and 1944. Recent ^{14}C-AMS dating of the three bows gave dendrochronologically corrected ages of around 4,000 years (Bellwald, 1992). These remarkable findings confirm that glacier length variation and overall glacier mass change indeed occur simultaneously if considered at the secular time scale; the use of simple steady-state approaches is thus justified for climatically relevant time intervals. Most essentially, however, the recent archaeological findings from melting ice in saddle configurations indeed prove that the extent of glaciers and permafrost in the Alps may be more reduced today than ever before during the Upper Holocene (see Figure 5.6).

CONCLUSIONS

Besides the extreme differences in the length of the time intervals considered, detailed interpretation of the observations remains difficult and somewhat uncertain. In general, however, alpine glacier and permafrost signals of recent warming trends constitute some of the clearest evidence available concerning past and ongoing changes in the climate system. The glacier changes in particular indicate that the secular changes in surface energy balance may well be in accordance with the estimated anthropogenic greenhouse forcing. In fact, the evidence strongly indicates that the situation is now evolving at a high and accelerating rate beyond the range of (natural) Holocene variability.

REFERENCES

Bellwald, W. (1992) 'Drei spätneolithisch/frühbronzezeitliche Pfeilbogen aus dem Gletschereis am Lötschenpass', *Archäologie der Schweiz*, 15 (4): 166–71.

Chen, J. and Funk, M. (1990) 'Mass balance of Rhonegletscher during 1882/83–1986/87', *Journal of Glaciology*, 36 (123): 199–209.

Finsterwalder, R. and Rentsch, H. (1980) 'Zur Höhenänderung von Ostalpengletschern im Zeitraum 1969–1979', *Zeitschrift für Gletscherkunde und Glazialgeologie*, 16 (1): 111–15.

Frenzel, B., Pécsi, M. and Velichko, A.A. (1992) *Atlas of palaeoclimates and palaeoenvironments of the northern hemisphere – late Pleistocene/Holocene*, Hungarian Academy of Sciences and Gustav Fischer Verlag, Budapest and Stuttgart.

Funk, M. (1985) 'Räumliche Verteilung der Massenbilanz auf dem Rhonegletscher und ihre Beziehung zu Klimaelementen', *Zürcher Geographische Schriften*, 24.

Gamper, M. and Suter, J. (1982) 'Postglaziale Klimageschichte der Alpen', *Geographica Helvetica*, 37 (2): 105–14.

Greuell, W. (1992) 'Hintereisferner, Austria: mass balance reconstruction and numerical modelling of historical length variation', *Journal of Glaciology*, 38 (129): 233–44.

Haeberli, W. (1985) *Creep of Mountain Permafrost – Internal Structure and Flow of Alpine Rock Glaciers*, Mitteilungen der Versuchsanstalt für Wasserbau, Hydrologie und Glaziologie der ETH Zürich 77.

—— (1990) 'Glacier and permafrost signals of 20th-century warming. Symposium on Ice and Climate, Seattle 1989', *Annals of Glaciology* 14: 99–101.

—— (1991) *Alpengletscher im Treibhaus der Erde*, Regio Basiliensis (Sonderband Deutscher Geographentag 1991), 32 (2): 59–72.

Haeberli, W. and Funk, M. (1991) 'Borehole temperatures at the Colle Gnifetti coredrilling site (Monte Rosa, Swiss Alps)', *Journal of Glaciology*, 37 (125): 37–46.

Haeberli, W. and Schmid, W. (1988) 'Aerophotogrammetrical monitoring of rock glaciers', *Proc. 5th International Conference on Permafrost*, Trondheim/Norway 1988, 764–9.

Haeberli, W., Hoelzle, M., Keller, F., Schmid, W., Vonder Mühll, D. and Wagner, S. (1993) 'Monitoring the long-term evolution of mountain permafrost in the Swiss Alps', *Proc. 6th International Conference on Permafrost*, Beijing/China 1993.

Haeberli, W., Huder, J., Keusen, H.-R., Pika, J. and Röthlisberger, H. (1988) 'Core drilling through rock glacier permafrost', *Proc. 5th International Conference on Permafrost*, Trondheim/Norway 1988, 937–42.

IAHS (ICSI)/UNESCO (1985) *Fluctuations of Glaciers 1975–1980* (ed. W. Haeberli), Paris.

IAHS (ICSI)/UNEP/UNESCO (1988) *Fluctuations of Glaciers 1980–1985* (ed. W. Haeberli and P. Müller), Paris.

—— (1989) *World Glacier Inventory – Status 1988* (ed. W. Haeberli, H. Bösch, K. Scherler, G. Østrem and C.C. Wallén), Nairobi.

IPCC (1992) *Climate Change 1992 – the Supplementary Report to the IPCC Scientific Assessment*, Cambridge University Press, Cambridge.

Johannesson, T., Raymond, C.F. and Waddington, E.D. (1989) 'Time-scale for adjustment of glaciers to changes in mass balance', *Journal of Glaciology*, 35 (121): 355–69.

Kasser, P., Aellen, M. and Siegenthaler, H. (1986) *Die Gletscher der Schweizer Alpen 1977/78 und 1978/79*, Glaziologisches Jahrbuch (99, 100, Bericht).

King, L., Gorbunov, A.P. and Evin, M. (1992) 'Prospecting and mapping of mountain permafrost and associated phenomena', in *International Workshop on Permafrost and Periglacial Environments in Mountain Areas*, Interlaken/Switzerland 1992. *Permafrost and Periglacial Processes*, 3 (2): 73–81.

Kuhn, M. (1981) *Climate and Glaciers*, IAHS Publication 131: 3–20.

—— (1990) 'Energieaustausch Atmosphäre – Schnee und Eis', in *Schnee, Eis und Wasser der Alpen in einer wärmeren Atmosphäre*, Mitteilungen der Versuchsanstalt für Wasserbau, Hydrologie und Glaziologie der ETH Zürich, 108: 21–32.

Kruss, Ph. (1983) 'Climate change in East Africa: a numerical simulation from the 100 years of terminus record at Lewis Glacier, Mount Kenya', *Zeitschrift für Gletscherkunde und Glazialgeologie*, 19 (1): 43–60.

Maisch, M. (1992) *Die Gletscher Graubündens – Rekonstruktion und Auswertung der Gletscher und deren Veränderungen seit dem Hochstand von 1850 im Gebiet der östlichen Schweizer Alpen (Bündnerland und angrenzende Regionen)*, Geographisches Institut der Universität Zürich.

Müller, F., Caflisch, T. and Müller, G. (1976) *Firn und Eis der Schweizer Alpen. Gletscherinventar*, Publ. no. 57 des Geographischen Instituts ETH Zürich.

Nye, J.F. (1960) 'The response of glaciers and ice-sheets to seasonal and climatic changes', *Proceedings of the Royal Society of London*, Series A, 256: 559–84.

Oerlemans, J. (1988) 'Simulation of historic glacier variations with a simple climate-glacier model', *Journal of Glaciology*, 34 (118): 333–41.

—— and Fortuin, J.P.F. (1992) 'Sensitivity of glaciers and small ice caps to greenhouse warming', *Science*, 258: 115–18.

Ohmura, A. (1988) 'Role of glaciers in a climatic change', in *Schnee, Eis und Wasser alpiner Gletscher* (Festschrift H. Röthlisberger). Mitteilungen der Versuchsanstalt für Wasserbau, Hydrologie und Glaziologie der ETH Zürich, 94: 109–26.

Patzelt, G. and Aellen, M. (1990) 'Gletscher', in *Schnee, Eis und Wasser der Alpen in einer wärmeren Atmosphäre*, Mitteilungen der Versuchsanstalt für Wasserbau, Hydrologie und Glaziologie der ETH Zürich, 108: 49–69.

Robin, G. de Q. (1983) *The climatic record in polar ice-sheets*, Cambridge University Press, Cambridge.

Scheuer, J. (1992) *Quantifizierung des Gletscherschwundes seit der kleinen Eiszeit im südlichen Wallis*, Diploma thesis, University of Trier (unpublished).

VAW (1993) 'Greenhouse gases, isotopes and trace elements in glaciers as climate evidence for the Holocene – report on the ESF/EPC workshop, Zürich, 27–28 October 1992', *VAW Arbeitsheft*.

Vonder Mühll, D. (1992) 'Evidence of intrapermafrost groundwater flow beneath an active rock glacier in the Swiss Alps', in *International Workshop on Permafrost and Periglacial Environments in Mountain Areas*, Interlaken/Switzerland 1992. *Permafrost and Periglacial Processes* 3 (2): 169–73.

—— (1993) *Geophysikalische Untersuchungen im Permafrost des Oberengadins*, Dissertation no. 10107, ETH Zürich.

Vonder Mühll, D. and Haeberli, W. (1990) 'Thermal characteristics of the permafrost within an active rock glacier (Murtèl/Corvatsch, Grisons, Swiss Alps)', *Journal of Glaciology*, 36 (123): 151–8.

Vonder Mühll, D. and Holub, P. (1992) 'Borehole logging in alpine permafrost, Upper Engadin, Swiss Alps', in *International Workshop on Permafrost and Periglacial Environments in Mountain Areas*, Interlaken/Switzerland 1992. *Permafrost and Periglacial Processes*, 3 (2): 125–32.

Vonder Mühll, D. and Schmid, W. (1993) 'Geophysical and photogrammetrical investigations of rock glacier Muragl I, Engadin, Swiss Alps', *Proc. 6th International Conference on Permafrost*, Beijing/China 1993.

WGMS (1991) *Glacier mass balance bulletin no. 1* (ed. W. Haeberli and E. Herren), World Glacier Monitoring Service, ETH Zürich.

—— (1993) *Glacier mass balance bulletin no. 2* (ed. W. Haeberli, E. Herren and M. Hoelzle), World Glacier Monitoring Service, ETH Zürich.

Wood, F.B. (1990) 'Monitoring global climate change: the case of greenhouse warming', *Bulletin of the American Meteorological Society*, 71 (1): 42–52.

Zumbühl. H.J. and Holzhauser, H. (1988) 'Alpengletscher in der Kleinen Eiszeit', *Die Alpen*, 64 (3).

6

MONITORING SNOW COVER VARIATIONS IN THE ALPS USING THE ALPINE SNOW COVER ANALYSIS SYSTEM (ASCAS)

M.F. Baumgartner and G. Apfl

INTRODUCTION

The changes in snowfall patterns in the Alps during the last few years have widely raised public awareness to the possible consequences of global climate change. There are several scientific reasons why snow is, in fact, a significant factor in this context: first of all, snow plays an important role in the hydrologic cycle and can be affected significantly by climate change. It is likely that temperatures will increase over the next decades, influencing the snow cover patterns in alpine regions. Because of the different physical properties of the snow cover compared to non-snow-covered surfaces, the energy balance is changing with an increase in temperature and with a consecutive decrease of the snow cover, resulting in a feedback to the atmosphere. Additionally, changes in the snow situation can have serious economic consequences:

1 Melt-water in the Alps is stored in reservoirs and is used for electricity production (60 per cent of the electricity in Switzerland is produced by water power).
2 Tourism during winter time, in alpine regions an important economic factor, can be severely affected by such changes. Furthermore, the influence of changes of the snow cover duration on vegetation has to be studied carefully.

Recordings of alpine meteorological stations show a later start of snowfall and snow accumulation but no considerable change in snow depths. First comparisons of quick-look photos of NOAA–AVHRR (National Oceanic and Atmospheric Administration – Advanced Very High Resolution Radiometer) data have shown that after the mid-eighties, snow distribution in the

Alps and the pre-alpine lowlands has changed drastically. This means that the same amount of precipitation is measured as before but the snow-rain level has been changed both spatially and temporally. The significant changes of snowfall patterns led to the idea of monitoring such changes on a long-term basis. The Alpine Snow Cover Analysis System (ASCAS) has been designed, therefore, to monitor the full range of the Alps. For detailed, quantitative calculations, five basins in Austria, France and Switzerland were selected representing different climatic regions (western vs. central vs. eastern, and northern vs. southern alpine climate).

Spatial and temporal snow cover variations are difficult to monitor using point measurements of climate stations. Satellite remote sensing represents an excellent tool for this purpose, if a satellite system with a high temporal resolution is available. Today, NOAA–AVHRR data are the only useful data for large-scale monitoring tasks because of its temporal resolution of 12 hours and its large field-of-view (110°). A receiving station built in-house (Baumgartner and Fuhrer, 1991) for NOAA–AVHRR–HRPT (High Resolution Picture Transmission) data gives guaranteed access to remote sensing data and, therefore, allows a permanent and actual monitoring of the alpine snow cover. Additionally, data have been archived since 1981, making a comparison of the actual with the 'historical' snow cover patterns possible.

TECHNICAL BACKGROUND OF ASCAS

Monitoring and quantifying snow cover variations is carried out using the ASCAS system. ASCAS integrates commercial and user-developed software on a single microcomputer-based system including digital satellite image processing, geographic information systems (GIS), data base management systems, and hydrologic/climatologic modelling. A new approach in such an integrated system is the application of a hydrologic/climatologic modelling module. The combination of these software modules into an integrated system sounds rather trivial, but, as it has been shown by Ehlers *et al.* (1989), serious problems exist because of the basically different character of satellite-image (raster) data and of vector-oriented data in a GIS systems. This means that (1) the users have to deal with not only one single data format but with a variety of different formats, and (2) that the exchange of data between the different modules is complicated.

Since the practical aspect of snow cover monitoring is dominant in this project, we focus with ASCAS not on a theoretical development of a 'totally integrated' system as proposed by Ehlers *et al.* (1989) but on the integration of commercial and user-developed software modules for operational applications. With ASCAS we are able to show that even commercial software can be used in research and only project-specific software modules have to be developed by the user. For flexible adaptions to user-specific needs, a software tool-kit (or the source code) must be available with commercial software. The integration of the different software modules into ASCAS and connecting them with interfaces leads to a greater ease of the interpreter's work, that is, to work with such a system configuration is less time-consuming and, therefore, economically more attractive.

To make ASCAS interesting for other users (for example, for hydro-power stations), it is implemented on a single microcomputer-based system (Baumgartner and Rango, 1991). All software modules are implemented on the same system using conventional hardware components including a high-resolution colour monitor, a digitizing tablet, a high-capacity hard disk (1.2 Gb), a cartridge tape drive, and a colour printer. The only specialized component is the image memory (frame buffer) for displaying and processing satellite data with a resolution of 1024 x 1024 pixels and 32 Bits per pixel (8 Bits for the four planes: red, green, blue and graphics overlay).

In designing ASCAS, the following guidelines had to be taken into account:

- Input and processing of remote sensing data for monitoring snow cover variations using NOAA–AVHRR data (image processing module).
- Input and manipulation of climate (temperature, precipitation, and snow depth, etc.) and hydrologic data (runoff, etc.) on a daily basis using a data base management system (DBMS module).
- Input of topographic information, digitized from topographic maps (GIS module).
- Transfer of snow cover maps to the GIS module.
- Transfer of DBMS data to the GIS module.
- Spatial and temporal analyses (GIS module).
- Transfer of DBMS as well as of GIS data, including the results of the analyses, to the Snowmelt Runoff Model (SRM) (Martinec et al., 1983).
- Snowmelt runoff simulations, snow volume calculations, climate change simulations (SRM module).
- Transfer of SRM-simulation results to the GIS or DBMS module.

ANALYSES AND MODELLING WITH ASCAS

In the following section, we give an overview of possible calculations using the ASCAS system including the processing of satellite data for deriving snow cover maps, GIS snow cover analyses, and SRM modelling in three test basins in the Swiss Alps.

Snow cover monitoring

Satellite image processing is, for the most part, still interpreter-controlled. Such semi-automated procedures give accurate results but they are time-consuming if time series evaluations are carried out (Baumgartner, 1990a and 1990b).

In a first step, NOAA–AVHRR data are transferred band by band from the archives of the receiving station to the image processing system. Originally, AVHRR data are received in a 10-bit format. For display reasons, they are reduced to 8 bits (most of the frame buffers today operate on a 8-bit basis) using a user-developed software module offering several possibilities of bit reduction. Following this procedure, the single bands are copied on to one file (band sequential) adding a software-dependent, system-internal header to the image file. This header gives specific information about the raw data and the actual situation of the data set (for example, coordinates of subframe, processing steps performed, etc.).

Secondly, the satellite data are displayed on the RGB monitor via the frame buffer. The frame buffer allows the interpreter to enhance the contrast of the image using look-up tables. We found that the combination of AVHRR bands 2 (near infrared), and 1 (visible) in the red, green, and blue layer of the frame buffer gives the best representation for the interpretation of snow cover variations. Using the middle or far infrared bands proved to be very difficult in alpine terrain due to temperature inversions and extreme differences of emitted radiation based on the exposure and slope angles.

A third procedure is the determination of the snow-covered area (see, for example, Baumgartner, 1990a). Most commonly, classification techniques by supervised learning are used. These techniques are based on the statistics of samples in the satellite data for several predetermined categories. Based on the probability-density function of the different categories, the feature space is subdivided using algorithms as maximum likelihood, minimum distance or mahalanobis distance classifiers, classifying each pixel into its specific category. The final result of such a procedure is a thematic map differentiating, for example, between snow-covered and non-snow-covered areas.

In this project the following intermediate categories were distinguished:

- water
- agricultural land
- forests
- urban area
- transition zone (snow/non-snow)
- dry snow
- wet snow

For each of these categories, several samples are selected in an iterative process based on scatterplots (two-dimensional histograms). Best results were obtained by classifying AVHRR bands 1 and 2 using the minimum distance or mahalanobis distance classifier. The accuracy can be checked by the interpreter very easily by displaying the results on the image processing system. A fading or split-screen function allows the comparison of the original and of the classified satellite image. It should be noted that estimating the accuracy of a classification requires the interpreter's skill and knowledge of the basin; otherwise, extensive fieldwork is necessary. Furthermore, the categories are summarized into two main classes, that is, snow and non-snow.

In a fourth step, the classified data are related to a reference coordinate system. In this project, an Albers conical, equal-area projection was used for geocoding. This projection system minimizes distortions in east–west direction which is important for monitoring the full range of the Alps. The transformation is based on the ground-control-point approach taking into account corrections for skew and for panoramic distortions (due to AVHRR's large field-of-view of 110°).

These four processing steps are repeated for each satellite image during the period of interest, e.g. the ablation period (in this project, as is discussed below, 42 images were analysed). Since all data are geocoded they can be superimposed on to a multivariate data set, ready to transfer to the GIS system. The interface between the image processing and the GIS module was commercially available and is operational but rather complex to handle. First, the classified snow cover maps are vectorized and reformatted to the USGS/DLG (digital line graph) intermediate format. Secondly, the DLG format is transformed into the BNA (Atlas ASCII boundary) intermediate format. Due to the internal organization of the GIS, the BNA format has to be converted into a graphics intermediate format and finally into the GIS format.

GIS analyses

Topologic and attribute data are integrated into the GIS system. Topologic information such as rivers, lakes (including reservoirs), elevation lines (with 500 m equidistance), basin boundaries, and locations of climatic stations and stream gauges is digitized from topographic maps. Since each of the countries has its own reference coordinate system and its own data formats, ASCAS was designed to transform the different data sets on to a common reference system and to handle all the different data formats. For each geographic layer (e.g., lakes), a meta-attribute file was added giving topology-specific information. These files do not contain data (such as climate and hydrologic data) but they give detailed information on where the data are stored. It was decided to store all these climatological (daily temperature minimum, temperature maximum, precipitation, snow depths, etc.) and hydrologic (daily runoff) data in an external data base management system. In this manner, the GIS is not overloaded and GIS analyses are facilitated resulting in a reduction of CPU time.

After the integration of these data, advanced GIS analyses can be carried out:

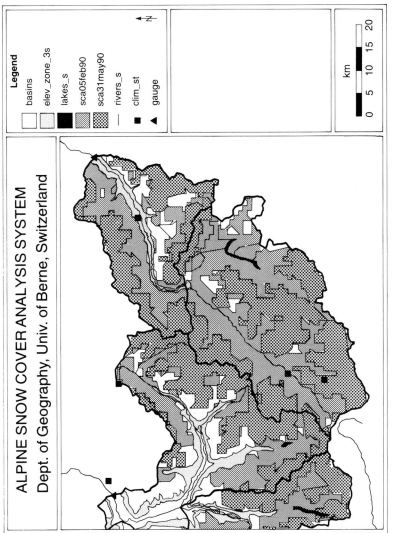

Figure 6.1 Intersection of the layer elevation zone and snow cover on two dates (5 February 1990, 31 May 1990) for the Inn–Martina Basin (Switzerland).

- Based on the layer elevation lines, elevation zones of 500 m equidistance were derived.
- Based on the layer elevation zones and basin boundaries, elevation zones per basin and sub-basin were derived by intersecting these two layers.
- By intersecting the layers elevation zone and snow cover of a specific date, the snow coverage (in per cent) per elevation zone and basin was determined (Figure 6.1). These results serve as input to the SRM module or can, additionally, be intersected with the layer aspects (e.g. north-east, south-east, south-west, north-west), used in radiation computations (Figure 6.2).
- Using a time series of snow cover layers, the temporal variation of the snowline was derived (either for the full basin or for a specific elevation zone) representing the characteristic accumulation and ablation patterns of a basin. If snow cover layers over several years are available, snow cover or ablation probability charts can be produced.
- An interesting feature is the overlay of snow cover layers with the elevation model resulting in elevation-dependent snow cover variation plots. It can be shown that during the winter of 1990 in three Swiss basins (Rhine–Felsberg, Ticino–Bellinzona (Figure 6.3), the snowline never reached further down than 1,600 m a.s.l. which is significantly higher than some years ago when even the pre-alpine lowlands (600 m a.s.l.) were snow covered for at least three to four weeks.

Hydrologic and climatological modelling

For modelling snowmelt runoff and the influence of climate change on to runoff, the Snowmelt Runoff Model (SRM) is used (Martinec et al., 1983). SRM is based on the degree-day method and calculates the snowmelt runoff on a daily basis. Input variables are daily values for temperature (minimum and maximum), precipitation and snow cover. SRM is one of the few models allowing the input of snow cover data derived from satellites. Extensive testing by WMO (1986) has shown the effectiveness and accuracy of SRM: the model was considered best in most of the tests. Due to the simple degree-day method, where temperature stands as index for the energy balance, the model runs with low CPU times; computing time is another important consideration for an operational use of a model.

The input of the variables stems from two sources within ASCAS: temperature, precipitation, and runoff from the DBMS module; and snow cover per elevation zone from the GIS module.

Using SRM, snowmelt runoff simulations for the ablation period 1990 have been carried out in three Swiss basins representing different climatic regions:

- Rhine–Felsberg basin. Influence of oceanic climate (westerly winds).

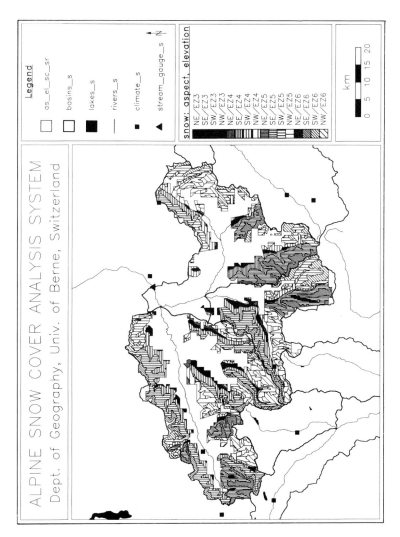

Figure 6.2 Intersection of the layers elevation zones (EZ3, EZ4, EZ5, EZ6), aspects (NE, SE, SW, NW), and snow coverage (5 February 1990) for the Rhine–Felsberg Basin (Switzerland).

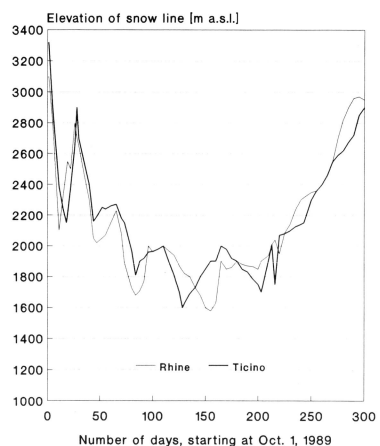

Figure 6.3 Overlay of a time series of snow cover data with the elevation model showing the snowline variations (starting 1 October 1989) in the Rhine–Felsberg and the Ticino–Bellinzona Basins (Switzerland).

- Ticino–Bellinzona. Mediterranean influence.
- Inn–Martina. Inner-alpine, dry valley (continental).

Figure 6.4 gives an example for the snowmelt runoff simulations. In general, the simulated runoff corresponds well to the measured runoff. The accuracy of the simulations was between 3 per cent and 7 per cent (comparison of simulation and measured runoff over the full snowmelt period). These results can be considered satisfactory since estimates of the accuracy made by the water-power stations do not reach these values. The main reason for the inaccuracies in our simulations is the manipulation of reservoirs by the water-power stations for electricity production. For a comparison with the simulated runoff, the natural runoff has to be known. This must,

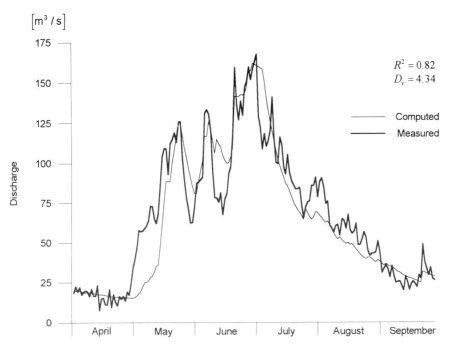

$R^2 = 0.82$
$D_v = 4.34$

——— Computed
——— Measured

Figure 6.4 Snowmelt runoff simulation for the Inn–Martina Basin (1990) using SRM.

therefore, be reconstructed by eliminating the influence of all major reservoirs in the basin. It is obvious that it was not possible to eliminate all these effects. Another reason for certain inaccuracies is heavy rainfall occurring during the snowmelt season.

For simulating the snowmelt runoff, the model needs the snow volume for 1 April. Therefore, the model calculates the snowmelt depth which is, consecutively, related to the snow cover. The results can be extracted from the model by plotting the snowmelt depths vs. the snow-covered area for a specific basin and elevation zone. It can be shown that between the three basins, significant differences in snow volume exist: in the elevation zone of 1,500 m–2,500 m a.s.l., the Inn valley had in 1990 less snow than the other two basins which both showed almost the same volume. In the next higher elevation zone (2,500 m–3,500 m a.s.l.), even the Rhine and Ticino basin showed a clear difference in snow volume. The Ticino basin is often influenced by the Mediterranean climate bringing a large amount of precipitation to the southern part of the Alps which falls in higher elevations mainly as snow. All the results fulfilled the expectations we had from regional climatology.

The SRM model includes a submodule for calculating the influence of temperature and precipitation changes on to snow cover, snow volume, and

117

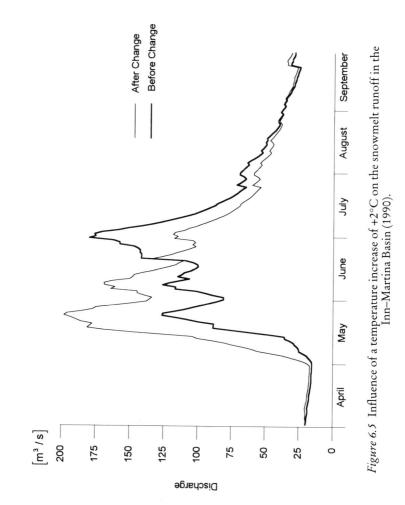

Figure 6.5 Influence of a temperature increase of +2°C on the snowmelt runoff in the Inn–Martina Basin (1990).

on to runoff (Rango, 1992). In our project, a scenario for a temperature increase of +2°C was calculated for all three basins. For these calculations, no changes in the variable precipitation and in the parameters were included since tests of such changes have shown less effect on the hydrograph than the effects caused by the higher temperatures. It can be shown, that with this scenario, SRM produces a markedly changed hydrograph (Figure 6.5). The snowmelt runoff occurs in all basins about two to three weeks earlier than under present conditions. Additionally, the ablation period is shorter, that is, the snowmelt is faster and produces, therefore, more runoff in a shorter time. Calculations in the USA (Rango and van Katwijk, 1990) have shown similar results. These calculations are a hint for the effects climate change could have on snow cover and on snowmelt runoff. For hydro-power stations this would imply at least a change in the storage and production schedule. If the snowline remains at the elevation pointed out above, or if it even ascends, major losses of storage water stemming from snowmelt runoff have to be expected. For winter tourism in the Alps, the ski resorts at lower elevations have to expect a shorter season (artificial snow cannot be produced at these elevations because of the warmer temperatures). Furthermore, the influence on vegetation has to be mentioned.

SUMMARY AND OUTLOOK

The paper showed how the Alpine Snow Cover Analysis System (ASCAS) can be used for monitoring snow cover variations in the Alps. In three different alpine basins, snow cover variations were monitored during the winter of 1990 using NOAA–AVHRR data. Spatial analyses of the snow cover and of topographic information were carried out on a GIS system. The results of the image processing and the GIS analyses were transferred to the Snowmelt Runoff Model (SRM) for snow volume and snowmelt runoff calculations in all three basins. The results have shown that during the winter of 1990, the snowline was never (for more than two or three days) below 1,600 m a.s.l. which means a significant difference compared to the winters before 1985. Differences in the snow volume between the three basins under investigation reflect the different climatic situations. With the climate change module of SRM, the influence of temperature variations on the snow cover ablation and on the snowmelt runoff were discussed.

The results presented here have shown that ASCAS is a valuable tool for monitoring the snow cover variations in the Alps as well as for simulating snowmelt runoff and effects of climate change. The simulations here served only as a test for ASCAS. Consequently, ASCAS will be used for further calculations: (1) Additional basins such as the Durance (France) and the Salzach (Austria) basins will be included. (2) To get an impression of the past and present snow cover situation in these basins, several ablation periods will be taken into account. (3) Climatic and economic impact studies are planned.

REFERENCES

Baumgartner, Michael F. (1990a) 'Snowmelt runoff simulations based on snow cover mapping using digital Landsat–MSS and NOAA–AVHRR data', Technical Report HL-16, US Dept of Agriculture, Agricultural Research Service, Hydrology Laboratory, Beltsville, MD.

—— (1990b) 'Snow cover mapping and snowmelt runoff simulations on micro-computers', in *Remote Sensing and the Earth's Environment*, ESA SP-301: 77–85.

Baumgartner, Michael F. and Fuhrer, M. (1991) 'A Swiss AVHRR and Meteosat receiving station', *Proc. 5th European AVHRR Users Meeting* held in Tromso, Norway, 23–33.

Baumgartner, Michael F. and Rango, A. (1991) 'Snow cover mapping using micro-computer image processing systems', *Nordic Hydrology*, 22: 193–210.

Ehlers, M., Edwards, G. and Bedard, Y. (1989) 'Integration of remote sensing with geographic information systems: a necessary revolution', *Photogrammetric Engineering and Remote Sensing*, 55 (11), November: 1619–27.

Martinec, J., Rango, A., and Major, E. (1983) *The Snowmelt Runoff Model (SRM) User's Manual*, NASA Reference Publ. 1100, Scientific and Technical Information Branch.

Rango, A. (1992) 'Worldwide testing of the Snowmelt Runoff Model with applications for predicting the effects of climate change', *Nordic Hydrology*, 23: 155–72.

Rango, A. and van Katwijk, V. (1990) 'Climate change effects on the snowmelt hydrology of western North American mountain basins', *IEEE Transactions on Geoscience and Remote Sensing*, GE-38 (5): 970–4.

WMO (1986) *Intercomparison of models of snowmelt runoff*, Operational Hydrology Report no. 23, World Meteorological Organization, Geneva.

7

EFFECTS OF MESOSCALE VEGETATION DISTRIBUTIONS IN MOUNTAINOUS TERRAIN ON LOCAL CLIMATE

R. A. Pielke, T. J. Lee, T. G. F. Kittel, T. N. Chase, J. M. Cram, and J. S. Baron

INTRODUCTION

Even a casual observer from an aircraft will note the varied landscape of mountainous terrain. These variations in land surface include the terrain features themselves as well as patchiness from different vegetation types, surface geology, urbanization, etc. There are two major questions related to climate system dynamics that need to be addressed concerning this landscape heterogeneity:

(i) how would a large-scale climate change influence this landscape, and
(ii) does landscape spatial and temporal structure influence the larger-scale climate?

In this paper, we examine aspects of both questions.

GOALS OF COLR – A RESEARCH PROJECT TO STUDY POTENTIAL CLIMATE CHANGE IN A MOUNTAINOUS REGION

The Colorado Rockies (COLR) Global Change Program is a five-year research effort supported by the US National Park Service with three major, connected goals. We will evaluate important mesoscale climate change scenarios for the Colorado Rocky Mountains, explore the consequences of changing climate on mountain aquatic and terrestrial ecosystems, and address the important feedbacks from land surface processes to mesoscale climate. This will be accomplished with a combination of field experiments in Rocky Mountain National Park and nearby Niwot Ridge, coupled with the

Regional Atmospheric Modeling System (RAMS) developed at Colorado State University (Pielke *et al.*, 1992) and a land surface process data and simulation system, RHESSys (Regional HydroEcological Simulation System, Band *et al.*, 1993), that combines a set of remote sensing and Geographic Information System (GIS) techniques with integrated hydrological and ecological models.

Critical to the Colorado Rockies research effort is the assumption that the RAMS and RHESSys models adequately simplify complex atmospheric, hydrologic, and ecologic processes to where we can use them as tools for increasing understanding of two-way interactions between climate and land surface processes (Pielke *et al.*, 1993a). Are responses of components of climate, hydrologic, biogeochemical, and ecologic systems significantly modified by these interactions? What are the feedbacks of surface processes to mesoscale and larger-scale atmospheric dynamics? How does feedback of surface processes to the atmosphere vary as the specificity and complexity of surface and atmospheric models are altered? How much complexity do the models require in order to retain the important interactions and dynamics that can be observed? These are questions that will be vital to address in order to evaluate the potential effects of increasing levels of atmospheric greenhouse gases and land-use change on hydrologic, climate, and ecosystem processes (i.e., the environmental vulnerability).

Land surface processes are explicitly defined as combined hydrologic and ecosystem processes. While the addition of land surface characteristics improves the performance of mesoscale atmospheric models, up to the present these surface boundary conditions are usually fixed over time or have a prescribed annual cycle, such that vegetation or hydrologic structure cannot respond to climatic excursions away from assigned boundaries. Similar circumstances are often found in land surface process models; climate is prescribed and fixed in order for the model to focus instead on feedbacks and interactions between hydrologic and ecosystem processes. We know, however, that such simplifying constraints are artificial. Changes in regional or global precipitation and temperature patterns would have dramatic long-term consequences to terrestrial ecosystems and regional-scale hydrology (Rango and van Katwijk, 1990; Revelle and Waggoner, 1983; Running and Nemani, 1991; Poiani and Johnson, 1991) that, in turn, will influence mesoscale climate. Possible changes in ecosystem function that affect physical and chemical climate include alteration of flux rates and storage terms for water and nutrients (including carbon), trace gas exchanges, and changing vegetation structure (Ojima *et al.*, 1991, 1992; Schimel *et al.*, 1990, 1991).

MESOSCALE CLIMATE MODELLING RESULTS RELATED TO COLR

The Regional Atmospheric Modeling System (RAMS) developed at Colorado State University is being used to investigate the coupling of landscape processes with weather and climate. RAMS is described in Tremback *et al.* (1986), Cotton *et al.* (1988), Tripoli and Cotton (1989a, b), Schmidt and Cotton (1990), Cram *et al.* (1992a, b), Lee *et al.* (1993), and Pielke *et al.* (1992). Two examples of its use so far in COLR to investigate the impact of vegetation and its spatial distribution on mesoscale climate are presented – one for the Rocky Mountain National Park area and one for the adjacent Central Great Plains of the United States.

Central Great Plains results: sensitivity to vegetation characterization

Based on the work of Deardorff (1978), McCumber (1980) and McCumber and Pielke (1981) developed a soil-vegetation model which was later adopted and modified by Avissar and Mahrer (1982, 1988). In 1989, this vegetation model was installed into a version of the RAMS. It was completely restructured in 1991 as the Land Ecosystem-Atmosphere Feedback (LEAF) model (Lee, 1992; Lee *et al.*, 1993).

LEAF uses an elevated canopy structure, which is similar to that used by BATS (Dickinson *et al.*, 1986) and SiB (Sellers *et al.*, 1986), and incorporates three aerodynamic resistance functions and one stomatal conductance function. The stomatal conductance function was originally developed by Avissar *et al.* (1985) where environmental controls (e.g., solar radiation, ambient temperature, vapour pressure deficit, and soil water potential) on stomatal closure were considered.

The canopy-air-to-atmosphere aerodynamic resistance function was constructed according to surface-layer similarity theory, the canopy-to-canopy-air aerodynamic resistance function followed Goudriaan (1977), and the soil-to-canopy-air aerodynamic resistance function was based on Shuttleworth and Wallace (1985). The energy combination theory by Shuttleworth and Wallace provides an estimation of below-the-canopy energy transport which is believed to be important for sparse canopies. Primary parameters used in the model are leaf area index (LAI), maximum stomatal conductance, and albedo. Secondary parameters are displacement height and environmental controls on stomatal function.

LEAF has been tested (Lee *et al.*, 1992) against FIFE87 and FIFE89 observational data (the acronym FIFE refers to the First ISLSCP Field Experiment; Sellers *et al.*, 1988). It is being used, along with data supplied by the United States Geological Survey (USGS), to study the vegetation impact on the atmospheric boundary layer and convective storms for the Central Great Plains of United States (Lee, 1992, as reported in the following text).

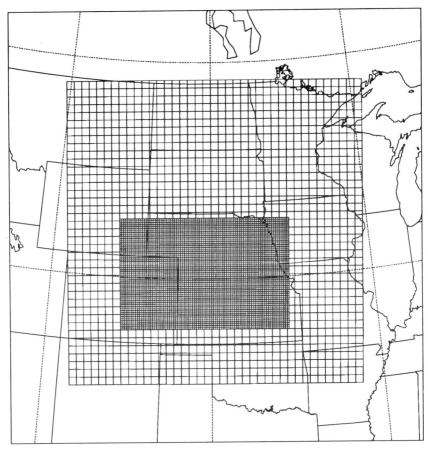

Figure 7.1 Computational grids. The outer mesh shows a 40 × 40 grid with 40 km
grid increment. The finer 90 × 58 grid has a grid increment of 10 km.

A nested grid system as shown in Figure 7.1 was designed for the simulation
of surface evapotranspiration over the area. The outer (coarse) grid has a grid
increment of 40 km and covers the Northern and Central Great Plains. The
inner (fine) grid has a 10 km grid increment and covers eastern Colorado,
Nebraska, and Kansas, including the immediate eastern flanks of the Rocky
Mountains. A study day of 6 June 1990 was chosen because a well organized
mesoscale convective complex developed near the Nebraska–Kansas border.
One outflow boundary travelled westward and became stationary in eastern
Colorado. Although synoptic forcing was strong (see Figure 7.2), the
strength of vegetation forcing on mesoscale circulation was explored by
evaluating differences between simulations with and without LEAF.

Advanced Very High Resolution Radiometer (AVHRR) derived data sets
were used in the study. Among these data, the Normalized Difference

124

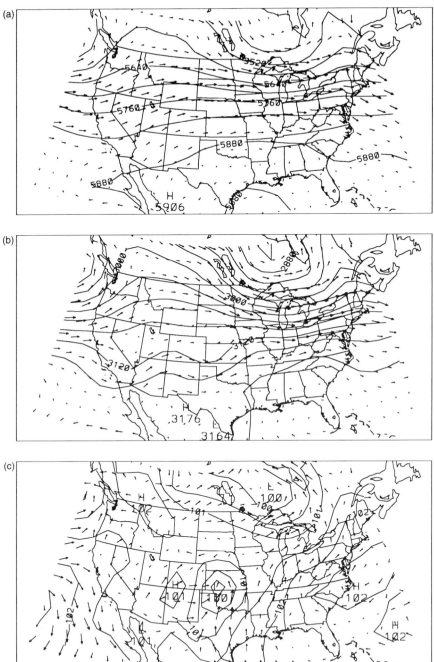

Figure 7.2 50 kpa height (m) (a), 70 kpa height (m) (b), and surface pressure (kpa) analyses (c) at 1,200 UTC on 6 June 1990.

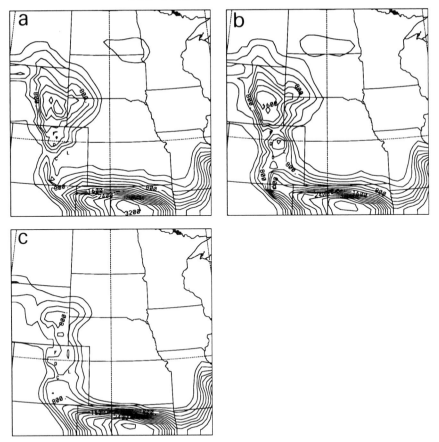

Figure 7.3 RAMS simulated convective available potential energy (CAPE m²/s⁻²), assuming an air parcel at the surface is lifted upward, at 0000 UTC 7 June 1990 for three different surface covers: (a) simulation with current vegetation; (b) simulation with uniform short grass replacing agricultural areas; and (c) simulation with bare soil. Locations marked on the maps are Ft. Collins (F), Denver (D), Limon (L), and Colorado Springs (C).

Vegetation Index (NDVI) and land cover characteristics (Loveland *et al.*, 1991) were most valuable in determining vegetation type and potential evapotranspiration. The high resolution land cover characteristics data base was also used in determining percentage coverage of land, vegetation, and water bodies. A digital elevation model (DEM) was utilized to provide terrain forcing.

Figure 7.3 shows results at 0000 UTC (1800 LST for the central time zone) from simulations with LEAF (Figures 7.3 a, b) and without LEAF (Figure 7.3c; the latter experiment assumes bare ground everywhere). Experiments with LEAF used the AVHRR-based current characterization of the land-scape (Figure 7.3a) or were run with a uniform short grassland replacing agricultural land to better represent pre-European settlement conditions (Figure 7.3b). The fields shown are convective available potential energy (CAPE) which is a measure of the expected intensity of thunderstorms. In eastern Colorado, the values of CAPE were about twice as large when vegetation was included (Figures 7.3b versus 7.3c). Moreover, the man-modified pattern of land cover resulted in a very different pattern of CAPE (Figure 7.3a) in this region than, according to the model, would occur with a uniform landscape.

Also, when vegetation was included, the simulated onset of cumulus convection lagged behind the results without vegetation. This is due to stronger heating over bare soil than over vegetated surfaces. However, the maximum condensate is higher in the simulation with transpiring vegetation because of the larger value of CAPE; also the water vapour mixing ratio near the surface is higher throughout the computational domain when vegetation is considered. Additional intercomparison details are found in Lee (1992).

Rocky Mountain National Park results: sensitivity to potential vegetation distribution change

A numerical modelling experiment was performed using RAMS to test the effect of changing the elevation of lower and upper treelines (the lower boundary between grasslands on the plains and conifer forests in the foothills and the upper boundary between conifer forests and alpine tundra) on local weather circulations in the mountain-plains region of northern Colorado.

The simulation options in this study included a nonhydrostatic formulation, a surface layer and soil model (Tremback and Kessler, 1985), a radiation parameterization (Chen and Cotton, 1983a, b), a convective parameterization based solely on the prediction of vertical variances (Weissbluth, 1991), and explicit microphysics.

The model domain includes most of the continental US and consists of three nested grids (Figure 7.4) centred over the northern Colorado Front Range and Rocky Mountain National Park. The largest grid has a 125 km spacing with 38 x 30 horizontal points. The second grid is spaced at 25 km with 47 x 37 points, and the finest grid includes 50 x 42 points with a 6.25 km grid spacing.

There are 28 vertical levels reaching to approximately 19 km. The model was initialized from ECMWF (European Centre for Medium-range Weather Forecasting) data for 26 July 1985 at 1200 UTC. This was during a period of summer mid-continental monsoon conditions where high pressure over the western US minimized dynamically forced convection.

Two model simulations were designed to test the effect of changing treelines (presumably a potential effect of global climate change although it

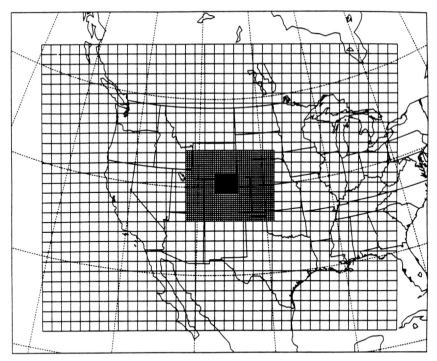

Figure 7.4 Nested model grids centred over the northern Colorado Front Range and Rocky Mountain National Park. The largest grid has 38 × 30 points with a 125 km grid increment. The second grid has 47 × 37 points with a grid increment of 25 km. The finest grid has 50 × 42 points and a 6.25 km grid increment.

Table 7.1 Vegetation albedo and roughness length values (in metres) and elevations distributions used in control (CON) and altered vegetation distribution (WARM) simulations

Region	Albedo	Roughness	CON	WARM
Alpine tundra	0.2	0.01 m	> 3050 m	> 3660 m
Forest	0.1	1.0 m	> 1982 m	> 2745 m
			< 3050 m	< 3660 m
Grasslands	0.3	0.05 m	< 1982 m	< 2745 m

could be caused by direct human intervention) on regional-scale circulations. Three different vegetation areas – tundra, forest, and grassland – were defined in the model by the albedo and roughness lengths shown in Table 7.1.

Vegetation effects were not included in any other manner in the model simulations reported in this section. The control simulation (CON) set the treelines at their approximate current levels, while in the second simulation

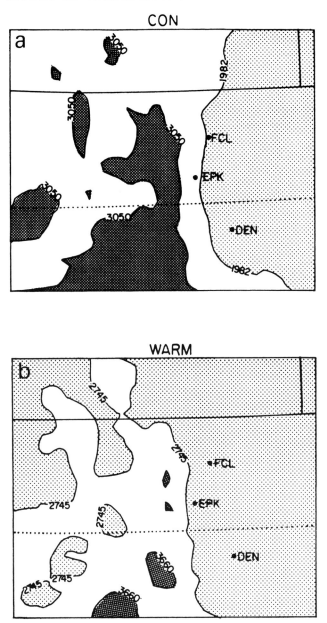

Figure 7.5 Vegetation patterns in Grid 3 (see Figure 7.4) used in (a) CON (control) and (b) WARM simulations. The light grey areas are grasslands, the white is forest, and the dark grey is alpine tundra. The specific values of albedo and roughness length corresponding to each simulations are given in Table 7.1. Complete topography is shown in Figure 7.6.

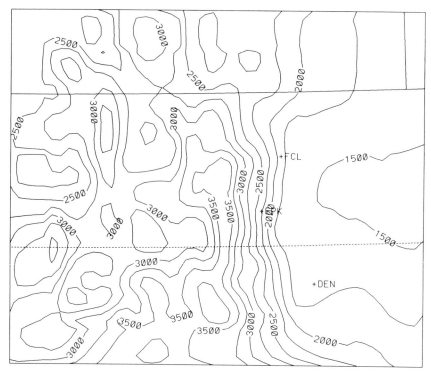

Figure 7.6 Topography in Grid 3. Contours are at 250 m intervals.

(WARM) the treelines were moved higher to account for a hypothesized regional warming effect. Figure 7.5a and b shows the different vegetation areas in the two simulations. Figure 7.6 gives the topography for this region.

The WARM simulation had a larger area of lower albedo (the conifer forest area), which was also shifted to a higher elevation than in the CON simulation. Surfaces of lower albedo will heat up more quickly. However, the lower albedo areas in this situation are also associated with a much greater roughness length, which implies increased mixing above the surface which leads to cooling of the surface. The net effect in the model was cooler surfaces associated with forest areas and warmer surfaces associated with the alpine tundra. The WARM simulation thus ended up with overall lower surface temperatures. Although it might be expected that this would lead to a decrease in the intensity of vertical circulations in the WARM simulation, or that warming in the alpine zone would intensify convection in that region, there was no discernible systematic trend in differences in the magnitudes of the vertical velocities in the two simulations. Although the magnitudes were not consistently stronger or weaker, differences in the spatial pattern of vertical velocity did occur.

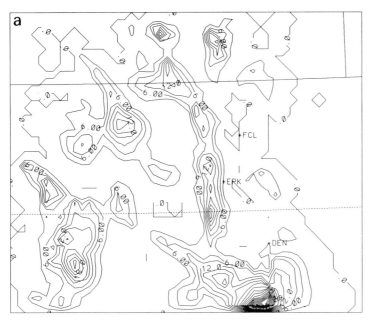

Figure 7.7a Total precipitation after 12 hours on the finest grid (Grid 3) for the CON simulation. Contour interval is 3 mm.

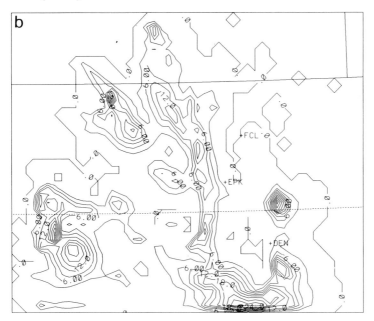

Figure 7.7b Total precipitation after 12 hours on the finest grid (Grid 3) for the WARM simulation. Contour interval is 3 mm.

131

The pattern of rainfall also differed (Figure 7.7a and b), although the terrain was clearly still the dominant influence. These simulations suggest that a change in land surface characteristics, such as might be expected from regional warming, can lead to significant and perhaps unexpected changes in regional-scale climate. This arises because the interacting processes controlling regional dynamics are nonlinear in nature and can initiate both positive and negative feedbacks.

CONCLUSIONS

In simulations for the Central Plains, inclusion of a detailed model of biophysical processes (LEAF) and realistic geography of land cover characteristics demonstrated the effect of vegetation and its spatial heterogeneity on the development of convective systems. In mountainous regions, the influence of vegetation might be expected to be swamped by terrain effects.

However, simulations showed sensitivity of Colorado Front Range mesoscale processes to vegetation distribution and suggest that even for regions with strong topographic forcing, vegetation has a significant impact on weather and climate.

The potential for changes in vegetation distribution under changing climate is high in mountain regions because of sharp spatial gradients in climatic factors controlling vegetation. This sensitivity to climate and the demonstrated effect of vegetation on climate emphasizes the need to evaluate the coupled response of climate, ecosystems, and surface hydrology to altered global climate in mountain regions. Development of fully linked models of land surface and climatic processes are required for such assessments (Pielke *et al.*, 1993a).

Furthermore, 3-D simulations with two-way interactive atmosphere-land process models could effectively evaluate the role of climatic and hydrologic linkages between mountains and adjacent lowlands in their joint response to global climate change.

ACKNOWLEDGEMENTS

This work was supported by the National Science Foundation under grant #ATM–8915265, the National Park Service Global Change Program under contracts \#0479–8–8001, Amendment 91–02, #CA–1268–2–9004–TO–03, #CA–1268–2–90CSU–06, and the United States Geological Survey, Department of the Interior, under Assistance No. 14–08–0001–A0929 and a grant from the Department of Energy Theoretical Ecology Program. Dallas McDonald ably completed the preparation of this manuscript. Portions of this work were previously reported in Pielke *et al.*, 1993b.

REFERENCES

Avissar, R. and Mahrer, Y. (1982) 'Verification study of a numerical greenhouse microclimate model', *Trans. Amer. Soc. Agric. Eng.*, 25 (6): 1711–20.

—— and —— (1988) 'Mapping frost-sensitive areas with a three-dimensional local-scale numerical model. Part I: Physical and numerical aspects', *J. Appl. Meteor.*, 27: 400–13.

Avissar, R., Avissar, P., Mahrer, Y. and Bravado, B.A. (1985) 'A model to simulate response of plant stomata to environmental conditions', *Agric. For. Meteor.*, 34: 21–9.

Band, L.E., Patterson, P., Nemani, R. and Running, S.W. (1993) 'Forest ecosystem processes at the watershed scale: incorporating hillslope hydrology', *Agric. Forest Meteor.*, 63: 93–126.

Chen, C. and Cotton, W.R. (1983a) 'A one-dimensional simulation of the stratocumulus-capped mixed layer', *Bound.-Layer Meteor.*, 25: 289–321.

—— and —— (1983b) 'Numerical experiments with a one-dimensional higher order turbulence model: Simulation of the Wangara Day 33 case', *Bound.-Layer Meteor.*, 25: 375–404.

Cotton, W.R., Tremback, C.J. and Walko, R.L. (1988) 'CSU RAMS – A cloud model goes regional', *Proc. NCAR Workshop on Limited-Area Modeling Intercomparison*, 15-18 Nov., NCAR, Boulder, CO, 202–11.

Cram, J.M., Pielke, R.A. and Cotton, W.R. (1992a) 'Numerical simulation and analysis of a prefrontal squall line. Part I: Observations and basic simulation results', *J. Atmos. Sci.*, 49: 189–208.

——, —— and —— (1992b) 'Numerical simulation and analysis of prefrontal squall line. Part II: Propagation of the squall line as an internal gravity wave', *J. Atmos. Sci.*, 49: 209–25.

Deardorff, J.W. (1978) 'Efficient prediction of ground surface temperature and moisture, with inclusion of a layer of vegetation', *J. Geophys. Res.*, 83 (C4): 1889–903.

Dickinson, R.E., Henderson-Sellers, A., Kennedy, P.J. and Wilson, M.F. (1986) 'Biosphere-atmosphere transfer scheme for the NCAR Community Climate Model', *Technical Report NCAR/TN–275 +STR*, NCAR, Boulder, CO.

Goudriaan, J. (1977) *Crop Micrometeorology: A Simulation Study*, Pudoc, Wageningen, The Netherlands.

Lee, T.J. (1992) 'The impact of vegetation on atmospheric boundary layer and convective storms', Ph.D. Dissertation, Department of Atmospheric Science, Colorado State University, Fort Collins, CO.

Lee, T.J., Pielke, R.A. and Mielke, Jr, P. (1992) 'Optimizing Land Ecosystem-Atmosphere Feedback model using MRBP', *12th Conference on Probability and Statistics in Atmospheric Sciences*, 22-26 June 1992, AMS, Toronto.

Lee, T.J., Pielke, R.A., Kittel, T.G.F. and Weaver, J.F. (1993) 'Atmospheric modeling and its spatial representation of land surface characteristics', in M. Goodchild, B. Parks and L.T. Steyaert (eds), *Environmental Modeling with GIS*, Oxford University Press, Oxford, pp. 108–22.

Loveland, T.R., Merchant, J.W., Ohlen, D.O. and Brown, J.F. (1991) 'Development of a land-cover characteristics database for the conterminous US', *Photo. Eng. Rem. Sens.*, 57: 1453–63.

McCumber, M.C. (1980) 'A numerical simulation of the influence of heat and moisture fluxes upon mesoscale circulations', Ph.D. Dissertation, University of Virginia.

McCumber, M.C. and Pielke, R.A. (1981) 'Simulation of the effects of surface fluxes

133

of heat and moisture in a mesoscale numerical model – Part 1: Soil layer', *J. Geophys. Res.*, 86: 9929–38.

Ojima, D.S., Kittel, T.G.F., Rosswall, T. and Walker, B.H. (1991) 'Critical issues for understanding global change effects on terrestrial ecosystems', *Ecological Applications*, 1: 316–25.

Ojima, D.S., Kittel, T.G.F., Schimel, D.S., Wessman, C.A., Curtiss, B., Archer, S., Brown, V.B. and Parton, W.J. (1992) 'Global arid and semi-arid ecosystems: linkage between process models and remote sensing', in R. Williamson (ed.), *International Geoscience and Remote Sensing Symposium*, IGARSS '92. IEEE #92CH3041–1, New York, 1027–9.

Pielke, R.A., Cotton, W.R., Walko, R.L., Tremback, C.J., Nicholls, M.E., Moran, M.D., Wesley, D.A., Lee, T.J. and Copeland, J.H. (1992) 'A comprehensive meteorological modeling system – RAMS', *Meteor. Atmos. Phys.*, 49: 69–91.

Pielke, R.A., Schimel, D.S., Lee, T.J., Kittel, T.G.F. and Zeng, X. (1993a) 'Atmosphere-terrestrial ecosystem interactions: implications for coupled modeling', *Ecological Modelling*, 67: 5–18.

Pielke, R.A., Lee, T.J., Kittel, T.G.F., Cram, J.M., Chase, T.N., Dalu, G.A. and Baron, J.S. (1993b) 'The effect of mesoscale vegetation on the hydrologic cycle and regional and global climate', Preprints, *Conference on Hydroclimatology: Land-Surface/Atmosphere Interactions on Global and Regional Scales*, American Meteorological Society, 17–22 January 1993, Anaheim, California, 82–7.

Poiani, K.A. and Johnson, W.C. (1991) 'Global warming and prairie wetlands: potential consequences for a waterfowl habitat', *Bioscience*, 41: 611–18.

Rango, A., and van Katwijk, V.F. (1990) 'Climate change effects on snowmelt hydrology of western North American mountain basins', *IEEE Transactions on Geoscience and Remote Sensing*, 28: 970–4.

Revelle R.R. and Waggoner, P.E. (1983) *Effects of a Carbon-dioxide-induced Climatic Change on Water Supplies in the Western United States*, National Academy of Sciences, National Academy Press, Washington, DC.

Running, S.W., and Nemani, R.R. (1991) 'Regional hydrologic and carbon balance responses of forests resulting from potential climate change', *Climatic Change*, 19: 349–68.

Schimel, D.S., Parton, W.J., Kittel, T.G.F., Ojima, D.S. and Cole, C.V. (1990) 'Grassland biogeochemistry: links to atmospheric processes', *Climatic Change*, 17: 13–25.

Schimel, D.S., Kittel, T.G.F. and Parton, W.J. (1991) 'Terrestrial biogeochemical cycles: global interactions with the atmosphere and hydrology', *Tellus*, 43AB: 188–203.

Schmidt, J.M. and Cotton, W.R. (1990) 'Interactions between upper and lower tropospheric gravity waves on squall line structure and maintenance', *J. Atmos. Sci.*, 47: 1205–22.

Sellers, P.J., Mintz, Y., Sud, Y.C. and Dalcher, A. (1986) 'A simple biosphere model (SiB) for use within general circulation models', *J. Atmos. Sci.*, 43: 505–31.

Sellers, P.J., Hall, F.G., Asrar, G., Strebel, D.E. and Murphy, R.E. (1988) 'The First ISLSCP Field Experiment (FIFE)', *Bull. Amer. Meteor. Soc.*, 69: 22–7.

Shuttleworth, W.J. and Wallace, J.S. (1985) 'Evaporation from sparse crops – an energy combination theory', *Quart. J. Roy. Meteor. Soc.*, 111: 839–55.

Tremback, C.J. and Kessler, R. (1985) 'A surface temperature and moisture parameterization for use in mesoscale numerical models', Preprints, *7th Conference on Numerical Weather Prediction*, 17–20 June 1985, Montreal, Canada.

Tremback, C.J., Tripoli, G.J., Arritt, R., Cotton, W.R. and Pielke, R.A. (1986) 'The Regional Atmospheric Modeling System', in P. Zannetti (ed.), *Proc. Inter. Conf.*

Development and Application of Computer Techniques to Environmental Studies, November, Los Angeles, California, Computational Mechanics Publications, Boston, 601–7.

Tripoli, G. and Cotton, W.R. (1989a) 'A numerical study of an observed orogenic mesoscale convective system. Part 1: Simulated genesis and comparison with observations', *Mon. Wea. Rev.*, 117: 273–304.

—— and —— (1989b) 'A numerical study of an observed orogenic mesoscale convective system. Part 2: Analysis of governing dynamics', *Mon. Wea. Rev.*, 117: 305–28.

Weissbluth, M.J., (1991) 'Convective parameterization in mesoscale models', Ph.D. Dissertation, Department of Atmospheric Science, Colorado State University.

8

CLIMATE SCENARIOS FOR MOUNTAIN REGIONS

An overview of possible approaches

M. Beniston

INTRODUCTION

Mitigation and adaptation strategies to counteract possible consequences of abrupt climate change in mountain regions require climatological information at high spatial and temporal resolution. Unfortunately, present-day simulation techniques for predicting climate change on a regional scale are by no means satisfactory, and few policy options can be prepared on the basis of currently available information on future climate scenarios in mountain regions.

A number of options exist today, however, which may help improve the quality of climate data used in impacts assessments and economic decision making. Among these, one can identify statistical techniques of downscaling from synoptic to local atmospheric scales, coupling of mesoscale models to general circulation models, palaeoclimatic analogues, and geographical analogues. It is obvious that any meaningful climate prediction for mountain regions – and indeed for any area which is less than a continental scale – needs to take into account not only those processes which occur in the region of interest but also global-scale processes which will inevitably impact upon regional-scale climates. It is this coupling of scales and the associated scientific disciplines which form the essence of the regionalization of climate scenarios.

REGIONAL CLIMATE PREDICTIONS: PROBLEMS OF SCALE

Regional climate predicitions are not sufficiently reliable when using present-day General Circulation Models (GCMs). One reason for this is related to space and time scales. Figure 8.1 illustrates some of the temporal scales inherent to the climate system. It can be seen that, according to the component of climate considered, response time to a particular forcing can

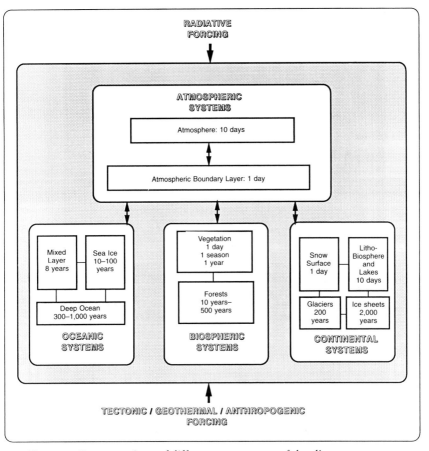

Figure 8.1 Response times of different components of the climate system to a
particular forcing (adapted from Saltzmann, 1983).

vary from a few hours to several centuries. The direct coupling of different
elements of the system in an integrated climate model is not feasible in view
of this situation, and even the expected increase in computing power in
coming years will not alleviate this problem to any significant extent.

General Circulation Models are those which attempt to incorporate as
many elements of the climate system as possible. They are among the largest
and most demanding operational applications in terms of computing
resources. They typically solve large sets of equations at up to several
hundred thousand grid points, and these computations must be repeated 50
or more times per simulated day in order to represent the temporal evolution
of the system with enough accuracy. Because the GCMs are producing
results over a network of grid points distributed in three dimensions over the
globe, computer time and space requirements are extremely large, and much

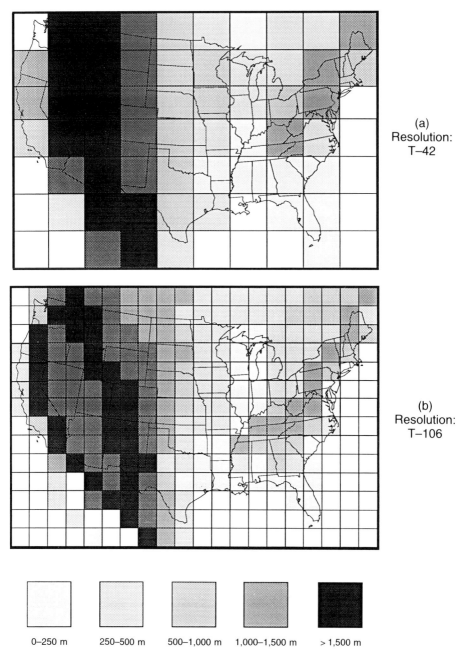

(a)
Resolution:
T–42

(b)
Resolution:
T–106

0–250 m   250–500 m   500–1,000 m   1,000–1,500 m   > 1,500 m

Figure 8.2 Orography of the United States as seen by a spectral GCM at a T–42 resolution (approx. 500 km horizontal grid) (a) and orography of the United States as seen by a spectral GCM at a T–106 resolution (approx. 120 km horizontal grid) (b).

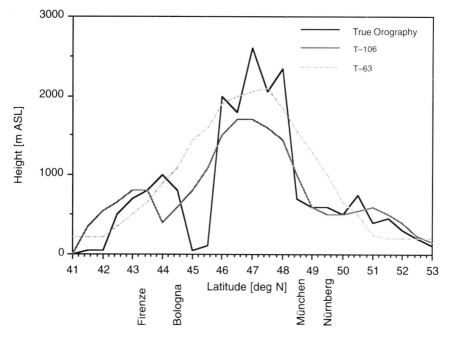

Figure 8.3 Cross-section across the Alps, showing true orography, and that resolved by spectral models at T–63 and T–106 resolutions.

of the physics representing feedbacks within the climate system needs to be parameterized, often in an oversimplified manner. As already mentioned, the coupling of such elements within a single modelling system is still in its infancy and is by no means a trivial matter. As an example of computing resource requirements, an advanced model such as the Max-Planck-Institute T–106 spectral model (wave number truncation 106, corresponding to a spatial resolution of approximately 120 km over the globe) requires 270 hours on a supercomputer of the CRAY generation to simulate a one-year atmospheric cycle. The data generated for the one-year period is about 16 Gbytes (L. Bengtsson, personnal communication, 1991).

Despite the impressive computational resources used for climate modelling, GCMs have a far too low resolution to provide any meaningful data on the regional scale; the 'regional' experiments undertaken during the IPCC process (Houghton *et al.*, 1990) were disappointing at best. Figure 8.2 illustrates the nature of the problem: a typical GCM attempting long-range (decade to century scale) will run at much lower resolution than the T–106 model mentioned above. Typical simulations are made with T–21 or T–42 models, the latter corresponding to a 4.5° × 4.5° mesh (roughly 500 × 500 km). A GCM at T–42 resolution would therefore 'see' the United States as in Figure 8.2a, in which the shading represents average elevation within a

500 km square. At such resolution, the Rockies stretch from the Pacific Coast to the Mississippi Basin in an unsubtle manner. At T–106 resolution, however (Figure 8.2b), it is possible to see more detail; in particular, the Sierra Nevada and Rocky Mountains are separated by the high deserts of the SW United States, and the Appalachians are somewhat better resolved. Atmospheric flows would certainly be more realistic than at lower resolutions, but this would be at the expense of a 16-fold increase in computer time and space requirements compared to a T–42 simulation.

If alpine climates are to be investigated, then the situation is even worse than for the Rockies, which are a bigger mountain chain. Figure 8.3 gives a cross-section from Germany to Italy across the Alps for the true orography and that perceived by GCMs with T–106 and T–63 resolution. As would be expected, the higher resolution model takes into account the true elevation in a more realistic manner than the T–63 model, even though in both cases only the gross features of the terrain are resolved.

In an overview paper in 'Hierarchy theory and global change', O'Neill (1988) points out that in many environmental studies, the scale of observation determines the level of interest, and that a subdivision of a physical or ecological process into different scales is useful for a particular class of problems, and 'no single division stands out as fundamental. There certainly seems no good reason to force all problems into a single framework' (O'Neill, 1988: 32). Applying the hierarchy principle to ecology, O'Neill furthermore states (ibid.: p. 33) that 'the theory recommends focusing on a single level but the appropriate level should be based on the problem at hand. It is not useful to force a new problem area into the mold that was appropriate for other problems at other levels.'

This principle has been applied in the past to other scientific disciplines, and in the field of atmospheric physics partitioning of scales has taken place to investigate turbulence, mesoscale processes, synoptic and climatological-scale flows. A scale diagram is given in Figure 8.4, where the shaded areas indicate the typical scale levels for simulation models investigating processes occurring at the micro, meso, and macro scales. As most models have been developed for a single scale, it is difficult to transpose the physical parameterizations and the numerical principles associated with one level to another. As will be seen later, coupling of models at different scales is possible, especially from the larger to the smaller scale where the data from the larger-scale model are used as boundary conditions for the finer-scale model. This principle is much less obvious to apply in the opposite direction, although in the climate system certain processes at scales unresolved by GCMs may have climatic significance, as has been shown by Beniston and Perez Sanchez (1992). When treating climate as a chaotic system, however, it becomes apparent that continuous forcing (for example by GHGs) could flip the system to a state different from that of today. If such a bifurcation were to occur, then the rapid dynamics of processes occurring at smaller scales

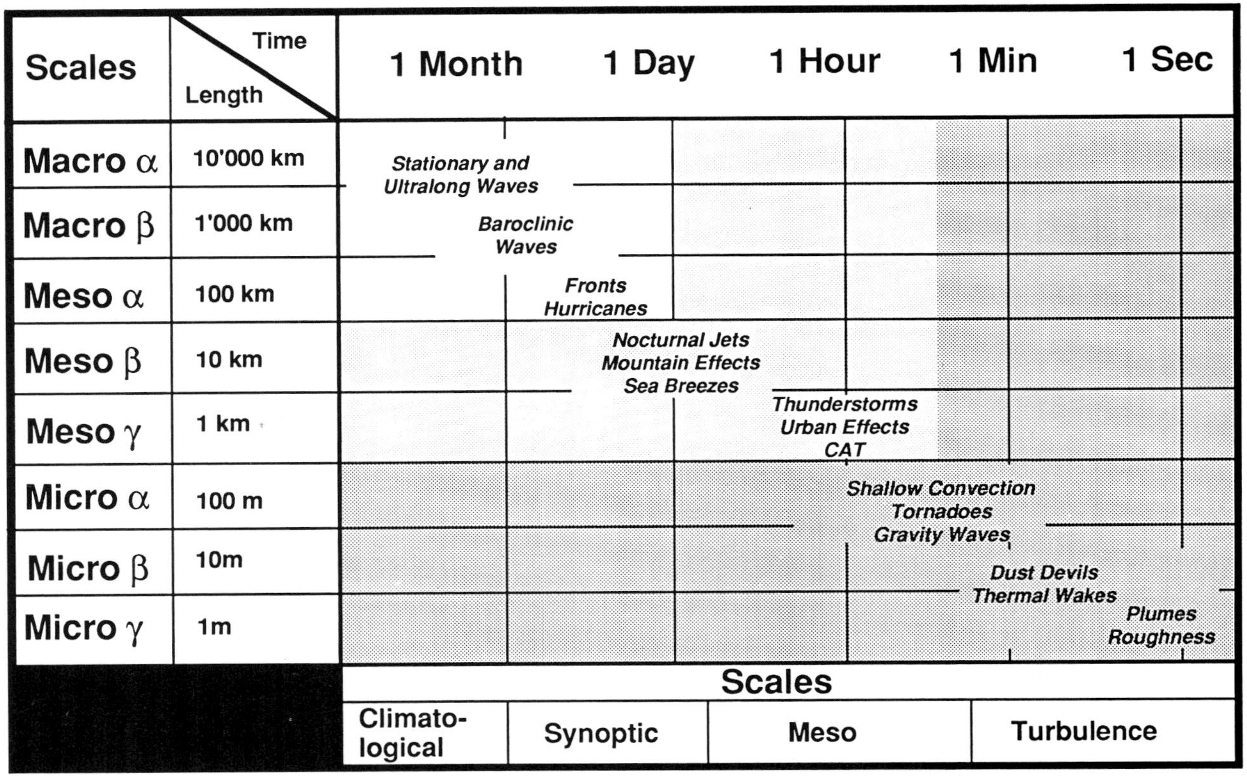

Scales	Length \ Time	1 Month	1 Day	1 Hour	1 Min	1 Sec
Macro α	10'000 km	Stationary and Ultralong Waves				
Macro β	1'000 km	Baroclinic Waves				
Meso α	100 km		Fronts, Hurricanes			
Meso β	10 km		Nocturnal Jets, Mountain Effects, Sea Breezes			
Meso γ	1 km			Thunderstorms, Urban Effects, CAT		
Micro α	100 m			Shallow Convection, Tornadoes, Gravity Waves		
Micro β	10m				Dust Devils, Thermal Wakes	
Micro γ	1m					Plumes, Roughness

Scales			
Climato-logical	Synoptic	Meso	Turbulence

Figure 8.4 Atmospheric scales and characteristic processes.

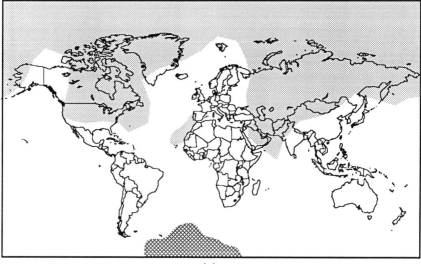

(a)

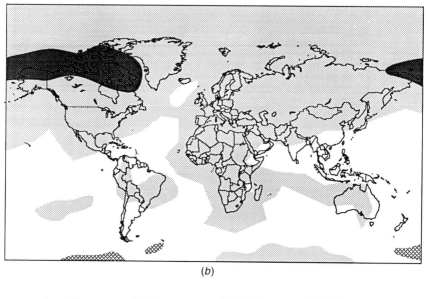

(b)

| | <0 C | | 0 - 2 C | | 2 - 4 C | | > 4 C |

Figure 8.5 Increases in annual mean surface temperature with respect to that of today, for an equilibrium double CO_2 simulation using the GISS GCM for the decades 1999–2008 (a) and 2019–28 (b).

would be largely responsible for moving the system into a new configuration, so that they need to be taken into account in some physically coherent manner. The understanding of the interaction between small-scale and larger-scale processes would help in the prediction of the occurrence of unstable responses to a given forcing. In addition to the non-trivial mathematics involved in coupling of models operating at different levels, the principal constraint is linked to computational resources: decadal to century scale simulations can only be undertaken at the expense of lower spatial resolution and limited coupling between various components of the climate system.

The foregoing discussion has shown that climate simulations in general, and regional climate estimates in particular, suffer from limits of time and space resolution imposed by computer power, and also by the complex, interdisciplinary nature of feedback mechanisms between the atmosphere, the oceans, the cryosphere, and the biosphere.

CLIMATE SCENARIOS FOR MOUNTAIN REGIONS

Figure 8.5 illustrates a simulated result for climate in the next century; the two figures show the near-surface average annual temperatures as computed by the Goddard Institute of Space Sciences model (GISS, New York, USA) based on a start-up time in 1958 and predictions for the decades 1999–2008 (Figure 8.5a) and 2019–2028 (Figure 8.5b). The essential feature in these two figures is the warming trend over continental areas, with enhanced greenhouse warming occurring in northern latitudes in Figure 8.5b. The range of uncertainty is so large that for Europe as a whole, including the Alps, the warming is expected to be between 2° and 4°C. Indeed, this amplitude of temperature is valid over most of North America, Africa, and Asia as well, so that the information provided by this low spatial resolution, long timescale simulation is of little help to scientists who wish to investigate ecological, hydrological, or economic response to climate change in a particular region of interest.

Palaeoclimatic analogues

Climate is known to have fluctuated in response to natural causes in the past, and the analysis of proxy data – based on tree rings, ice-cores, lake and ocean sediments – has allowed an insight into the complex inter-relations between mean atmospheric temperatures and GHG concentrations, as well as the response of ice, water, and living organisms to changing climates. A number of warm epochs have been recorded in the past, with temperatures well above that of today, as reported by Budyko and Izrael (1987); the authors argue that these warm climates could be used as analogues of a future warm climate. Three warm periods stand out in the past, namely the climatic optimum of the Pliocene, which occurred between 3.3 and 4.3 million years ago, the

Eemian interglacial optimum around 130,000 years ago, and the more recent mid-Holocene optimum which was recorded at 5,000 to 6,000 years before present (BP).

The most recent warm period was characterized by a climate with summer temperatures 3–4 °C warmer than present in high latitudes; increased precipitation and higher lake levels in subtropical and high latitudes occurred, whereas mid-latitude precipitation and hydrological levels experienced decreases. Because mid-Holocene summer temperatures are in the range of 2–4 °C higher than present, this period is often used as an analogue to climate in the first half of the twenty-first century. However, the transition to this climatic optimum certainly took far longer than what is expected from GHG forcing for the next century, so that the *consequences* of the warmer climate may certainly be different for ecosystems in view of the difficulties of adaptation of certain species to new environmental conditions. In terms of rates of change, the only period in the past where climate change was as abrupt as for GHG forcing is the so-called Younger Dryas period, which followed the last deglaciation and occurred around 10,500 BP. The Younger Dryas was an abrupt reversal of the general warming trend at the end of the glaciation and lasted 200–500 years.

Because of a long-standing Swiss tradition in palaeoclimatology and palaeoecology, numerous studies have been undertaken in the Alps using proxy data from lakes, ice-cores, and tree rings, to reconstruct alpine climates and environmental conditions which occurred during the Holocene. Records prior to the last glaciation are difficult to interpret because of the significant environmental changes which occurred through the movement of glaciers in the alpine valleys, which displaced or destroyed geological and biological evidence of older climates.

Work by Ammann and Lotter (1989), Lotter (1989), and Wegmüller and Lotter (1990) have used pollen stratigraphy in Swiss lakes to obtain high time resolution reconstructions of climate in the alpine region and estimates of vegetation behaviour in response to the climate changes which occurred from 1,000 to 12,000 BP. Burga (1990) notes that during the Holocene optimum, the timberline reached its highest extension at 2,300 m above sea level, and permanent snowlines were located above the 3,100 m level. Peat formation in the Aosta Valley reached a peak around 6,500 years BP in what was then a warmer, drier climate than present. Similar climates predicted for coming decades could result in such cryospheric and biospheric conditions, but only if these systems can adapt rapidly to a new equilibrium climate.

The reconstruction of palaeoclimate and palaeoenvironments inferred from a wide range of proxy data is of prime value in underlining the natural variability of climate; it does not, however, provide a direct analogy for the future, because in the last centuries mankind has changed his environment far more rapidly and radically than any natural cause. Where species may have migrated in response to past climate changes, the same species would have great difficulties in avoiding the man-made barriers to migration, such as cities and highways.

Difficulties in identifying 'short-lived' climate events such as the Younger Dryas in the proxy data render complex the study of climate–environment links in rapidly changing climates. The resolution required of proxy data to attempt some form of decadal timescale climate analogue study is such that very little reliable data is available to investigate the profound environmental changes in the face of abrupt climate change. It is perhaps also erroneous to attempt analogue studies on the basis of climates whose *temperature and precipitation regimes* may have been similar to what might occur in the twenty-first century, but which reached those levels of temperature and moisture through internal or external forcings other than by increases in GHGs. There is indeed *no* parallel to man-induced climate change in the climate record, and unfortunately palaeoclimatology will provide few of the answers to problems of future regional climate change at the required space and time scales.

Mathematical approaches to scale hierarchy

There exist some possibilities of applying mathematical models of variation to problems of interlinks between scales. These include harmonic analysis where high-frequency variations can be superimposed on lower frequencies. Harmonic analysis allows time-series fluctuations of different amplitudes and frequencies to be combined. Cliff and Ord (1981) have laid the theoretical groundwork for the use of harmonic analysis in scale problems, although the application to incomplete data sets remains problematic.

A more 'exotic' approach in the form of Markov theory to the scale problem can be of use as a first approximation to changes taking place from one state to another when only the probabilities of transitions can be quantified. The modelling of state variations as a Markov process leads to a series of embedded Markov chains which can be analysed to assess the nature of coupling between scales. Few studies using Markov model techniques have been applied to the regionalization of climate predictions.

The use of fractals to reproduce patterns at smaller and smaller scales is a relatively new development which could be applied to the problem of climate scales. Indeed, Mandelbrot (1983) has suggested that all the heterogeneity observed in natural systems could be represented by fractal geometry, although the application of the technique to geophysical problems is very much in its infancy.

Statistical downscaling in the EPOCH-FUTURALP programme

As a specific example of a downscaling procedure, the FUTURALP project of the European Program on Climate and Natural Hazards (EPOCH) is of interest to the present discussion. FUTURALP is aimed at assessing the response of alpine ecosystems, particularly plant and bird species, to global warming. The scales needed for ecosystem modelling range from the entire alpine arc down to 250 m squares as used by some of the Geographic Information Systems applied to vegetation modelling. In view of the diverse requirements of FUTURALP, a number of strategies have been adopted.

The first short-term solution consists in the generation of analogue scenarios based on historical instrumental records. A comparison of the characteristics of warmer years with those of colder years or periods may be used to infer changes in regional climatology under regionally, hemispherically, or globally elevated mean temperatures (e.g. Wigley *et al.*, 1980; Williams, 1980; Pittock and Salinger, 1982; Kellogg, 1982; Jaeger and Kellogg, 1983; Lough *et al.*; 1984; Palutikof *et al.*; 1984). The main limitation of such scenarios is that they reflect boundary conditions for the climate system different from those expected in the future under increased greenhouse-gas concentrations; in addition they do not give any information on transient climate change as it might occur within the course of the next century. Furthermore, these scenarios are based on relatively small climatic variations which may not be reliably extrapolated due to the highly nonlinear nature of the climate system.

The second short-term solution is based on an interpretation of the results of global climate models, in particular the outputs of GCMs. Here it has been decided to focus on the two cases of simulated changes under a doubling of equivalent CO_2-concentrations (equilibrium response) and under the IPCC-Scenario A ('Business-as-Usual', transient climate change) for future greenhouse-gas emissions (Houghton *et al.*, 1990). However, as has been previously explained, these models operate on a global scale so that confidence in simulated changes on a regional scale is low (Wilson and Mitchell, 1987; Santer *et al.*, 1990).

In order to be able to generate more realistic and internally consistent scenarios, the application of empirical-statistical methods (Kim *et al.*, 1984; Wilks, 1989; Karl *et al.*, 1990; Wigley *et al.*, 1990; Von Storch *et al.*, 1991) has been selected as a longer-term strategy. Compared to purely empirical or analogue scenarios, statistical scenario generation has two main advantages: first, the statistical methods allow description of the evolution of climate variables within time (transient climate reponse to man-induced changes), and second, the climate scenarios may be objectively linked to GCM outputs. In particular, statistical models designed to relate global and local climatology to each other are, at least within the range of climatic changes covered by the data used for model development, suited to bridge the gap between the scales of GCM-simulations and the needs of regional ecosystem studies.

The method estimates by means of Canonical Correlation Analysis, a linear regression model relating a small number of Empirical Orthogonal

Functions (EOFs; e.g. Kutzbach, 1967) of large-scale sea-level pressure deviations from their long-term monthly (or seasonal) mean fields to the first few EOFs of anomalies in monthly (or seasonal) statistics of weather elements. The weather elements considered are daily mean, minimum and maximum temperature, temperature amplitude, precipitation sum, mean windspeed, relative humidity and relative sunshine duration. In this manner, data of use to ecosystem modellers other than only mean temperature or precipitation can be generated.

The main results of FUTURALP scenario generation are based on simulations by the T–21 GCM of the Max-Planck-Institute (Hamburg). Results from the model indicate a general weakening of the circulation over the Atlantic, because future climate will be characterized by a smaller Equator-to-Pole temperature contrast (Cubasch et al., 1992). As a consequence, a general weakening of the Westerlies may be expected, resulting in more marked decreases in precipitation in the western Alps compared to the eastern Alps.

Albedo-temperature feedbacks will be affected by global warming. As a result of higher mean temperatures in winter months, less precipitation is likely to fall as snow, resulting in a lowered surface albedo and smaller radiative energy losses into the atmosphere. Therefore, warming is assumed to be larger in areas with longer contemporary snow-cover duration, the latter being associated to the degree of continentality (Schüepp and Schirmer, 1977).

Soil moisture–temperature feedback (Mitchell et al., 1990), according to the analysis, will undergo significant transformation due to reduced evaporation and thus reduced latent heat flux into the atmosphere under the assumed precipitation decrease in summer. Consequently, warming in areas with contemporary wet summers is predicted to be stronger compared to the warming occurring in areas which are already dry under the current climate.

Once a statistical relationship has been established, it can be validated by applying it to observed sea-level pressure variations not used for model estimation, or it can be used for the derivation of transient scenarios based on the sea-level pressure simulated by a GCM under different conditions of GHG forcing.

It should be noted here that the downscaling component of the FUTUR-ALP programme has the objective of proposing climate *scenarios* and not *predictions*, that is, a scenario is a possible or plausible picture of future climate trends with no measure as to the probability of its actually occurring; this would be more the objective of a prediction. However, a scenario can provide the necessary guidelines for a wide range of impact studies.

Once future scenarios have been established, it would be of interest to investigate which Alpine regions *today* experience climatic conditions close to those which the scenarios are suggesting for *tomorrow*, as outlined by

Beniston and Price (1992). One plausible consequence of global warming could be that the cooler northern Alps may have a more Mediterranean-type climate as climatic belts shift northwards. It may then be possible to describe the distribution of the chosen vegetation species in today's warmer alpine regions, in order to estimate the rate and extent of migration of the species along different altitudinal levels in today's cooler Alps when they become as warm as the southern Alps – with the assumption that the vegetation and fauna can *adapt* to abrupt climate change, and that they will find at higher altitude the soil type and other necessary environmental conditions to ensure the continued existence of the species. This approach can be described as a 'geographical analogue'.

Nested global-mesoscale model approach

Pioneering work has been undertaken by Giorgi *et al.* (1990) and Giorgi and Mearns (1991) in coupling a mesoscale model to a GCM, in order to improve the spatial resolution of climate variables. In an overview paper, Giorgi and Mearns (1991) come to the same basic conclusions as in this paper concerning approaches to regional climate scenarios, that is, that they can be generated through empirical approaches, palaeoclimatic analogues, semi-empirical approaches in which GCM predictions are used as a basis for taking into account mesoscale responses to climatic change (as described in the previous subsection), and coupled modelling approaches which are more physically based than empirically based.

In the modelling approach, Giorgi *et al.* (1990) have applied the National Center for Atmospheric Research (NCAR, Boulder, Colorado, USA) Community Climate Model (CCM) and the medium resolution MM4 mesoscale model to attempt coupled climatic simulations over the continental United States and over western Europe. Long-term climatology has been simulated with a coarse-resolution (R–15, i.e., a $7.5° \times 4.5°$ longitude–latitude resolution) version of the CCM, which allows the salient features of large-scale circulations, major storm tracks, and large-scale precipitation patterns to be captured over the region of interest. Giorgi *et al.* (1990) note, however, that because of the coarseness of the CCM resolution, certain climatologically relevant features such as cyclogenesis in the western Mediterranean were not highlighted in the GCM experiments. The NCAR team has then used the R–15 data to provide lateral boundary conditions for a 70 km grid resolution version of the MM4 limited area model; the higher resolution allows the broad topographic features of the region to be taken into account, and ten-year January climatologies have been simulated with the nested approach. Among the significant results of the numerical experiments, certain storm cases resulting from cyclogenesis in the Mediterranean and in the lee of the Alps were well simulated. Regional definition of climatic variables, such as precipitation, temperature, cloudiness, or snow cover are far better taken into account by the higher.resolution model than by the coarse CCM, and the preliminary results obtained by the authors seem to compare favourably with observational data.

The nested model approach is now being adapted to the alpine regions through a collaborative approach between Switzerland (Department of Geography at ETH-Zürich), the Max-Planck-Institute in Hamburg (MPI) and NCAR (Beniston et al., 1993). The coupled approach uses the MPI T–106 GCM (approx. 120 km horizontal resolution) and a version of the NCAR MM4 designed to operate down to a 10 km grid resolution, and the system is currently being tested over the Alps. It is hoped that this order-of-magnitude increase in resolution compared to the original NCAR experiments will yield climatological data on scales closer to those required by impact modellers than those available in the past. The method is, of course, extremely resource-intensive from a computational point of view.

One criticism of the nested model approach is that the coupling between scales is one-way only, that is, forcing takes place from the large scale to the small scale (see also the discussion on scaling problems in the introduction to this paper). No feedback is possible from cryospheric, hydrospheric, or biospheric processes occurring on the smaller scales, despite the fact that response of these systems to climatic forcings may alter synoptic patterns to a considerable degree. The nested model approach in its present form can only be used to provide a high-resolution 'snapshot' of regional scale dynamic and thermodynamic patterns over a given region as a function of large-scale forcings.

CONCLUSIONS

The paper has reviewed some of the possible consequences of abrupt climate change on mountain environments and economies, and has argued the need for accurate climate scenarios on the scales required by impact modellers.

Predictions of future climates are carried out principally by General Circulation Models, which operate at present on spatial scales which are far too crude to be used directly for regional impact assessment studies. Because of the inevitable trade-off between spatial resolution and long time integrations imposed by computer resource limitations, GCMs will have difficulty in resolving the mountain areas of interest to the impact modellers. Indeed, in low-resolution spectral models (T–21 or T–42 scales), mountain chains such as the Alps are not even 'perceived' by such models and the physical influence of mountain barriers is taken into account through parameterizations such as gravity-wave drag. In the face of the scaling problem, a number of methods can be applied to extract information from global models and scale these down to the region of interest. Among the techniques available are mathematical procedures such as harmonic analysis and fractal geometry, statistical downscaling, or the coupling of different types of numerical model.

None of the techniques will provide more than a crude scenario for a region of interest, but the refinement of the downscaling methods may reduce the margins of error in future.

Palaeoclimatic studies offer only some limited information on the links between climate change and environmental response, partly because man-made changes to the present-day environment leave little room for analogue studies, and partly because the rates of change expected from GHG warming have occurred too infrequently in the past to offer any conclusive evidence of what may happen to forests, natural ecosystems, or hydrological regimes in the next decades.

Waiting for high-resolution GCM climate simulations in coming years as new generations of supercomputers come on to the market is probably not an optimal solution. Whereas higher resolution has led to marked improvements in forecasting skills for 5–10 days, a corresponding improvement for *climatological-scale* problems cannot be expected, because in the climate system scale is not the only dominant feature of the problem to be solved. The feedbacks from other components of climate, in particular the oceans, the cryosphere, and the biosphere, will also play a determining role in the evolution of climate forced by increased GHG concentrations. Despite improvements in recent years in the coupling of different systems to atmospheric GCMs, there are still many uncertainties and unknowns in the simulation of coupled systems. In addition, climate is an inherently chaotic system, so that a measure of unpredictability will always need to be kept in mind when analysing GCM climate statistics, whatever the resolution of the climate model. Regional climate predictions cannot be solved solely by increased computer power and higher resolution models. They will progress through a better integration of ocean, ice, and biological processes and feedback mechanisms with the atmosphere. This implies major investments in interdisciplinary research, which in turn requires breaking through the traditional barriers of the different disciplines, in order to achieve a convergent approach to research in climate and global change.

REFERENCES

Ammann, B. and Lotter, A.F. (1989) 'Late-glacial radiocarbon and palynostratigraphy on the Swiss Plateau', *Boreas*, 18: 109–26.

Beniston, M. and Perez-Sanchez, J. (1992) 'An example of climate-relevant processes unresolved by present-day General Circulation Models', *Env. Conserv.*, 19: 165–9.

Beniston, M. and Price, M. (1992) 'Climate scenarios for the alpine regions: a collaborative effort between the Swiss National Climate Program and the International Center for Alpine Environments', *Env. Conserv.*, 19: 360–3.

Beniston, M., Ohmura, A., Wild, M., Tschuck, P., Marinucci, M., Bengtsson, L., Schlese, U., Esch, M., Giorgi, F. and Bernasconi, A. (1993) 'Coupled simulations of global and regional climate in Switzerland', *Supercomputing Switzerland*, 1: 80–6.

Budyko, M.I. and Izrael, Y. (eds) (1987) *Anthropogenic Climate Change*, Hydro-meteoizdat, Leningrad (in Russian).

Burga, C. (1990) 'Vegetation history and paleoclimatology', *Quart. Bull. of the Zürich Nat. Sci. Soc.*, 135: 17–30.

Cliff, A.D. and Ord, J.K. (1981) *Spatial Autocorrelation*, Pion, London.

Cubasch, U., Hasselmann, K., Höck, H., Maier-Reimer, E., Mikolajewicz, U., Santer, B.D. and Sausen, R. (1992) 'Time-dependent greenhouse warming computations with a coupled atmosphere–ocean model', *Climate Dynamics*, 8: 55–69.

Giorgi, F., Marinucci, M.R. and Visconti, G. (1990) 'Application of a limited area model nested in a general circulation model to regional climate simulation over Europe', *J. Geophys. Res.*, 95: 18413–31.

Giorgi, F. and Mearns, L.O. (1991) 'Approaches to the simulation of regional climate change: a review', *Rev. Geophys*, 29: 191–216.

Houghton, J.T., Jenkins, G.J. and Ephraums, J.J. (eds) (1990) *Climate Change – The IPCC Scientific Assessment,* Cambridge University Press, Cambridge.

Jaeger, J, and Kellogg, W.W. (1983) 'Anomalies in temperature and rainfall during warm arctic seasons', *Climatic Change*, 5: 34–60.

Karl, T. R., Wang, W.C., Schlesinger, M.E., Knight , R.W. and Portman, D. (1990) 'A method relating General Circulation Model simulated climate to the observed local climate, Part I: Seasonal statistics', *J. Clim.*, 1: 1053–79

Kellogg, W.W. (1982) 'Precipitation trends on a warmer earth', in R.A. Reck and J.R. Hummel (eds), *Interpretation of Climate and Photochemical Models, Ozone and Temperature Measurements*, AIP conference proceedings, New York, 35–45.

Kim, J.W., Chang, J.T., Baker, N.L., Wilks, D.S. and Gates, W.L. (1984) 'The statistical problem of climate inversion: determination of the relationship between local and large-scale climate', *Mon. Wea. Rev.*, 113: 2069–77.

Kutzbach, J.E. (1967) 'Empirical eigenvectors of sea-level pressure, surface temperature and precipitation complexes over North America', *J. Appl. Met.*, 6: 791–802.

Lotter, A.F. (1989) 'Evidence of annual layering in Holocene sediments of Soppensee, Switzerland', *Aquatic Sciences*, 51: 19–30.

Lough, J.M., Wigley, T.M.L. and Palutikof, J.P. (1984) 'Climate and climate impact scenarios for Europe in a warmer world', *J. Clim. and Appl. Met.*, 23: 1673–84.

Mandelbrot, B.B. (1983) *The Fractal Geometry of Nature*, W.H. Freeman & Co., New York.

Mitchell, J.F.B., Manabe, S., Meleshko, V. and Tokioka, T. (1990) 'Equilibrium climate change – and its implications for the future', in J.T. Houghton, G.J. Jenkins and J.J. Ephraums (eds), *Climate Change – The IPCC Scientific Assessment*, Cambridge University Press, Cambridge, pp. 139–73.

O'Neill, R.V. (1988) 'Hierarchy theory and global change', in T. Rosswall, R.G. Woodmansee and P.G. Risser (eds), *Scales and Global Change*, J. Wiley & Sons, London.

Palutikof, J.P., Wigley, T.M.L. and Lough, J.M. (1984) 'Seasonal climatic scenarios for Europe and North America in a high CO_2, warmer world', *Climatic Research Unit Publication*, University of East Anglia, Norwich.

Pittock, A.B. and Salinger, M.J. (1982) 'Towards regional scenarios for a CO_2-warmed earth', *Climatic Change*, 5: 23–40.

Saltzmann, B. (ed.) (1983) *Theory of Climate*, series *Advances in Geophysics*, vol. 25, Academic Press, New York.

Santer, B.D., Wigley, T.M.L., Schlesinger, M.E. and Mitchell, J.F.B. (1990) 'Developing climate scenarios from equilibrium GCM-results', *Max-Planck-Institut für Meteorologie Report No. 47*, Hamburg.

Schüepp, M. and Schirmer, H. (1977) 'Climates of central Europe', in C.C. Wallén (ed.), *Climates of Central and Southern Europe*, World Survey of Climatology, Elsevier, Amsterdam-Oxford-New York, 3–74.

Von Storch, H., Zorita, E. and Cubasch, U. (1991) 'Downscaling of global climate change estimates to regional scales: an application to Iberian rainfall in Wintertime', *Max-Planck Institut für Meteorologie Report No. 64*, Hamburg.

Wegmüller, S. and Lotter, A.F. (1990) 'Palynostratigraphic investigations of the lateglacial and Holocene vegetation history of the northwestern calcareous Swiss Prealps', *Botan. Helv.*, 100: 37–73.

Wigley, T.M.L., Jones, P.D. and Kelly, P.M. (1980) 'Scenario for a warm, high CO_2-world', *Nature*, 283: 17–21.

Wigley, T.M.L., Jones, P.D., Briffa, K.R. and Smith, G. (1990) 'Obtaining sub-grid scale information from coarse-resolution general circulation model output', *J. Geophys. Res.*, 95: 1943–53.

Wilks, D.S. (1989) 'Statistical specification of local surface weather elements from large-scale information', *Theor. Appl. Climatol.*, 119–34.

Williams, J. (1980) 'Anomalies in temperature and rainfall during warm arctic seasons as a guide to the formulation of climate scenarios', *Climatic Change*, 3: 249–66

Wilson, C.A. and Mitchell, J.F.B. (1987) 'Simulated climate and CO_2-induced climate change over Western Europe', *Climatic Change*, 10: 11–42.

Part II

IMPACTS OF CLIMATE CHANGE ON VEGETATION

Observations, modelling, networks

9

IMPACT OF ATMOSPHERIC CHANGES ON HIGH MOUNTAIN VEGETATION

C. Körner

INTRODUCTION

On a global scale, plant life at high elevations is primarily constrained by direct and indirect effects of low temperature and perhaps also by reduced partial pressure of CO_2. Other atmospheric influences, such as increased radiation, high wind speeds or insufficient water supply may come into play on a regional scale, but exert no globally uniform characteristics of high mountain systems (for climatological data see Lüdi, 1938; Fliri, 1975; Lauscher, 1977; Körner and Larcher, 1988; Barry, 1992).

Plants respond to these features of mountain climate through a number of morphological and physiological adjustments such as stunted growth forms and small leaves, low thermal requirements for basic life functions, and reproductive strategies that avoid the risk associated with early life phases. In this review I will briefly summarize the current knowledge about five aspects of mountain plant life that relate to the climate change issue:

- thermal acclimation of primary life processes,
- growth strategies and developmental processes,
- water relations,
- mineral nutrition, and
- carbon dioxide effects.

An exhaustive review of the literature would have to encounter about 1,500 publications. For the present purpose, only some key points will be highlighted here, and the reader is referred to previous, more specialized treatments of the subject (for example, Billings, 1974; Billings and Mooney, 1968; Bliss, 1971, 1985; Larcher, 1980; Körner and Larcher, 1988; Körner *et al.*, 1989a). The term 'alpine' will be used strictly in the phytogeographic sense, namely for the zone between the upper treeline and the snowline.

155

THERMAL ACCLIMATION OF PHOTOSYNTHESIS AND RESPIRATION IN MOUNTAIN PLANTS

Contrary to what is commonly believed, photosynthetic assimilation of CO_2 in mountain plants is not particularly restricted by low temperatures during the growing period. This is a consequence of the combined effect of perfect physiological acclimation and microclimatic pecularities in narrow vegetation. Most mountain plants are able effectively to utilize periods of high photon flux density which are also the periods of greatest radiation-driven warming of the plant canopy (Cernusca, 1976; Cernusca and Seeber, 1981; Grace, 1988; Körner and Cochrane, 1983; Körner et al., 1983). Canopy warming in the narrow alpine vegetation is a consequence of aerodynamic uncoupling, whereas needle temperatures in upright treeline trees are more closely coupled to ambient temperatures (Grace, 1988).

Temperature optima for CO_2 uptake in alpine plants reflect this microclimatic peculiarity, and are not very different from those in lowland plants (Körner and Diemer, 1987). Overall, the actual limitation of annual CO_2 uptake by suboptimal daytime temperatures is in the order of 6 to 7 per cent of the potential uptake (Körner, 1982) which is similar to what is reported for lowland plants (Küppers, 1982) and reflects the physiological adjustment for the utilization of warm periods that usually coincide with high radiation. In contrast, low temperatures during periods of low solar radiation and during the night seriously restrict the utilization and investment of photoassimilates (Körner and Pelaez, 1989).

In the physiological range rates of CO_2 losses by mitochondrial respiration follow the well known exponential temperature response characteristics of all enzymatic reactions (Precht et al., 1973). However, it is well documented that plants from cold climates have adjusted their respiratory temperature response so that rates of CO_2 loss (and oxygen consumption) are greater than those in warm acclimated plants when both groups of plants are briefly exposed to the same warm experimental temperature. These differences disappear partly or completely when plants are grown for longer periods under similar temperatures (Christophersen, 1973; Körner and Larcher, 1988).

Thus, there is no justification for assuming that climate warming will enhance respiratory losses of plant tissue. Although widely ignored, it has been known for over 60 years (Stocker, 1935; see also Hiesey and Milner, 1965; Pearcy et al., 1987) that plant tissues from arctic environments exhibit respiratory losses similar to those in tropical rainforest plants when both are studied under their real life conditions. However, we know little about the time constants of thermal acclimation, in particular the differential responses of above- and below-ground organs that are exposed to widely contrasting thermal conditions. At least in the temperate zone assimilation and dissimilation of CO_2 by plants, although of key importance to plant growth, are the least likely points of action where climate warming would become effective.

GROWTH STRATEGIES AND DEVELOPMENTAL PROCESSES

It is widely acknowledged that plant biomass production is largely controlled by the way in which assimilates are invested by plants, and by the timing and duration of life phases such as bud burst, leaf maturation, flowering, or senescence. On an annual basis the duration of the growing season is a key component of biomass production of vegetation, but whether an individual plant species may profit from a lengthening of the snowfree period largely depends on its genotypic life rhythm and its reproductive strategy.

Mountain plants grow more slowly but have equal or higher photo-synthetic capacity than lowland plants (Körner and Diemer, 1987; Körner and Pelaez, 1989). It has been shown that this has partly to do with the fact that most alpine plants produce only one leaf generation (Diemer et al., 1992), and have somewhat greater below-ground carbon investments (Körner and Renhardt, 1987). The greater the 'non-green' plant compartment and the longer the period during which photosynthesis is halted due to snow cover or wintertime dormancy, the greater the difficulty of achieving a long-term positive carbon balance.

The subalpine treeline is a typical carbon balance controlled ecotone at which the life form with the smallest photosynthetically active biomass fraction (leaf mass ratio only 1–4 per cent of total biomass) is replaced by life forms with greater leaf mass ratio such as shrubs (10–15 per cent) and forbs (20–25 per cent, Körner, 1993). This phenomenon is observed at all latitudes. Locally, additional constraints may come into play such as winter desiccation (see below), but these have only moderate overall impact. The reason why temperate and boreal zone treelines with their short growing season are situated at similar isotherms of the warmest month as in subtropical and tropical treelines (Wardle, 1974) may have to do with the fact that the cold season in the temperate and boreal zone simply does not 'count' because respiratory losses during this part of the year are comparatively small (Tranquillini, 1979).

Thus, if climate warming becomes a reality and the growing seasons become warmer and longer, key questions will be whether alpine plants (and treeline trees) will profit from direct effects of higher temperatures, or from the increasing length of the snowfree period, or by a combination of both. In either case the net profit will depend on the extent of thermal acclimation and developmental constraints of plants. If direct effects of temperature come into play it will primarily be via their influence on tissue development (cell division and differentiation), and night-time temperatures will be more important than daytime temperatures (Körner and Pelaez, 1989; Körner, 1992).

The most important point seems to be the photoperiodic control of development. Heide and co-workers have clearly shown that photoperiod constraints in cold climates may be strong and of overriding importance compared to moderate warming effects (e.g., Heide, 1974, 1985; Solhaug, 1991). High-altitude provenances of forest tree species have been shown to maintain part of their developmental rhythms even when grown in low-altitude nurseries (Holzer, 1967, 1975). From Holzer's work with *Picea abies* it appears that bud burst in spring is less (or not) photoperiod dependent, while the formation of winter buds and the associated cessation of growth in late summer or autumn are exclusively under photoperiodic control.

From work in progress (S. Prock and C. Körner, unpublished) it appears that photoperiodic control of the development in alpine and arctic plants is quite variable. Cross-continental transplant experiments revealed at least two types of behaviour: while most species shed leaves in autumn independently of weather (photoperiod-sensitive species or ecotypes that are unable to profit from extended season length, for instance *Ranunculus glacialis*), others are photoperiod-insensitive and continue to grow new leaves as long as temperatures permit (for instance Rosaceae species from the genera Geum and Potentilla). For most native species we do not know the precise photoperiodic requirements and their interaction with temperature, thus making it difficult to estimate plant migration based on temperature scenarios alone (Ozenda and Borel, 1990). If migrations of whole vegetation types are considered, differential photoperiodic sensitivities of species will lead to new species compositions and altered species dominance.

WATER RELATIONS

Plants in the alpine zone of high mountain systems are not particularly limited by water supply in most parts of the world, since precipitation tends to increase while the vapour pressure deficit and the length of the snowfree season decrease. Exceptions are some continental mountain regions (Central Asia – Izmailova, 1977), sheltered valleys in the core of larger mountain complexes (Valais/Alps – Richard, 1985), exposed ridges (Oberbauer and Billings, 1981) or leeward slopes of N–S stretching Cordilleras (Andes – Geyger, 1985; Gonzalez, 1985). However, even in these areas little support was found, after careful study, for the notion that alpine plants are particularly stressed by water shortage. Hot, south exposed scree slopes often supposed to be dry habitats, actually have been shown to be places of particular good water supply (Pisek and Cartellieri, 1941). Stunted growth forms and leaf dieback in mountain plants are often mistaken as water stress symptoms, while they may result from shortage in mineral nutrients, wind sheering, or normal developmental processes, as exemplified by most alpine tussock graminoids (*Carex curvula* in the Alps, *Carex tristis* in the Caucasus, Chionochloa sp. in New Zealand, Kobresia sp. in the Rocky Mountains).

For the temperate zone a simple estimate illustrates the meteorological change that would be required to produce significant water stress in alpine vegetation. The length of the season in the alpine belt in the Alps (*c.* 2,500 m) lasts about three to four months, a period during which evapotranspiration accumulates to about 300 mm per year (Körner *et al.*, 1980, 1989b). Soils are wet at the beginning of the season. Summer rainfall during this period may vary between 300 and 400 mm. Assuming a worst case scenario with only 200 mm summer rainfall (although increasing rather than decreasing precipitation is predicted), the soil water pool would have to contribute about 100 mm of easily available moisture, which requires a depth of fine soils of 250 to 350 mm. Most alpine soils under closed vegetation are deeper and/or roots penetrate into deeper layers of weathered substrate. Thus, water supply to leaves will rarely exert an existential problem to alpine plants, but may locally and/or periodically reduce their activity.

However, periodic water shortage in the top soil may induce indirect water stress effects. The top 5 cm of alpine soils are regularly drying out during periods of bright weather in summer. Since microbial and fine root activity is concentrated in this rather narrow, warm soil layer (Schinner and Gstraunthaler, 1981; Klug-Pümpel, 1989), its desiccation impairs decomposition and mineral recycling and thus, soil nutrient availability. Frequently observed late summer browning of alpine mats is partly associated with this nutrient shortage. Fertilized patches, remain 'green spots' in the landscape without any water addition (unpublished observations).

On the other hand, many alpine soils are badly aerated because of surplus of water, also impairing plant nutrition and litter decomposition. With increasing rainfall anoxia situations may become more serious, while decreasing precipitation would lead to better soil aeration in some places. Thus, if changes in water supply were to occur during summer, they would become effective largely by influencing top soil processes and plant nutrition rather than via any direct impact on plant water status.

A special form of water stress may occur in evergreen species in later winter, when soils may still be frozen, while evaporative driving forces are already strong. This 'winter desiccation' of leaves may lead to shoot damage in certain species, particularly in isolated young conifers (Michaelis, 1934; Pisek and Larcher, 1954; Tranquillini, 1976; Baig and Tranquillini, 1980), but most treelines of the world are in regions where such stress phenomena are meteorologically impossible (Wardle, 1981; McCracken *et al.*, 1985; Grace, 1990; Perkins *et al.*, 1991). There are reports that deciduous trees at the treeline may also undergo winter desiccation, but drought damage is even less likely (e.g. Tranquillini and Plank, 1989). Thus, winter desiccation is relevant only in certain regions and species and is not a general problem for mountain plants including treeline trees. The relatively balanced water status in mountain plants, is underlined by increased stomatal frequency, high abundance of stomata on the upper leaf side, high leaf water potential, high maximum leaf diffusive conductance and little diurnal reduction of stomatal opening (Körner and Mayr, 1981; Körner *et al.*, 1986; Körner and Cochrane, 1985).

MINERAL NUTRITION

There is a persistent controversy among plant ecologists about what 'nutrient limitation' means in an ecological context. Adopting horticultural reasoning, almost all natural plant communities of the world are nutrient limited, since they will grow more if fertilized. In an ecological context the definition of limitation is not straightforward since the continued addition of nutrients will almost always lead to a new type of ecosystem, with new species gaining dominance, while others will disappear, hence, the former, supposedly 'limited' system will become extinct. It is now well established that nutrient addition to wild plants weakens their stress tolerance and can cause developmental cycles to become outphased with seasonal climatic trends. Such effects have been demonstrated also for alpine plants (Körner, 1984). The current increase of wet deposition of soluble nitrogen compounds (Psenner and Nickus, 1986) is thus enhancing the overall risk potential.

A global comparison of the nutritional status of plants from high altitudes has shown that the functioning of leaves is unlikely to be limited by mineral nutrient concentrations (Körner, 1989). Since most alpine plants are inherently slow growers, any enhancement of nutrient supply will stimulate the few, potentially fast growing species in a community, and can cause changes in vegetation structure more dramatically than by any other environmental change. Among woody species (dwarf shrubs), deciduous species are likely to be more responsive than evergreen species.

Besides mineral nutrition, there are many other interactions between mountain plants and their soils. The most important one is the mechanical stabilizing effect by plants. Any change in the plant cover will affect soil stability. When slow growing species with their extensive root systems and below-ground stems are replaced by fast growing species, which invest more in above-ground structures, soil stability on slopes will decline.

CO₂ EFFECTS

The altitudinal reduction of total pressure leads also to reduced partial pressures of CO_2 and oxygen. Alpine plants of the temperate zone live with *c.* 75 per cent, and those of the highest ranging subtropical plants with 50 per cent of the partial pressure of CO_2 that prevails at lowlands (the mixing ratio of CO_2 in air does not change with altitude). While numerous adaptive features to oxygen depletion have been found in animals and humans living in high altitudes, there is no clear evidence that plants 'acclimate' to the reduced availability of their basic photosynthetic resource. However, measurements of photosynthesis have shown higher rates in alpine versus lowland plants, whatever partial pressure was simulated (Körner and Diemer, 1987). Stable carbon isotope ratios confirm these findings on a global scale, with mountain plant tissues being less depleted in the heavy ^{13}C than tissues of lowland plants (Körner et al., 1988). A causal interpretation (Körner et al., 1991) indicated that low temperature and a greater length in the diffusion path in the generally thicker leaves of mountain plants are most likely explanations. Whatever the reason may be, mountain plants on an average tend to fix more CO_2 per unit leaf area than lowland plants.

Whether rising levels of ambient CO_2 will ameliorate CO_2 shortage in high altitude plants is not yet clear. Growth chamber experiments have shown a moderate downward adjustment of photosynthesis in mountain plant species exposed to CO_2-enriched atmospheres that only partially offset CO_2 enrichment (Körner and Diemer, unpublished). This adjustment was relatively smaller than in comparable species from low altitudes, indicating a possibly greater 'profit' in high altitude species. Whether such a greater carbon gain per unit of leaf area will translate into more biomass accumulation will depend on certain climatic constraints that presently seem to exert major limitations to alpine plant growth (e.g. low night temperatures during summer). Field experiments are under way in the Central Swiss Alps to test a possible CO_2 stimulation of growth. First results from 1992 indicate that plants, at least initially, increase investments in reserve pools rather than in additional structural tissue. Furthermore it seems that small additions of fertilizer exert greater responses than a treatment with twice the present CO_2 concentration.

There is a possibility that plants in general, and mountain plants in particular, already exhibit some growth stimulation by the current 27 per cent higher CO_2 level compared to the middle of the nineteenth century. Increased width of tree rings in subalpine forests has been interpreted as such a response (Pinus aristata, California, La Marche et al., 1984), but later climatological consideration indicated confounding trends in the local moisture regime. In other areas such trends are possibly related to massive increases of nitrogen deposition (Picea abies, Vosges Mountains, Becker, 1991). The general difficulty of separating a CO_2 signal from other environmental influences on tree-ring formation was illustrated by Kienast and Luxmoore (1988). A recently completed survey in subalpine Pinus cembra of the Tyrolean Alps (Nicolussi, Bortenschlager and Körner, unpublished) appears to be devoid of such biases and indeed indicates a stimulation of radial stem growth in accordance with elevated CO_2. With respect to forest stability it will be important to investigate possible changes in wood chemistry and wood mechanics, since there is a possibility that CO_2 fertilization alters the cellulose/lignin ratio (Aber et al., 1990).

CONCLUSION

Plants selected over geological periods to cope with the adverse life conditions at high altitudes show a number of features that give rise to doubts as to whether our traditional concept of 'limitation' is applicable to natural situations. Most alpine species would be substantially more stressed under lowland growth conditions, and many of them die within few weeks when transplanted to low altitudes. Experimental manipulations indicate that inherent slow growth, efficient utilization of resources, high stress tolerance and strong genetical control over phenology converge into a life strategy whose advantages fade when environmental conditions 'improve' from a human perspective. Rising temperatures, longer season-length, increased nitrogen supply and enhanced CO_2 levels alone or in combination will eliminate some of those determinants that favour current plant communities and eliminate others. The natural consequences of such changes are plant migrations. The key question in future vegetation development particularly in high mountains is whether such migrations can track the speed of environmental change (Körner, 1992). Palaeoecological evidence for treeline migration indicates half times for the establishment of a new mature treeline on the order of 200 years. No evidence exists for such migration rates in the majority of alpine plants that predominantly propagate vegetatively. Clonal expansion growth in one of the dominant alpine species, *Carex curvula* amounts to 0.5 mm per year (Grabherr *et al.*, 1978), rates definitely too slow to track expected isotherm elevations of 300 m altitude within the next 50 to 100 years. If such scenarios become reality, vegetation boundaries will become labile zones with largely unpredictable transitory behaviour.

REFERENCES

Aber, J.D., Melillo, J.M. and McClaugherty, C.A. (1990) 'Predicting long-term patterns of mass loss, nitrogen dynamics, and soil organic matter formation from initial fine litter chemistry in temperate forest ecosystems', *Can. J. Bot.*, 68: 2201–8.

Baig, M.N. and Tranquillini, W. (1980) 'The effects of wind and temperature on cuticular transpiration of Picea abies and Pinus cembra and their significance in desiccation damage at the alpine treeline', *Oecologia*, 47: 252–6.

Barry, R.G. (1992) 'Mountain climatology and past and potential future climatic changes in mountain region: a review', *Mountain Research and Development*, 12: 71–86.

Becker, M. (1991) 'Incidence des conditions climatiques, édaphiques et sylvicoles sur la croissance et la santé des forêts', in *Les recherches en France sur le dépérissement des forêts*, Programme DEFORPA, 2ème rapport, ENGREF, Nancy, pp. 25–41.

Billings, W.D. (1974) 'Arctic and alpine vegetation: plant adaptations to cold summer climates', in J.D. Ives and R.G. Barry (eds), *Arctic and Alpine Environments*, Methuen, London, pp. 403–43.

Billings, W.D. and Mooney, H.A. (1968) 'The ecology of arctic and alpine plants', *Biol. Rev.*, 43: 481–529.

Bliss, L.C. (1971) 'Arctic and alpine plant life cycles', *Ann. Rev. Ecol. Systematics*, 2: 405–38.

—— (1985) 'Alpine plant life', in B.F. Chabot and H.A. Mooney (eds), *Physiological Ecology of North American Plant Communities*, Chapman & Hall, London and New York, pp. 41–65.

Cernusca, A. (1976) 'Bestandesstruktur, Bioklima und Energiehaushalt von alpinen Zwergstrauchbeständen', *Oecol. Plant.*, 11: 71–102.

Cernusca, A. and Seeber, M.C. (1981) 'Canopy structure, microclimate and the energy budget in different alpine plant communities', in J. Grace, E.D. Ford and P.G. Jarvis (eds), *Plants and their Atmospheric Environment*, 21st Symp. Brit. Ecol. Soc., Blackwell Scientific Publishers, Oxford, London, Edinburgh, pp. 75–81.

Christophersen, J. (1973) 'Basic aspects of temperature action on microorganisms', in H. Precht, J. Christophersen, H. Hensel and W. Larcher (eds), *Temperature and Life*, Springer, Berlin, Heidelberg, pp. 3–59.

Diemer, M., Körner, C. and Prock, S. (1992) 'Leaf life spans in wild perennial herbaceous plants: a survey and attempts at a functional interpretation', *Oecologia*, 89: 10–16.

Fliri, F. (1975) *Das Klima der Alpen im Raume von Tirol*, Universitätsverlag Wagner, Innsbruck, München.

Geyger, E. (1985) 'Untersuchungen zum Wasserhaushalt der Vegetation im nord-westargentinischen Andenhochland', *Dissertationes Botanicae*, 88, J. Cramer, Berlin, Stuttgart, p. 176.

Gonzalez, J.A. (1985) 'El potencial agua en algunas plantas de altura y el problema del stress hidrico en alta montana', *Lilloa*, 36: 167–72.

Grabherr, G., Mahr, E. and Reisigl, H. (1978) 'Nettoprimärproduktion und Repro-duktion in einem Krummseggenrasen (Caricetum curvulae) der Ötztaler Alpen, Tirol', *Oecol. Plant.*, 13: 227–51.

Grace, J. (1988) 'The functional significance of short stature in montane vegetation', in M.J.A. Werger, P.J.M. Van der Aart, H.J. During and J.T.A.Verhoeven (eds), *Plant Form and Vegetation Structure*, SPB Academic Publishers, The Hague, The Netherlands, pp. 201–9.

—— (1990) 'Cuticular water loss unlikely to explain tree-line in Scotland', *Oecologia*, 84: 64–8.

Heide, O.M. (1974) 'Growth and dormancy in Norway spruce ecotypes (Picea abies). I. Interaction of photoperiod and temperature', *Physiol Plant*, 30: 1–12.

—— (1985) 'Physiological aspects of climatic adaptation in plants with special reference to high-latitude environments', in A. Kaurin, O. Junttila and J. Nilsen (eds), *Plant Production in the North*, Norweg University Press, Oslo, pp. 1–22.

Hiesey, W.M. and Milner, H.W. (1965) 'Physiology of ecological races and species', *Ann. Rev. Plant Physiol.*, 16: 203–16.

Holzer,K. (1967) 'Das Wachstum des Baumes in seiner Anpassung an zunehmende Seehöhe', *Mitt. Forstl. Bundeversuchsanst*, Wien 75: 427–45.

—— (1975) 'Zur Identifizierung von Fichtenherkünften (Picea abies (L.) Karst.)', *Silvae Genertica*, 24: 169–75.

Izmailova, N.N. (1977) 'Wasserhaushalt kryophiler Polsterpflanzen im östlichen Pamir (russ)', *Ekologia* (Akad nauk SSSR) 2: 17–22.

Kienast, F. and Luxmoore, R.J. (1988) 'Tree-ring analysis and conifer growth responses to increased atmospheric CO_2 levels', *Oecologia*, 76: 487–95.

Klug-Pümpel, B. (1989) 'Phytomasse und Nettoproduktion naturnaher und anthro-pogen beeinflusster alpiner Pflanzengemeinschaften in den Hohen Tauern', in A. Cernusca (ed.), *Struktur und Funktion von Graslandökosystemen im Nationalpark*

Hohe Tauern, Veröffentl. Österr. MaB-Progr 13: 331–55, Wagner, Innsbruck.

Körner, C. (1982) 'CO$_2$ exchange in the alpine sedge Carex curvula as influenced by canopy structure, light and temperature', *Oecologia*, 53: 98–104.

—— (1984) 'Auswirkungen von Mineraldünger auf alpine Zwergsträucher', *Verhandl. Ges. Ökol.*, 12: 123–36.

—— (1989) 'The nutritional status of plants from high altitudes. A worldwide comparison', *Oecologia*, 81: 379–91.

—— (1992) 'Responses of alpine vegetation to global climate change', *CATENA Suppl.*, 22: 85–96.

—— (1993) 'Scaling from species to vegetation: the usefulness of functional groups', in E.D. Schulze and H.A. Mooney (eds), *Biodiversity and Ecosystem Function*, Springer, Berlin, Heidelberg, pp. 117–40.

Körner, C. and Cochrane, P. (1983) 'Influence of plant physiognomy on leaf temperature on clear midsummer days in the Snowy Mountains, south-eastern Australia', *Acta Oecol., Oecol. Plant.*, 4: 117–24.

—— and —— (1985) 'Stomatal responses and water relations of Eucalyptus pauciflora in summer along an elevational gradient', *Oecologia*, 66: 443–55.

Körner, C. and Diemer, M. (1987) 'In situ photosynthetic responses to light, temperature and carbon dioxide in herbaceous plants from low and high altitude', *Functional Ecol.*, 1: 179–94.

Körner, C. and Larcher, W. (1988) 'Plant life in cold climates', in S.F. Long and F.I. Woodward (eds), *Plants and Temperature*, Symposium of the Society for Experimental Biology, 42: 25–57, The Company of Biology Ltd, Cambridge.

Körner, C. and Mayr, R. (1981) 'Stomatal behaviour in alpine plant communities between 600 and 2600 metres above sea level', in J. Grace, E.D. Ford and P.G. Jarvis (eds), *Plants and their Atmospheric Environment*, Blackwell, Oxford, London, pp. 205–18.

Körner, C. and Pelaez Menendez-Riedl, S. (1989) 'The significance of developmental aspects in plant growth analysis', in H. Lambers *et al.* (eds), *Causes and Consequences of Variation in Growth Rate and Productivity of Higher Plants*, SPB Academic Publishers, The Hague, pp. 141–57.

Körner, C. and Renhardt, U. (1987) 'Dry matter partitioning and root length /leaf area ratios in herbaceous perennial plants with diverse altitudinal distribution', *Oecologia*, 74: 411–18.

Körner, C., Allison, A. and Hilscher, H. (1983) 'Altitudinal variation in leaf diffusive conductance and leaf anatomy in heliophytes of montane New Guinea and their interrelation with microclimate', *Flora*, 174: 91–135.

Körner, C., Bannister, P. and Mark, A.F. (1986) 'Altitudinal variation in stomatal conductance, nitrogen content and leaf anatomy in different plant life forms in New Zealand', *Oecologia*, 69: 577–88.

Körner, C., Farquhar, G.D. and Roksandic, Z. (1988) 'A global survey of carbon isotope discrimination in plants from high altitude', *Oecologia*, 74: 623–32.

Körner, C., Farquhar, G.D. and Wong, S.C. (1991) 'Carbon isotope discrimination by plants follows latitudinal and altitudinal trends', *Oecologia*, 88: 30–40.

Körner, C., Neumayer, M., Pelaez Menendez-Riedl, S. and Smeets-Scheel, A. (1989a) 'Functional morphology of mountain plants', *Flora*, 182: 353–83.

Körner, C., Wieser, G. and Cernusca, A. (1989b) 'Der Wasserhaushalt waldfreier Gebiete in den österreichischen Alpen zwischen 600 und 2600 m Höhe', in A. Cernusca (ed.), *Struktur und Funktion von Graslandökosystemen im Nationalpark Hohe Tauern*, Veröffentl. Österr. MaB-Progr. 13: 119–53, Wagner, Innsbruck.

Körner, C., Wieser, G. and Guggenberger, H. (1980) *Der Wasserhaushalt eines alpinen*

Rasens in den Zentralalpen, Veröffentl. Österr. MaB-Hochgebirgsprogramm Hohe Tauern 3: Untersuchungen an alpinen Böden in den Hohen Tauern 1974–1978. Stoffdynamik und Wasserhaushalt: 243–64.

Küppers, M. (1982) *Kohlenstoffhaushalt, Wasserhaushalt, Wachstum und Wuchsform von Holzgewächsen im Konkurrenzgefüge eines Heckenstandortes*, Dissertation, University of Bayreuth, 222 S.

LaMarche, V.C., Graybill, D.A., Fritts, H.C. and Rose, M.R. (1984) 'Increasing atmospheric carbon dioxide: tree-ring evidence for growth enhancement in natural vegetation', *Science*, 225: 1019–21.

Larcher, W. (1980) *Klimastress im Gebirge – Adaptationstraining und Selektionsfilter für Pflanzen*, Rheinisch-Westfälische Akad. Wiss. Vorträge N 291: 48–88.

Lauscher, F. (1977) *Ergebnisse der Beobachtungen an den nordchilenischen Hochgebirgsstationen Collahuasi und Chuquicamata*, Jahresbericht des Sonnblickvereines für die Jahre 1976–1977: 43–67.

Lüdi, W. (1938) 'Mikroklimatische Untersuchungen an einem Vegatationsprofil in den Alpen von Davos III', *Ber Geobot Forschungsinst Rübel*, Zürich, pp 29–49.

McCracken, I.J., Wardle, P., Benecke, U. and Buxton, R.P. (1985) 'Winter water relations of tree foliage at timberline in New Zealand and Switzerland', in H. Turner and W. Tranquillini (eds), *Establishment and Tending of Subalpine Forests: Research and Management*, Proc. 3rd IUFRO Workshop, pp. 1.07–00, 1984. Ber. Eidg. Anst. forstl. Versuchswes. 270: 85–93.

Michaelis, P. (1934) 'Ökologische Studien an der alpinen Baumgrenze. V. Osmotischer Wert und Wassergehalt während des Winters in den verschiedenen Höhenlagen', *JB Wiss. Bot.* 80: 336.

Oberbauer, S.F. and Billings, W.D. (1981) 'Drought tolerance and water use by plants along an alpine topographic gradient', *Oecologia*, 50: 325–31.

Ozenda, P. and Borel, J.L. (1990) 'The possible responses of vegetation to a global climatic change. Scenarios for western Europe, with special reference to the Alps', in M.M. Boer and R.S. De Groot (eds), *Landscape-ecological Impact of Climatic Change*, Proceedings of European Conference, Lunteren, The Netherlands, 3–7 December 1989, IOS Press, Amsterdam, pp. 221–49.

Pearcy, R.W., Bjorkman, O., Caldwell, M.M., Keeley, J.E., Monson, R.K. and Strain, B.R. (1987) 'Carbon gain by plants in natural environments', *BioScience*, 37: 21–9.

Perkins, T.D., Adams, G.T. and Klein, R.M. (1991) 'Desiccation or freezing? Mechanisms of winter injury to red spruce foliage', *Amer. J. Botany*, 78: 1207–17.

Pisek, A. and Cartellieri, E. (1941) 'Der Wasserverbrauch einiger Pflanzenvereine', *JB Wiss. Bot.*, 90: 282–91.

Pisek, A. and Larcher, W. (1954) 'Zusammenhang zwischen Austrocknungsresistenz und Frosthärte bei Immergrünen', *Protoplasma*, 44: 30–45.

Precht, H., Christophersen, J., Hensel, H. and Larcher, W. (1973) *Temperature and Life*, Springer, Berlin, Heidelberg, New York.

Psenner, R. and Nickus, U. (1986) 'Snow chemistry of a glacier in the Central Eastern Alps (Hintereisferner, Tyrol, Austria)', *Z Gletscherkunde Glazialgeologie*, 22: 1–18.

Richard, J.L. (1985) 'Pelouses xerophiles alpines des environs de Zermatt (Valais, Suisse)', *Bot. Helvetica*, 95: 193–211.

Schinner, F. and Gstraunthaler, G. (1981) 'Adaptation of microbial activities to the environmental conditions in alpine soils', *Oecologia*, 50: 113–16.

Solhaug, K.A. (1991) 'Long day stimulation of dry matter production in Poa alpina along a latitudinal gradient in Norway', *Holarctic Ecol.*, 14: 161–8.

Stocker, O. (1935) 'Assimilation und Atmung westjavanischer Tropenbäume', *Planta*, 24: 402–45.

Tranquillini, W. (1976) 'Water relations and alpine timberline', in O.L. Lange, L. Kappen and E.D. Schulze (eds), *Ecological Studies. Analysis and Synthesis*, Springer Verlag, Berlin, Heidelberg, New York, pp. 473–91.

—— (1979) 'Physiological ecology of the alpine timberline. Tree existence at high altitudes with special reference to the European Alps', *Ecological Studies*, 31, Springer, Berlin, Heidelberg.

Tranquillini, W. and Plank, A. (1989) 'Ökophysiologische Untersuchungen an Rotbuchen (Fagus sylvatica L.) in verschiedenen Höhenlagen Nord- und Südtirols', *Centralblatt f d ges Forstwesen*, 106: 225–46.

Wardle, P. (1974) 'Alpine timberlines', in J.D. Ives and R.G. Barry (eds), *Arctic and Alpine Environments*, Methuen, London, pp. 372–402.

—— (1981) 'Winter desiccation of conifer needles simulated by artificial freezing', *Arctic Alpine Res.*, 13: 419–23.

10

LONG-TERM VEGETATION CHANGE IN MOUNTAIN ENVIRONMENTS

Palaeoecological insights into modern vegetation dynamics

L.J. Graumlich

INTRODUCTION

The spectre of human-induced alteration of atmospheric composition and associated changes in climate has focused attention on how species, communities, and ecosystems respond to climate change (Solomon and Shugart, 1993). Such concern has prompted a wide range of research from short-term experiments which examine specific controls over the responses of individual tissues or plant parts, to the modelling of whole systems which predict ecological response at different levels of biological organization and spatio-temporal scales (Ehleringer and Field, 1993). While long-term, whole system experimental programmes in a broad range of biomes would be the ideal complement to efforts to understand the physiological response of individual plants, such idealism is constrained by funding and logistics. An alternative source of information about how species, communities, and ecosystems respond to environmental change is the palaeoecological record. Palaeoecology offers insights into the nature of climate–vegetation interactions that derive from the well-documented response of plant communities to environmental changes of the past: a record that is equivalent to natural, if unplanned, experiments (Davis, 1989). Since these natural experiments typically include conditions not observed in the twentieth century, the palaeo-record documents a substantially broader range of biotic responses to environmental variations than can be obtained from observational data.

The spatial and temporal resolution of palaeoecological data sets has increased in recent decades so that relatively detailed histories of vegetation change are available for much of North America and Europe. In addition, comparisons of records of past vegetation dynamics to palaeoclimatic simulations by general circulation models have improved the understanding of the role of climate in governing past vegetation change. Two major findings of palaeo-research have importance to investigations of the effects of future climate change on the Earth's biota. These are: (1) the large-scale controls over climate can vary independently, producing sustained periods of past climate and vegetation with no present-day analogues; and (2) short-term climatic controls, although poorly understood, have led to rapid environmental changes and substantial changes in tree distributions.

In this paper, patterns of montane forest response to climatic variation at two temporal scales are discussed: the Late Quaternary and the last millennium. Examples illustrate the wide range of potential responses of montane forests to climatic variation and emphasize opportunities for applying palaeoecological findings to questions of modern ecological dynamics. While the paper draws largely on examples from North America, the conclusions are well-supported by parallel research results in Europe and Asia.

CLIMATIC VARIATION

Weather and climate vary on time scales from seconds to millions of years. The response of forest vegetation to climate is scale-dependent, ranging from physiological responses to relatively short-term (e.g. diurnal) weather events to population responses to long-term (e.g. 10^3 to 10^5 years) climatic oscillations leading to altered species distributions and community structure (Davis, 1986). We focus on two illustrative periods: (1) the last 18,000 years, which encompasses a global warming of 6°C from the end of the last glacial maximum to the warmest portion of the current interglacial; and (2) the last 1,000 years, a period of relatively small magnitude (*c.* 1.5°C) temperature oscillations.

Climatic change over the last glacial/interglacial cycle

Over the last two million years the Earth's climate has oscillated between glacial and interglacial conditions. Oxygen isotope records from deep sea sediments indicate that these oscillations have a quasi-periodicity of *c.* 100,000 years and that periods when ice-sheets dominate high latitudes are relatively long (*c.* 90,000 to 120,000 years) compared to periods with little ice-cover (*c.* 10,000 years) (Shackelton and Opdyke, 1973). The present interglacial (the Holocene) began *c.* 10,000 years ago, although ice-sheets did not decrease to their current size until *c.* 6,000 years ago. These long-term climate oscillations are governed by periodic changes in the earth's orbit, which modify the latitudinal and seasonal distribution of solar radiation at periods ranging from 10^3 to 10^5 years (Hays *et al.*, 1976). Long-term changes in concentration of atmospheric trace gases (i.e., carbon dioxide, methane) mimic long-term changes in ice volume. Atmospheric concentrations of carbon dioxide and methane are substantially higher during interglacial than glacial periods (*c.* 100 ppmv for carbon dioxide and *c.* 200 ppbv for methane). This observation has led to speculation as to the mechanisms by which trace gas variations may reinforce the control of glaciation by changes in solar radiation (Raynaud *et al.*, 1993).

The factors that govern the regional climate (i.e. solar insolation, atmospheric trace gas and aerosol concentration, sea surface temperatures, ice volume) vary at different timescales and, as a result, their combined effects generate unique climatic states through time (COHMAP, 1988). These unique climatic states, often referred to as 'no-analogue' climates, have combinations of the seasonal temperature and precipitation with no counterparts in twentieth-century observational climatic data. An oft-cited example of no-analogue climate conditions is a rather long period encompassing the transition between the last glaciation and the present interglacial *c.* 12,000 to 6,000 years ago. During this period the seasonal cycle of insolation was amplified due to the greater tilt of the earth's axis and the occurrence in July of the perihelion (i.e. the point on the earth's orbit which is nearest to the sun). These orbital changes increased summer insolation in mid-to-high latitudes by about 8 per cent as compared to present, and decreased winter insolation by about 8 per cent. Increased summer insolation contributed to the wastage of the northern hemispheric ice-sheet, which had decreased to 25 per cent of its maximum extent by *c.* 9,000 years ago. The expression of the increased seasonal cycle differed regionally. For example, summers in Alaska, western Canada and Pacific Northwest were warmer than present, but summer temperatures in eastern North America remained cold due to the continued influence of the Laurentide ice-sheet on regional circulation. The most prominent aspect of summer climate in the southwestern United States was increased precipitation due to the intensification of monsoonal circulation (Spaulding and Graumlich, 1986). Periods of no-analogue climate such as these serve as natural experiments in which the response of vegetation to altered climatic regimes can be studied. Such 'experiments' are of particular importance given the likelihood that no-analogue climates will arise from an enhanced greenhouse effect.

Climatic change over the last millennium

Instrumental climatic observations as well as proxy climatic data from tree rings, high-resolution ice-cores, and glacial deposits indicate decadal and longer-term temperature variation over the last millennium. These variations are roughly synchronous over parts of the northern mid- to high latitudes, although the magnitude, seasonality, and timing of warm and cold intervals varies from region to region (Grove, 1988; Wigley, 1989 and references therein). Such temperature records indicate a general pattern of more frequent warm episodes from *c.* AD 1000 to 1300, more frequent cool episodes from *c.* AD 1300 to 1850, and a trend towards higher temperatures from AD 1850 to the present. The causes of short-term (i.e. decadal to centennial scale) climatic variation differ from those of long-term climatic variation and are surprisingly poorly understood. Three external factors are hypothesized to contribute to climatic variation on these time scales (Wigley, 1989). First, variations in solar radiation, associated with variations in the sun's radius and sunspot activity, affect the amount of incoming radiation to the earth's atmosphere. Second, sulphur-rich aerosols, derived from volcanic, industrial, and biological sources, are thought to reduce the earth's temperature by increasing atmospheric reflectivity. Finally, increased carbon dioxide and other trace gases affect the earth's climate by absorbing longwave radiation. Ascertaining the relative importance of each of these external factors in governing climate is complicated by the internal variability of the coupled atmosphere–ocean system. Simple global climate models demonstrate that the thermal inertia of the ocean as well as changes in the vertical (thermohaline) circulation of the ocean are sufficient to induce multi-decadal trends in an otherwise unforced climate system (Wigley and Raper, 1990). A more complete understanding of the patterns and causes of decadal-scale climatic variation is likely in the near future with the increasing global coverage of various palaeoclimatic indicators (Mosley-Thompson *et al.*, 1990).

VEGETATION RESPONSES

Changes in montane forests over the last glacial/interglacial cycle

The climatic changes marking the transition from the last glaciation to the current interglacial period caused major reorganization of forest vegetation worldwide. Counter-intuitively, high-latitude, boreal forests covered substantially less area during full-glacial times as compared to today. Conversely, the montane coniferous forest cover in southwestern North America expanded during full-glacial times. A summary of montane forest response to climatic changes, emphasizing results from western North America, serves to illustrate the linkages between global-scale climatic forcings and regional vegetation response.

The montane forests of western North America experienced massive reorganization in the transition from late glacial to modern climates. Vegetation changes in western North America were driven both by changes in thermal regimes and by changes in soil moisture supplies as influenced by changes in the position and strength of westerly surface winds (COHMAP, 1988). In the Pacific Northwest and northern Rocky Mountains, conditions were colder and drier during the late glacial and, as a result, coniferous forest species retreated to lower elevations and were, in general, far less abundant than during the ensuing interglacial period (Barnosky *et al.*, 1987; Whitlock,

1993). In the Southwest (south of *c.* 34°N latitude), conditions during the late glacial were colder and wetter than present due to a weakening of the Pacific subtropical high pressure system and a southward displacement and strengthening of the westerly surface winds (COHMAP, 1988). Under these more mesic conditions, pinyon pine and juniper woodlands were widespread in lowland areas that today support only desert scrub vegetation (Spaulding *et al.*, 1983). As in the case of the migrational histories of boreal forest taxa, species of the coniferous forests of western North America demonstrated highly individualistic behaviour with respect to elevational displacement (Spaulding *et al.*, 1983, Barnosky *et al.*, 1987). Certain tree species (e.g. bristlecone pine (*Pinus longaeva*) and limber pine (*Pinus flexilis*) in the Southwest; Engelmann spruce (*Picea engelmannii*) and lodgepole pine (*Pinus contorta*) in the Northwest) had broader altitudinal and latitudinal ranges in glacial times as compared to today. Other species (e.g., ponderosa pine (*Pinus ponderosa*), Colorado pinyon (*Pinus edulis*)) appear to have been restricted to small ranges in what is today southern Arizona and New Mexico (Van Devender *et al.*, 1987).

During the early Holocene (*c.* 9,000 to 6,000 years ago), the regional contrasts in biotic response to large-scale climatic forcings continued. In the Pacific Northwest and the northern Rocky Mountains, the increased summer solar insolation was associated with drier climatic conditions associated with an intensification of the eastern Pacific subtropical high (Whitlock, 1993). Effectively drier conditions and an associated increase in fire frequency resulted in an increase in drought-tolerant and fire-adapted species (e.g. lodgepole pine, Douglas-fir (*Pseudotsuga menziessii*), alder (*Alnus*)) and the occurrence of open parklands in areas of currently closed forest. Conversely, in the Southwest, increased summer solar insolation was associated with an intensification of summer, monsoonal rainfall (Thompson *et al.*, 1994). Where summer soil moisture was adequate, montane conifers either expanded their lower elevational ranges or maintained their late-glacial range. In central Colorado, for example, increased summer temperatures combined with increased summer moisture to cause both a downward expansion of lower treeline (Markgraf and Scott, 1981) as well as an upward expansion of upper treeline (Fall, 1988; Carrara *et al.*, 1984). Elsewhere in the Southwest, evidence for the persistence of montane conifers at relatively low elevations is abundant, although species vary in the pattern and timing of their retreat to higher elevations. Variation in the timing of upslope movement by coniferous species reflects both the individuality of species response to climatic trends and the heterogeneity of climatic regimes in the region (Thompson *et al.*, 1994 and references therein).

Table 10.1 Examples of twentieth-century increases in tree recruitment at or above timberline

Region	Species	Interpretation	Reference
North Quebec	*Picea glauca* *Larix laricina*	Altitudinal seed regeneration limit increased *c.* 100 m since nineteenth century at protected sites; increase of tree density near forest limit attributed to climatic warming	Payette and Filion, 1985 Marin and Payette, 1984
Coast Range, British Columbia	*Abies lasiocarpa* *Tsuga mertensiana*	Decreased snowfall and resulting longer growing seasons led to increased recruitment in subalpine meadows	Brink, 1959
Canadian Rockies	*Abies lasiocarpa* *Picea engelmannii*	Increased seedling establishment in meadows at timberline associated with warmer summer temperatures	Kearney, 1982
Swedish Scandes	*Picea abies*	Upright growth of previously stunted individuals in conjunction with increased regeneration above treeline is correlated with climatic warming at two periods of heavy snowfall	Kullman, 1986
Finland	*Pinus sylvestris*	Climate amelioration led to increased regeneration at and above treeline	Hustich, 1958
Siberia, Russia	*Boreal forest* spp.	Increased regeneration in response to climate amelioration	Gorchakovsky and Shiyatov, 1978
Khibin Mts, Russia	*Pinus* sp.	Increased regeneration in response to presumed climatic amelioration began 200 years ago, but accelerated significantly in the 1900s	Kozubov and Shaydurov, 1965 (in Bray, 1971)
South Island, New Zealand	*Nothofagus* spp.	Climate amelioration led to increased regeneration at and above treeline	Wardle, 1963 (in Bray, 1971)
Cascade Range, Washington and Oregon	*Abies lasiocarpa* *Tsuga mertensiana* *Abies amabilis*	Warmer growing seasons and reduced snowpack led to increased regeneration and invasion of subalpine meadows	Franklin *et al.*, 1971; Heikkinen, 1984; Agee and Smith, 1984

Olympic Mts, Washington	*Abies lasiocarpa* *Tsuga mertensiana*	Tree invasion of meadows during periods of warm growing season and reduced snowpack	Fonda and Bliss, 1969 Kuramoto and Bliss, 1970
Lemhi Mts, Idaho	*Abies lasiocarpa* *Picea engelmannii* *Pinus contorta* *Pseudotsuga menziessii*	Tree invasion of meadows during warm, dry periods of reduced snowpack in the early twentieth century	Butler, 1986
Sierra Nevada, California	*Pinus balfouriana*	Warmer growing seasons led to increased recruitment at and above treeline	Scuderi, 1987
White Mts, California and Nevada	*Pinus longaeva*	Warmer growing seasons led to increased recruitment at and above treeline	LaMarche, 1973
Uinta Mts, Utah	*Picea engelmannii*	Young trees advancing towards relict treeline	Major, pers. comm., in Bray, 1971
LaSal Mts, Utah	*Picea* sp.	Climate amelioration led to increased recruitment	Richmond, 1962
Rocky Mountains, Colorado	*Pinus aristata* *Picea engelmannii*	Warmer growing seasons led to establishment of seedlings above current treeline	Daly and Shankman, 1985; K. Yamaguchi, pers. comm., 1993

Changes in montane forests over the last millennium

Two broad classes of response have occurred in coniferous species in response to the temperature fluctuations of the last millennium. First, at several alpine and arctic treelines, range limits have shifted due to altered reproductive and establishment rates. Second, in areas where climatic limitations to growth have changed, established trees have undergone phenotypic adjustments to the altered climate.

While climatic variables may subtly alter rates of reproduction and establishment in temperate coniferous ecosystems, the role of climate in governing population processes is clearly seen at alpine and arctic ecotones. Tree establishment has increased in subalpine and treeline stands and young trees have established at elevations or latitudes beyond current treeline in a variety of settings worldwide (Table 10.1). One must be cautious in inferring from these data a global upward movement of treeline, for several reasons. First, the fate of the young trees in these stands is unclear in that a return to more severe climatic conditions may slow or halt recruitment and decrease survivorship. Second, ecotonal movement may result from several different climatic factors (e.g. increased summer temperature, decreased snowpack, increased soil moisture supplies, alteration of wind regime) as well as different anthropogenic factors (e.g. grazing, altered fire regime). Finally,

Table 10.2 Evidence of altitudinal treeline fluctuations over the last several hundred years, with emphasis on locations where relict stands are preserved above current treeline

Location	Interpretation	Dating	Reference
Siberia	Treeline higher 800–900 ^{14}C yr BP; treeline depressed 270–300 m after 600 ^{14}C yr BP; treeline advanced 200–250 ^{14}C yr BP	radiocarbon	Kozubov and Shaydurov, 1965 (in Bray, 1971)
Europe	Treeline descended *c.* 100–200 m from the fourteenth to seventeenth century	pollen stratigraphy	Firbas and Losert, 1949 (in Bray, 1971)
Alps	Treeline descended *c.* 70 m after AD 1300		Gams, 1937 (in Bray, 1971)
Canadian Rockies	Subfossil trees above current treeline date to late 1300s and early 1400s; dieback at these sites in late 1700s	radiocarbon, dendro-chronology	Luckman, 1994, and references therein
Yukon	*Picea* forest, 60 m higher than present treeline, buried under 1.5 m of volcanic ash	none	Rampton, 1969
Wyoming Range, Wyoming	Tree stumps 60 m above current treeline	none	Griggs, 1938
LaSal Mts, Utah	Relict forest 30 m above current treeline	none	Richmond, 1962
Big Snowy Range, Montana	Relict treeline 100 m higher than present	none	Bamberg and Major, 1968
Uinta Mts, Utah	*Picea engelmannii* relicts 80 m above current treeline	none	Major (pers. comm., in Bray, 1971)
Sierra Nevada, California	Relict treeline up to 100 m above present treeline	dendro-chronology	Lloyd and Graumlich, 1993; Scuderi, 1987
Kirkliston Range, New Zealand	Logs of *Podocarpus totara* 155 m above present treeline, dated 500–1450 ^{14}C yr BP	radiocarbon	Burnett, 1926 (in Raeside, 1948)
New Zealand	*Nothofagus fusca* charcoal 210 m above present treeline	none	Molloy *et al.*, 1963

ecotonal movement is not universally observed in all regions that have experienced a recent upward trend in temperatures. A particularly well-studied example of this is the lichen-spruce woodland of northern Quebec in which only sites that are well-protected from the damaging effects of wind exposure show growth responses to recent climatic warming (Payette et al., 1989). Further, rather than treeline advance in northern Quebec, radiocarbon dated spruce charcoal records indicate that lichen-spruce woodlands have fragmented and retracted to their present distribution because of catastrophic fires during the past several millennia (Payette and Gagnon, 1985). The work of Payette and colleagues clearly demonstrates the complexity of the problem of assessing ecosystem response or resiliency to climatic change and emphasizes the importance of multiple climatic factors and exogenous factors such as disturbance.

Trees respond to decadal scale climatic variation by systemic changes in carbon balance and nutrient status, which, in turn, lead to changes in either absolute growth rates or altered patterns of carbon allocation. Sustained changes in growth and productivity have been observed during the twentieth century in trees growing near latitudinal and altitudinal treeline in western North America (Garfinkel and Brubaker, 1980; LaMarche et al., 1984; Graumlich and Brubaker, 1986; Graumlich, 1991; Graybill and Idso, 1993; Luckman, 1994). Such long-term growth trends often mirror similar trends in growing-season temperature. While growing-season temperature is generally the most important climatic variable governing growth at latitudinal and altitudinal treeline, detailed analyses of year-to-year variability indicate that other climatic factors, such as soil moisture and depth of snowpack, interact with temperature in controlling growth at treeline sites in the Cascade Mountains and Sierra Nevada (Graumlich and Brubaker, 1986; Graumlich, 1991). Non-climatic factors, specifically direct CO_2 fertilization, have been linked to the observed increasing trends in growth of upper treeline bristlecone pine (LaMarche et al., 1984). However, a recent comparison of bristlecone pine growth in individuals with strip-bark (i.e. partial bark and cambium) vs. full bark (i.e. entire bark and cambium) indicates that the growth enhancement is confined to strip-bark individuals (Graybill and Idso, 1993). Further, analyses of growth trends of other subalpine conifers have failed to detect evidence of direct CO_2 fertilization (Kienast and Luxmoore, 1988; Graumlich, 1991). Direct enhancement of forest growth by CO_2 thus appears to be highly specific to strip-bark bristlecone pine.

The two climatic responses documented above, increased regeneration rates and increased growth rates in treeline stands, appear unique when considered in the context of twentieth-century observations. However, in the relatively rare circumstances in which particularly long-lived conifers provide growth or regeneration records that extend over the last millennium, the responses to recent climatic trends are not unique and have occurred in the past when temperatures were equivalent to late twentieth-century values. In the Sierra Nevada, growth of foxtail pine exceeded that of late twentieth-century values for several episodes in the past (1370–1440 and 1480–1580; Graumlich, 1991). Similarly, evidence for tree establishment above current treeline exists at several localities throughout the world (Table 10.2), indicating that the climatic variables critical for regeneration and survival have fluctuated significantly on the timescales of hundreds of years to millennia. Therefore, while the warm temperatures of the twentieth century may be anomalous with respect to the life span of most coniferous species, they may not be anomalous in the context of the recent history of a given forest stand.

CONCLUSIONS

The palaeoecological record of climatic variation and vegetation response summarized here demonstrates that the individualism of the behaviour of species with respect to physiological attributes (Chapin and Shaver, 1985) is mirrored in long-term and large-scale vegetation dynamics. On Quaternary timescales we see that the communities we study are relatively shortlived, temporary assemblages characterized by continuous flux (Davis, 1981, 1986). Over the last 1,000 years, we see that the growth of individuals and the dynamics of populations reflect complex interactions among multiple climatic variables. Regardless of assumptions one makes regarding future climatic scenarios, we can expect climate factors to continue to behave in a quasi-independent manner and, as a result, future climates will be characterized by combinations of temperature and precipitation that are not replicated on the modern landscape. Consequently, predictions of vegetation response to climatic changes must be based on a mechanistic understanding of the relationship between climatic variation and vegetation processes.

ACKNOWLEDGEMENTS

The comments of Linda B. Brubaker and Andrea H. Lloyd greatly improved the manuscript. Andrea H. Lloyd compiled the information contained in Tables 10.1 and 10.2.

REFERENCES

Agee, J.K. and Smith, L. (1984) 'Subalpine tree establishment after fire in the Olympic Mountains, Washington', *Ecology*, 65: 810–19.

Bamberg, S.A. and Major, J. (1968) 'Ecology of the vegetation and soils associated with calcareous parent materials in three alpine regions of Montana', *Ecological Monographs*, 38: 127–67.

Barnosky, C.W., Anderson, P.M. and Bartlein, P.J. (1987) 'The northwestern U.S. during deglaciation; vegetational history and paleoclimatic implications', in W.F. Ruddiman and H.E. Wright, Jr (eds), *North America and Adjacent Oceans During the Last Deglaciation*, series *The Geology of North America*, vol. K–3, Geological Society of North America, Boulder, Colorado, pp. 289–321.

Bray, J.R. (1971) 'Vegetational distribution, tree growth, and crop success in relation to recent climate change', *Advances in Ecological Research*, 7: 177–223.

Brink, V.C. (1959) 'A directional change in the subalpine forest-heath ecotone in Garibaldi Park, British Columbia', *Ecology*, 40: 10–16.

Butler, D.R. (1986) 'Conifer invasion of subalpine meadows, Central Lemhi Mountains, Idaho', *Northwest Science*, 60: 166–73.

Carrara, P.E., Mode, W.N., Meyer, R. and Robinson, S.W. (1984) 'Deglaciation and postglacial timberline in the San Juan Mountains, Colorado', *Quatenary Research*, 21: 42–56.

Chapin, F.S., III, and Shaver, G.R. (1985) 'Individualistic growth response of tundra plant species to environmental manipulations in the field', *Ecology*, 66: 564–76.

COHMAP Project Members (1988) 'Climatic changes of the last 18,000 years: observations and model simulations', *Science*, 241: 1043–51.

Daly, C. and Shankman, D. (1985) 'Seedling establishment by conifers above tree limit on Niwot Ridge, Colorado U.S.A.', *Arctic and Alpine Research*, 17: 389–400.

Davis, M.B. (1981) 'Quaternary history and the stability of forest communities', in D.C. West, H.H. Shugart and D.B. Botkin (eds), *Forest Succession: Concepts and Applications*, Springer-Verlag, New York, pp. 132–53.

—— (1986) 'Climatic instability, time lags, and community disequilibrium', in J. Diamond and T.J. Case (eds), *Community Ecology*, Harper and Row, New York, pp. 269–84.

—— (1989) 'Insights from paleoecology on global change', *Bulletin of the Ecological Society of America*, 70: 222–8.

Ehleringer, J.R. and Field, C.B. (1993) *Scaling Physiological Processes: Leaf to Globe*, Academic Press, New York.

Fall, P.L. (1988) 'Vegetation dynamics in the southern Rocky Mountains: Late Pleistocene and Holocene timberline fluctuations', unpublished Ph.D. dissertation, University of Arizona, Tucson.

Firbas, F. and Losert, H. (1949) 'Untersuchungen über die Entstehung der heutigen Waldstufen in der Sudeten', *Planta*, 36: 478–506.

Fonda, R.W. and Bliss, L.C. (1969) 'Forest vegetation of the montane and subalpine zones, Olympic Mountains, Washington', *Ecological Monographs*, 39: 271–96.

Franklin, J.F., Moir, W.H., Douglas, G.W. and Wiberg, C. (1971) 'Invasion of subalpine meadows by trees in the Cascade Range, Washington and Oregon', *Arctic and Alpine Research*, 3: 215–24.

Gams, H. (1937) 'Aus der Geschichteder Alpenwälder', *Z. dt. öst. Alpenver*, 68: 157–70.

Garfinkel, H.L. (1979) 'Climate-growth relationships in white spruce (Picea glauca [Moench] Voss) in the south central Brooks Range of Alaska', unpublished MS thesis, University of Washington, Seattle.

Garfinkel, H.L. and Brubaker, L.B. (1980) 'Modern climate–tree-ring relations and climatic reconstruction in sub-arctic Alaska', *Nature*, 286: 872–3.

Gorchakovsky, P.L. and Shiyatov, S.G. (1978) 'The upper forest limit in the mountains of the boreal zone of the USSR', *Arctic and Alpine Research*, 10: 349–63.

Graumlich, L.J. (1985) 'Long-term records of temperature and precipitation in the Pacific Northwest derived from tree rings', unpublished Ph.D. dissertation.

—— (1991) 'Subalpine tree growth, climate, and increasing CO_2: an assessment of recent growth trends', *Ecology*, 72: 1–11.

—— (1993) 'A 1,000-year record of temperature and precipitation in the Sierra Nevada', *Quaternary Research*, 39: 249–55.

Graumlich, L.J. and Brubaker, L.B. (1986) 'Reconstruction of annual temperature (1590–1979) for Longmire, Washington, derived from tree rings', *Quaternary Research*, 25: 223–34.

Graybill, D.A. and Idso, S.B. (1993) 'Detecting the aerial fertilization effect of atmospheric CO_2 enrichment in tree-ring chronologies', *Global Biogeochemical Cycles*, 7: 81–95.

Griggs, R.F. (1938) 'Timberlines in the northern Rocky Mountains', *Ecology*, 19: 548–64.

Hays, J.D., Imbrie, J. and Shackelton, N.J. (1976) 'Variations in the earth's orbit: pacemaker of the ice ages', *Science*, 194: 1121–32.

Heikkinen, O. (1984) 'Forest expansion in the subalpine zone during the past hundred years, Mount Baker, Washington, USA.', *Erdkunde*, 38: 194–202.

Hustich, I. (1958) 'On the recent expansion of the Scotch pine in northern Europe', *Fennia*, 82: 1–25.

Jones, P.D., Raper, S.C.B., Santer, B.D., Cherry, B.S.G., Goodess, C., Bradley, R.S., Diaz, H.F., Kelley, P.M. and Wigley, T.M.L. (1985) 'A gridpoint surface air temperature data set for the Northern Hemisphere: 1851–1984', *US DoE Tech. Rep. TR022*, US Dept Energy Carbon Dioxide Research Division, Washington, DC.

Kearney, M.S. (1982) 'Recent seedling establishment at timberline in Jasper National Park', *Canadian Journal of Botany*, 60: 2283–7.

Kienast, F. and Luxmoore, R.J. (1988) 'Tree-ring analysis and conifer growth responses to increased atmospheric CO_2 levels', *Oecologia*, 76: 487–95.

Kozubov, G.M. and Shaydurov, V.S. (1965) 'Vertical zonality in the Khibin mountains and fluctuations of the timberline', *Akademiia Nauk SSSR Izvestiia*, Ser. Geog 3: 101–4 (in Russian).

Kullman, L. (1986) 'Recent tree-limit history of Picea abies in the southern Swedish Scandes', *Canadian Journal of Forest Research*, 16: 761–77.

Kuramoto, R.T. and Bliss, L.C. (1970) 'Ecology of subalpine meadows in the Olympic Mountains, Washington', *Ecological Monographs*, 40: 317–45.

LaMarche, V.C., Jr (1973) 'Holocene climatic variations inferred from treeline fluctuations in the White Mountains, California', *Quaternary Research*, 3: 632–60.

LaMarche, V.C. Jr, Graybill, D.A., Fritts, H.C. and Rose, M.R. (1984) 'Increasing atmospheric carbon dioxide: tree-ring evidence for growth enhancement in natural vegetation', *Science*, 225: 1019–21.

Lloyd, A.H. and Graumlich, L.J. (1993) 'Late Holocene treeline fluctuations in the Southern Sierra Nevada', *Bulletin of the Ecological Society of America*, Supplement, 74: 334–5.

Luckman, B.H. (1994) 'Evidence for climatic conditions between *c.* AD 900–1300 in the southern Canadian Rockies', *Climatic Change*, in press.

Marin, A. and Payette, S. (1984) 'Expansion récente du Mélèze à la limite des forêts (Québec nordique)', *Canadian Journal of Botany*, 62: 1404–8.

Markgraf, V. and Scott, L. (1981) 'Lower timberline in central Colorado during the past 15,000 yr', *Geology*, 9: 231–4.

Molloy, B.P.J., Burrows, C.J., Cox, J.E., Johnston, J.A. and Wardle, P. (1963) 'Distribution of subfossil forest remains, Eastern South Island, New Zealand', *New Zealand Journal of Botany*, 1: 68–77.

Mosley-Thompson, E., Barron, E., Boyle, E., Burke, K., Crowley, T., Graumlich, L., Jacobson, G., Rind, D., Shen, G. and Stanley, S. (1990) 'Earth system history and modeling', in *'Research Priorities for the U.S. Global Change Research Program'*, Committee on Global Change, National Academy Press, Washington.

Payette, S. and Filion, L. (1985) 'White spruce expansion at the treeline and recent climate change', *Canadian Journal of Forest Research*, 15: 241–51.

Payette, S. and Gagnon, R. (1985) 'Late Holocene deforestation and tree regeneration in the forest-tundra of Quebec', *Nature*, 313: 570–2.

Payette, S., Filion, L., Delwaide, A. and Begin, C. (1989) 'Reconstruction of tree-line vegetation response to long-term climatic change', *Nature*, 341: 429–32.

Raeside, J.D. (1948) 'Some Post-glacial climatic changes in Canterbury and their effects on soil formation', *Transactions of the Royal Society of New Zealand*, 77: 153–71.

Rampton, V. (1969) 'Pleistocene geology of the Snag-Klutlan area of southwestern Yukon, Canada', Ph.D. thesis, University of Minnesota.

Raynaud, D., Jouzel, J., Barnola, J.M., Chappellaz, J., Delmas, R.J. and Lorius, C. (1993) 'The ice record of greenhouse gases', *Science*, 259: 926–34.

Richmond, G.L. (1962) 'Quaternary stratigraphy of the La Sal Mountains, Utah', *U.S. Geological Survey Professional Paper 324*.

Scuderi, L.A. (1987) 'Late-Holocene upper timberline variation in the southern Sierra Nevada', *Nature*, 325: 242–4.

Shackleton, N.J. and Opdyke, N.D. (1973) 'Oxygen isotope and paleomagnetic stratigraphy of equatorial Pacific core V28–238: oxygen isotope temperatures and ice volumes on a 10^5 year and 10^6 year scale', *Quaternary Research*, 3: 39–55.

Solomon, A.M. and Shugart, H.H. (1993) *Vegetation Dynamics and Global Change*, Chapman & Hall, New York.

Spaulding, W.G. and Graumlich, L.J. (1986) 'The last pluvial climatic episode in the deserts of southwestern North America', *Nature*, 319: 441–4.

Spaulding, W.G., Leopold, E.B. and Van Devander, T.R. (1983) 'Late Wisconsin paleoecology of the American Southwest', in S.C. Porter (ed.), *Late-Quaternary Environments of the United States. Vol. 1: The Late Pleistocene*, University of Minnesota Press, Minneapolis, pp. 259–93.

Thompson, R.S., Whitlock, C., Bartlein, P.J., Harrison, S.P. and Spaulding, W.G. (1994) 'Climatic changes in the western United States since 18,000 yr B.P.', in H.E. Wright, Jr, J.E. Kutzbach, T. Webb III, W.F. Ruddiman, F.A. Street-Perrot and P.J. Bartlein (eds), *Global Changes Since the Last Glacial Maximum*, in press. University of Minnesota Press, Minneapolis.

Van Devender, T.R., Thompson, R.S. and Betancourt, J.L. (1987) 'Vegetation history of the deserts of southwestern North America; the nature and timing of the Late Wisconsin-Holocene transition', in W.F. Ruddiman and H.E. Wright, Jr (eds), *North America and Adjacent Oceans During the Last Deglaciation*, series *The Geology of North America*, vol. K-3, Geological Society of America, Boulder, Colorado, pp. 323–52.

Wardle, P. (1963) 'The regeneration gap of New Zealand gymnosperms', *New Zealand Journal of Botany*, 1: 301–15.

Webb, T. III, Kutzbach, J.E. and Street-Perrot, F.A. (1985) '20,000 years of global climate change: palaeoclimatic research plan', in T.F. Malone and J.G. Roederer (eds), *Global Change*, Cambridge University Press, Cambridge, pp. 182–218.

Whitlock, C. (1993) 'Postglacial vegetation and climate of Grand Teton and southern Yellowstone National Parks', *Ecological Monographs*, 63: 173–98.

Wigley, T.M.L. (1989) 'Climatic variability on the 10–100-year time scale: observations and possible causes', in R. Bradley (ed.), *Global Changes of the Past*, UCAR, Office for Interdisciplinary Studies, Boulder, Colorado, pp. 83–101.

Wigley, T.M.L. and Raper, S.C. (1990) 'Natural variability of the climate system and detection of the greenhouse effect', *Nature*, 344: 324–7.

11

LATITUDINAL VARIATION IN THE POTENTIAL RESPONSE OF MOUNTAIN ECOSYSTEMS TO CLIMATIC CHANGE

P.N. Halpin

INTRODUCTION

The potential impact of future climatic changes on mountain ecosystems and nature reserves has become an increasingly important issue in the study of long-term biodiversity management and protection (McNeely, 1990; Peters and Darling, 1985). The possibility of rapid changes in the climatic habitats of montane ecosystems due to industrial emissions of radiative gases raises many difficult questions concerning the vulnerability of montane nature reserves to environmental disruption in the future. Proposed changes in global temperatures and local precipitation patterns could significantly alter the altitudinal ranges of important species within existing mountain nature reserves and create additional environmental stresses on already fragile mountain ecosystems. Until recently, generalized speculations concerning the possible future movement of ecoclimatic habitats within mountain nature reserves have been put forth with only limited reference to actual mountain sites or application of geographic modelling methods (Peters and Darling, 1985; Hunter *et al.*, 1988; Graham, 1988). However, current geographic analysis techniques now allow for the simulation of possible responses of both global and regional impacts of climate change on existing mountain nature systems (Halpin, 1992; Halpin and Secrett, in press; Leemans and Halpin, 1992; Halpin, 1994). This type of geographic analysis provides for the development and testing of more detailed hypotheses concerning global change and mountain environments.

In this brief overview, a pilot study is outlined which assesses the potential impacts of various future climate change scenarios on the global distribution of mountain vegetation regions, the possible impacts on important mountain nature reserve sites, and a regional case study of impacts on a tropical mountain area. A hypothetical analysis is also presented of the differences in

180

the sensitivity of mountain reserve areas to climatic change due the latitudinal position of the sites. In this pilot study, the importance of differences in the relationship of annual primary production and species ranges to topographic features at different latitudes is emphasized, and possible implications for future impacts on biodiversity management are suggested.

CLIMATIC CHANGE AND MOUNTAIN ECOSYSTEM CONSERVATION

The potential impacts of rapid climatic change on natural ecosystems and nature reserves has reinforced the need for the development of creative research techniques for the analysis of the potential movement of habitats within the confines of existing nature reserve boundaries. In the most basic form, rapid changes in climatic conditions coupled with ongoing changes in landscape fragmentation are seen to be two potentially interactive environmental forces. These combined forces could dramatically hinder the ability of natural resource managers to maintain viable mountain habitats and species populations in the future.

Peters and Darling (1985) outlined two basic hypotheses concerning potential climatic changes and nature reserves. The first general hypothesis defines the interaction of landscape fragmentation and species migration under changing climates. The basic premise of this hypothesis is that under changing land-use and climatic conditions, static nature reserve boundaries may fail to encompass the species ranges for which they were intended to protect. The explicit process outlined in this hypothesis is that the southern climatic ranges of species in the northern hemisphere would move poleward under proposed climate change scenarios in a direct response to increasing temperatures. The ability of species to track changing environmental conditions would be related to the rate of climatic change, the migratory potential of the species, competitive pressures between species, the availability of suitable habitats and physical obstacles in the path of migrating individuals (Smith and Tirpak, 1989; Solomon, 1986; Davis, 1989; Sedjo and Solomon, 1989; Hunter *et al.*, 1988; Graham, 1988; Schwartz, 1992).

Potential boundaries to the migration of species between reserve areas has spurred a wide interest in the geographical (GIS) analysis of protected area distributions and connectiveness in recent years (Mackintosh *et al.*, 1989; Scott *et al.*, 1987; Hudson, 1991). This type of analysis has led to the development of a variety of rule of thumb assumptions concerning the spatial connectiveness of nature reserves under changing climatic conditions. Noss (1992) states that corridors between reserve areas should be aligned upslope, coast–inland, and south–north in order to facilitate potential species movements under changing climatic conditions. This type of generalization has been coupled with other basic biogeographic theories to form the basis for numerous landscape level assessments now investigating the effects of fragmentation on natural resource management (Scott *et al.*, 1990; Hudson, 1991).

The second general hypothesis presented by Peters and Darling (1985) and others concerns the potential movement of the climatic ranges of species along altitudinal gradients. Under this general hypothesis, increases in local temperature would act to move the climatic habitats of species upslope, in a linear manner, where each successive altitudinal range would be replaced by the species habitat occupying the zone directly below (Figure 11.1). This conceptual model implies that the boundaries of present species climatic ranges will respond symmetrically to changes in temperature related to the adiabatic lapse rate (the loss of temperature with altitude) for a particular mountain side. The general biogeographical rule used to derived this conceptual model is attributed to 'Hopkins bioclimatic law' (Peters and Darling, 1985; MacArthur, 1972), which relates a 30°C change in temperature to a 500 m change in altitude. Under this general conceptual model, the expected impacts of climate change in mountainous nature reserves would include the loss of the coolest climatic habitats at the peaks of the mountains and the linear migration of all remaining habitats upslope.

These extremely generalized assumptions have been incorporated directly into much of the current literature on potential impacts of climatic change on biodiversity management and nature reserve systems (Smith and Tirpak, 1989; McNeely, 1990; IPCC, 1989; Noss, 1992). The general management conclusion derived from this species response theory is that nature reserves exhibiting large altitudinal ranges and topographic relief will offer the largest range of climatic habitats, and therefore will allow for the greatest amount of internal species movement under changing climates. In contrast, climatic shifts in areas of low altitudinal range would be expected to force species to migrate outside the boundaries of the reserve area.

This conceptual model has been used as the underpinning for the analysis of the altitudinal ranges of global nature reserve networks in an initial assessment of the risks posed to these nature reserves by changing climatic conditions. McNeely (1990) assessed the altitudinal ranges of more than 2,000 nature reserves greater than 1,000 hectares in size. In his study, McNeely states that 686 (approximately 30 per cent) of the sites assessed were found to exhibit more than 1,000 m of altitudinal range. Using the 3°C temperature change per 500 m of altitude rule of thumb, these reserves were expected to contain approximately 6°C of potential climatic range within their borders. This altitudinal temperature range was assumed to be within the expected limits of most future climate change scenarios and parks containing this range of altitude were expected to be able to withstand an upward shift of climatic habitats. Both predictive modelling of mountainous landscapes and palaeological records of past climatic changes suggest that changes in vegetation zones are not generally symmetrical in pattern along altitudinal gradients. The direct extrapolation of intuitive theories of

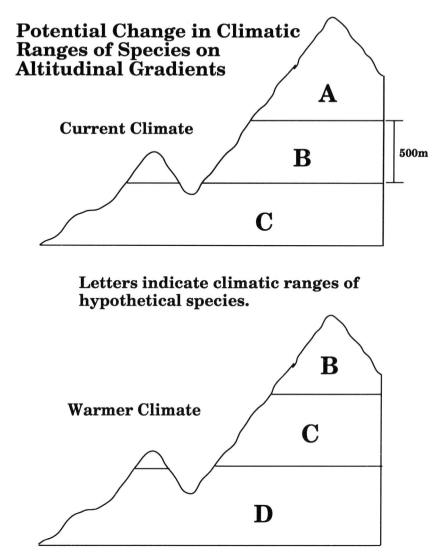

Potential Change in Climatic Ranges of Species on Altitudinal Gradients

Current Climate

A

B

C

500m

Letters indicate climatic ranges of hypothetical species.

Warmer Climate

B

C

D

Figure 11.1 A conceptual model of the potential movement of species climatic ranges along altitudinal gradients under a climatic change scenario. Movement of climatic ranges is based on a 500 m increase in altitude for each 3°C increase in temperature (adapted from Peters and Darling, 1985).

expected biogeographic response to management oriented assessments, without a geographic analysis of the spatial characteristics of potential climate changes on mountainous terrain can be misleading. While it is generally accepted that larger altitudinal gradients potentially offer a large number range of ecoclimatic habitats, the implicit assumption that changes in the climate will offer a similar number and spatial configuration of distinct climatic habitats along an altitudinal gradient is in question. Peters' and Darling's (1985) linear model of montane habitat response to climatic change was presented as a heuristic illustration, not a global model of responses for all mountain environments (Peters, personal communication).

METHODS OF IMPACT ASSESSMENT

In this pilot study, the global distribution of potential climatic impacts has been assessed for a wide range of nature reserves and altitudinal ranges outlined above; it has been found that the general assumptions used in previous analyses do not corespond to currently accepted biogeographical principles and that the simplistic rules of thumb used in past assessments may result in misleading conclusions (Halpin, 1992, 1994; Halpin and Secrett, in press).

First, all climate change scenarios are not spatially equivalent, therefore assessments of the expected impacts on mountain reserves to proposed climatic changes must be assessed in terms of the locations of mountain sites with respect to variation in the magnitude of the climate change forcing expected for specific locations. Second, all ecoclimatic habitats are not equivalent. For example, the potential climatic range or extent of a tropical montane forest and a northern alpine larch forest are significantly different, and any conceptual model which treats the responses of different climatic habitats as symmetrical features is destined to misrepresent the composition of altitudinal change for different latitudes and biogeographic regions. And third, the actual complexity of population level dynamics which would occur as individual species attempt to track shifting climates on to low fertility, high-slope alpine areas occupied by other persistent species is an extremely difficult problem to address with simple bioclimatic rules.

As a first step in the analysis of potential impacts of climate change on mountain ecosystems and nature reserves, a global grid model was used to establish current mean climatic conditions for the world. The 0.5° × 0.5° climatic grid GIS developed by Leemans and Cramer (1990) which contains information on monthly average temperature, precipitation, cloudiness, soil texture, soil water holding capacity, and elevation was used to develop an ecoclimatic classification (potential vegetation) for the world. The climatic grid GIS was then used to derive a Holdridge life zone class and an effective soil moisture budget for each grid cell (Holdridge, 1967; Emmanuel et al., 1985; Smith et al., 1990).

The Holdridge life zone model is a widely used global ecoclimatic classification system based on the relationship of current vegetation biomes to three general climatic parameters: annual temperature, annual precipitation and an estimated potential evapotranspiration presented on logarithmic axes (Holdridge, 1967). Establishing the climatic coordinates within the

Table 11.1 General Circulation Models used to construct climate change impact scenarios

General Circulation Model	Change in mean global temp. (°C)	Change in mean global precip. (%)
Oregon State University (OSU)[a]	2.84	7.8
Geophysical Fluid Dynamics Lab (GFDL)[b]	4.00	8.7
Goddard Institute for Space Studies (GISS)[c]	4.20	11.0
UK Meteorological Office (UKMO)[d]	5.20	15.0

Notes: [a] Schlesinger and Zhao (1988)
[b] Manabe and Wetherald (1987)
[c] Hansen *et al.* (1988)
[d] Mitchell (1983)

model defines the ecoclimatic life zone for a particular site. The model predicts ecoclimatic areas currently associated with major vegetation formations of the world, and does not directly model actual vegetation or land cover distribution.

In order to test the potential impacts of changing climatic conditions at a global scale, the differences in monthly temperature and precipitation for four General Circulation Models (GCMs) were geometrically interpolated and overlaid into a raster Geographic Information System (GIS) to develop data layers of new distributions of equilibrium ecoclimatic zones for each scenario (Emmanuel *et al.*, 1985; Smith *et al.*, 1990, 1992). The scenarios evaluated present a wide range of different climate change modelling experiments from different research institutions. Table 11.1 presents the general features of each of these climatic change scenarios. Under this modelling technique, areas which are shown to change ecoclimatic class under a climate change scenario represent areas of potential ecosystem stress, and should not be interpreted as direct changes in vegetation type to the new equilibrium state. Species populations are expected to respond individually to any potential changes in future environmental features under any scenario. Also, areas of urban land-use and agriculture are not expected to respond to changing climatic conditions through natural processes of ecosystem succession, and therefore are identified and excluded through GIS overlay methods from the continental scale analyses.

While all of the climate change scenarios tested present different spatial patterns and magnitudes of change, a few general trends can be observed. The highest percentage of change occurs in the high northern latitudes under all scenarios due to higher temperature forcing at these latitudes (Smith *et al.*, 1990, 1992). This feature becomes important when analysing potential

Table 11.2 Potential impacts on mountain nature reserves containing over 3,000 m
of altitudinal range

UKMO	OSU	GISS	GFDL
85.5%	64.4%	65.7%	51.3%

Table 11.3 Potential ecoclimatic change in mountain versus non-mountain
vegetation regions

	UKMO	OSU	GISS	GFDL
Mountain regions	84.7%	73.9%	64.5%	55.7%
Non-mountain regions	71.8%	60.8%	56.1%	49.4%

impacts on existing nature reserves, which have a numerical bias toward more northerly, developed countries (Halpin, 1992; Halpin and Secrett, in press; Smith *et al.*, 1990; Leemans and Halpin, 1992).

A selection of nature reserves containing more than 3,000 m altitudinal range was plotted and compared to the spatial distribution of climatic change under the four scenarios tested. Table 11.2 presents the percentage of these reserves which experience a change in ecoclimatic zone under each climate change scenario. As with the general distribution of important nature reserves, high mountain reserves received significant levels of impacts under all scenarios tested.

In order to assess the spatial distribution of climate change scenarios against global mountain ranges, a GIS data layer of mountain vegetation regions of the world was digitized and overlaid on the ecoclimatic change scenario data layers. As with the general distribution of nature reserves, global mountain systems received a higher percentage of aerial ecoclimatic change than terrestrial ecosystems in general. Table 11.3 depicts the percentage of area changing ecoclimatic zone under mountain and non-mountain environments. These elevated rates of ecoclimatic impact can be attributed to the position of the mountain regions with respect to the simulated climatic forcing, or alternatively, can be used as an argument to suggest that the climate models (GCMs) have a higher range of variance in representing the complex climate features of mountainous areas (Halpin, 1994).

While global scale GIS analysis of the potential ecological impacts of climatic change offers interesting insights into the distribution of impacts on entire systems of mountain reserve areas, high resolution, regional case studies must be conducted to develop an understanding of potential changes within individual mountain reserves. A case study of the potential regional impacts of climate change in Costa Rica illustrates a few key features of this type of analysis.

The spatial resolution of the global climate data sets outlined above is approximately 55 × 55 km grid cells at the equator in a raster based GIS system. This spatial resolution is sufficient to represent broad continental features of ecoclimatic transitions, but is entirely inappropriate for use at a regional scale, where questions concerning land management areas are being considered (Halpin and Secrett, in press). For example the entire country of Costa Rica is represented by only 20 grid cells in the global ecoclimatic data base. To conduct a more meaningful case study in this highly mountainous area, a regional climate, topography, soils, potential vegetation, vegetation cover, and land-use data base was developed at approximately a 400 × 400 m pixel resolution. This data set was better able to represent sufficiently the complex terrain and vegetation features of the region at a 1:500,000 map scale using 500 m elevation contours. A second, higher resolution analysis was conducted by collaborators at the Tropical Science Centre, in Costa Rica, through expert interpretation of climatic and topographic features at a 1:200,000 map scale using 100 m elevation contour intervals. This manually interpreted higher resolution analysis was digitized into a vector based GIS for later analysis (Tosi et al., 1992; Smith et al., in prep.; Secrett, 1992).

Elevations in Costa Rica range from sea level to over 3,800 m in less than 100 kilometres in the southern region of the country. Even though Costa Rica covers a relatively small area (51,000 km²), the distribution of complex climatic features in this country required the division of the country into five topographically distinct climatic regions for the development of sub-regional climatic change models. Separate lapse rates, sea-level temperatures, and precipitation regimes were interpolated from climate station data for each region. This base climate model was then modified to create two regional climate change sensitivity scenarios (Halpin et al., 1991; Kelly, 1991). The base climate mapping of life zones for Costa Rica has been related to specific vegetation patterns through substantial site level ground truthing over the last 25 years (Tosi, 1969; Holdridge et al., 1971; Sawyer and Lindsey, 1971). A moderate change scenario based on an increase of +2.5°C temperature and +10 per cent precipitation and a more extreme regional scenario depicting an increase of +3.6°C and +10 per cent precipitation were used to assess potential changes in tropical montane forest climate zones.

Figure 11.2a–b depicts areas of change in ecoclimatic zones for the country under current climate, the +2.5°C and the +3.6°C sensitivity test scenarios. A significant amount of change in ecoclimatic zonation occurs under both scenarios for the country with a 38 per cent change in zones for the 2.5°C scenario and 47 per cent change under the 3.6°C scenario (Halpin et al., 1991). Even larger areas of spatial change in ecoclimatic zones (43 per cent and 60 per cent respectively) were found under the higher resolution map interpretation analysis conducted by regional experts (Tosi et al., 1992).

(a)

(b)

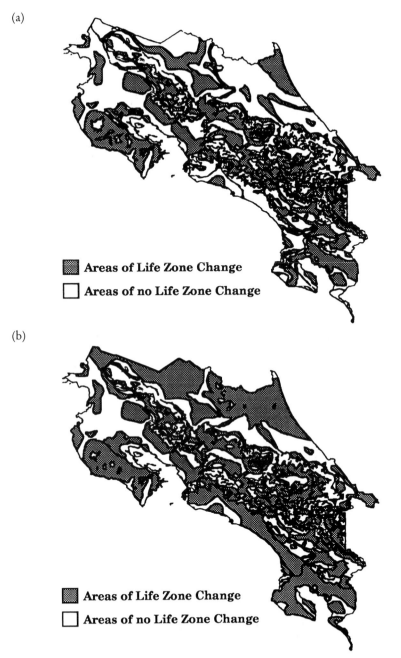

Figure 11.2 Provisional Holdridge life zones for Costa Rica under (a) +2.5°C +10 per cent precipitation sensitivity scenario; (b) + 3.6°C +10 per cent precipitation sensitivity scenario (modified from Tosi *et al.* 1992).

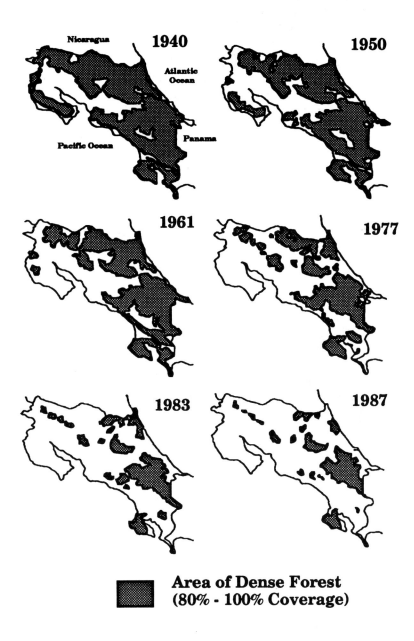

Figure 11.3 Areas of remaining natural forest in Costa Rica, 1940–1987.

Potential climatic change expresses only one factor of the two-part global change dilemma outlined in the introduction (Peters and Darling, 1985). Change in landscape fragmentation is the most noticeable and potent threat to mountain habitats in the humid tropical regions of the world at present (Myers, 1980; Sader and Joyce, 1989; Lugo, 1988). Figure 11.3 illustrates changes in forest cover for Costa Rica derived from field surveys, aerial photography, and satellite remote sensing (LANDSAT–TM) from 1940 to 1987. Areas of ecoclimatic zone change in non-agricultural (60–100 per cent natural vegetation cover) areas of the country, nature reserves and other special management areas were isolated in the analysis using standard GIS overlay techniques.

Under the 3.6°C scenario, an ecoclimatic threshold is reached where the distribution of warmer premontane climate zones expands up the slopes, lowering the total number of distinct ecoclimatic habitats from 9 to 6 in the southern mountainous region of La Amistad Biosphere Reserve. The reason for this non-linear change is actually quite simple to explain. The range, or extent of climate now associated with the lower elevation premontane forests is larger in terms of possible temperature and precipitation combinations than the cooler, narrower vegetation zones upslope. We just do not normally notice this potential altitudinal extent when the vegetation associated with these warmer ecoclimatic areas is distributed across the low foothill areas of the country. When the sea-level temperatures of the tropical mountain are increased past a threshold temperature range, the distribution of the lowland forest climates can be extended up a larger altitudinal gradient than the significantly narrower ranges of the cooler montane and alpine zones farther up the slopes (Halpin, 1992; Halpin, 1993; Halpin and Secrett, in press). The result is that instead of simply shifting all of the altitudinal zonations presently found on the mountain up one evenly spaced level, the size and composition of the resulting zones change dramatically. This result is in direct contrast to the present hypothesis that climatic habitats would respond in a regularly ordered pattern. To emphasize this point, the only time one could possibly expect changes in ecoclimatic zones to respond in a regular pattern would be if the climatic limits for each potential vegetation category were identical in their range of temperatures and precipitation responses. All generally accepted vegetation correlation models are based on delineation of differential climate spaces for different types of vegetation categories (Holdridge, 1967; Budyko, 1974; Whittaker, 1975; Box, 1981; Woodward, 1987; Stephenson, 1990; Prentice et al., 1992). Each of these approaches would predict very different patterns of mountain zonation responses to changing climates due to different interpretations of current vegetation climate ranges for each model, the only common feature that can be generally acknowledged between all models being that none of them will predict climatic zones moving in a symmetrical, staircase pattern upslope as is now commonly presumed.

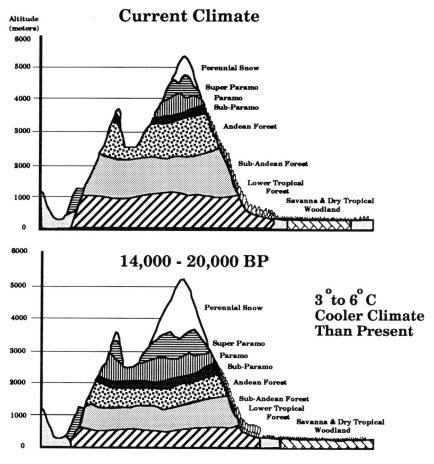

Figure 11.4 A palaeoecological reconstruction of vegetation zones of a tropical mountain site in Columbia from a 3–6°C cooler epoch (14,000–20,000 BP) to warmer, present climate conditions (adapted from Flenley, 1979).

In the tropics, there are a number of palaeoclimatic reconstructions of vegetation from fossil pollen records which can be viewed as qualitative climatic change analogies (Flenley, 1979; Vuilleumier and Monasterio, 1986; Peteet, 1987). Figure 11.4 represents a palaeovegetation reconstruction of a mountain in Columbia approximately 5° south from the Costa Rica case study area (adapted after Flenley, 1979). This palaeological record qualitatively illustrates the trend from a more narrowly compressed ecoclimatic zonation during a cooler epoch (14,000–20,000 BP) to broader zonations during a 3° to 6°C warmer (current climate) period. Similar spatial trends can be seen in various palaeoclimate reconstructions in the neotropical region (Flenley, 1979; Vuilleumier and Monasterio, 1986). This proposed differential expansion of ecoclimatic zonations along altitudinal gradients during different climatic epochs offers an intriguing problem for more rigorous analysis.

Comparative studies of mountain vegetation zonations of curent tropical mountains demonstrate significant variation in the width and distribution of similar structural vegetation types between mountain sites within the tropical latitudes (Gerrard, 1990). These differences can be attributed to generalized responses of the local vegetation to large-scale climatic characteristics of each region.

LATITUDINAL DIFFERENCES IN MOUNTAIN SENSITIVITY TO CHANGING CLIMATIC CONDITIONS

One of the first geographic questions to ask concerning potential differences in the response of mountain regions to changing climates is to do with differences in sensitivity of mountains at different latitudes to this type of change. In order to investigate the sensitivity of different mountain sites to climatic change based on their latitudinal position, three sites ranging from tropical to arctic latitudes were selected for a hypothetical analysis of ecoclimatic zonation changes. A 3,900 m hypothetical mountain with 100 m elevation intervals was digitized into a raster GIS and used to represent a typical mountain at each site. A single +3.6° temperature and +10 per cent precipitation climate change sensitivity scenario was imposed for all sites. This process was done to hold the topographic effect and magnitude of climatic forcing equal for the simulation. Actual climate data from each site approximating La Amistad Biosphere reserve in Costa Rica, the Sequoia-Kings Canyon Biosphere Reserve in California (US), and the Denali Biosphere Reserve in Alaska (US) were used to establish the baseline climates for each hypothetical mountain (Halpin, 1994).

The features of the wet tropical site change have already been outlined above. A loss of climatic zones occurred under the warming scenario and large bands of lower elevation tropical premontane forest climatic zones shifted upslope. The cool subalpine paramo climate zone was lost off the top of the mountain under this sensitivity test (Figure 11.5a).

The dry temperature mountain site produced a significantly different result, with a loss of two representative climate zones and the expansion of low and mid altitude climate ranges. While reduced in coverage, the high elevation nival (alpine tundra) climate zone at the peak of this temperate site was not lost under this scenario as would be expected under the accepted change paradigm (Figure 11.5b).

The cold arctic site begins with a very shallow range of subalpine and alpine forest and tundra sites which expand only slightly upslope under the same climate scenario. Unlike the conventional paradigm, the only loss in distinct ecoclimatic zones at the arctic site occurs near the base of the

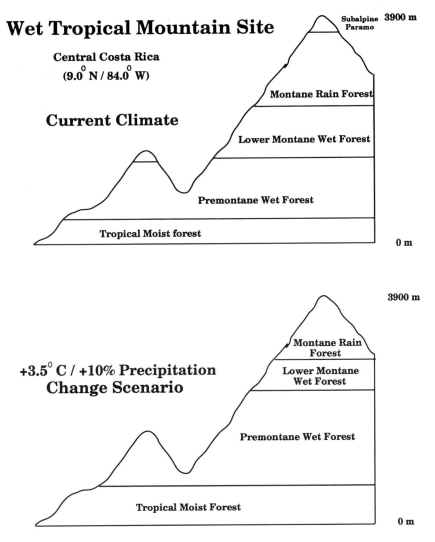

Figure 11.5a Current and changed ecoclimatic zonation for a hypothetical tropical climate mountain located in La Amistad Biosphere reserve, central Costa Rica.

mountain, instead of at the summit (Figure 11.5c).

Figure 11.6 illustrates the areas of ecoclimatic change for the three latitudinal sites under the same change scenario. This illustration does not represent a new generic paradigm for the expected changes in montane zones under changing climatic conditions, but instead is presented to caution us that changes in ecoclimatic zonations on altitudinal gradients cannot be explained by simple linear assumptions applied globally. Using extremely

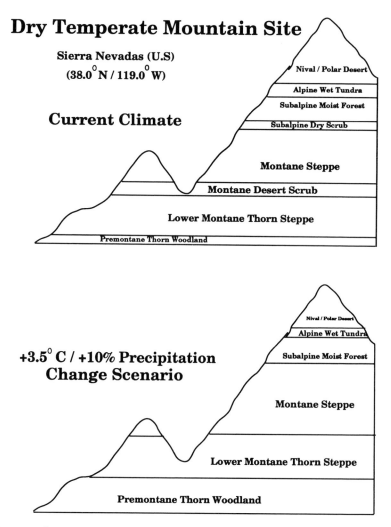

Figure 11.5b Current and changed ecoclimatic zonation for a hypothetical temperate climate mountain located in Sequoia-Kings Canyon Biosphere reserve, central California, USA.

straightforward climatic correlation models, in a spatially explicit GIS modelling framework, significantly more complex solutions than would generally be anticipated are derived. A new conceptual model based on assumptions of more complex reconfigurations of non-symmetrical ecoclimatic zones should replace current assumptions of symmetrical zonation change for mountain environments under changing environments (Halpin, 1992, 1994; Halpin and Secrett, in press).

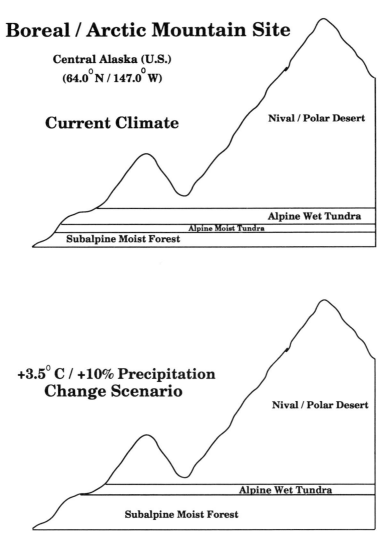

Figure 11.5c Current and changed ecoclimatic zonation for a hypothetical boreal/arctic climate mountain located in Denali Biosphere reserve, central Alaska, USA.

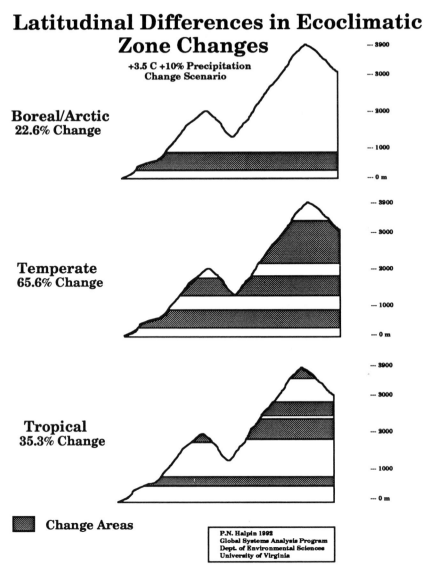

Latitudinal Differences in Ecoclimatic Zone Changes

+3.5 C +10% Precipitation Change Scenario

Boreal/Arctic
22.6% Change

Temperate
65.6% Change

Tropical
35.3% Change

Change Areas

P.N. Halpin 1992
Global Systems Analysis Program
Dept. of Environmental Sciences
University of Virginia

Figure 11.6 Areas of ecoclimatic change for hypothetical arctic, temperate, and tropical mountains.

OTHER LATITUDINAL DIFFERENCES IN MOUNTAIN ECOSYSTEMS

The potential changes in ecoclimatic zones along altitudinal gradients described above are related to general correlations between structural vegetation types and annual climatic characteristics. This type of correlation between ecoclimatic zones and vegetation structure are in general related to the amount and statification of standing biomass and the rates of primary production expected for a given site (Holdridge, 1967). Differences in species responses to climatic change for mountains at different latitudes is more difficult to generalize. However, a few studies demonstrate potentially important trends between the distribution of tree species and altitudinal gradients which may be very relevant to climate impact analysis. Three important potential differences in the latitudinal responses of mountain environments to changing climates are: first, the altitudinal position of critical climatic limits on vegetation life forms; second, the relative barriers mountains pose to species migration; and third, differences in the altitudinal ranges of tree species at different latitudes.

The climatic limits of vegetation life forms along altitudinal gradients have been explored extensively by numerous investigators (Wardle, 1974; Barry, 1979; Tranquillini, 1979; Arno, 1984; Stevens and Fox, 1991; Slayter and Noble, 1992). The altitudinal limits of closed forest timberline, tree limit and krummholz zones vary significantly with latitudinal position of the mountain site. A proposed empirical relationship between monthly climate and timberline has been established at the 10°C isotherm on a mountain slope for the warmest month of the year (Arno, 1984). When plotted across a large latitudinal range, this relationship exhibits a distinct latitudinal trend with timberlines occurring at lower elevations with distance from the equator.

The central consideration to be drawn from this relationship is that changes in the general climate for a site will alter the relative critical temperature regimes for mountain slopes at dramatically different altitudinal positions depending on the latitudinal position of the mountain. Conceptual models of potential impacts of climatic change must take into account the distinct differences in the spatial location of critical climatic features which affect ecotone boundaries. Once again, changes in the climatic mechanisms which control vegetation distributions on mountain gradients cannot be expected to occur in symmetrical patterns across all latitudes.

In an investigation of the ecological effects of mountains on species movements, Janzen (1967) explored the hypothesis that mountains in the tropics are potentially more effective barriers to species migration than higher latitude sites. This hypothesis is based on observed differences in the seasonal overlap of climates between sites along similar altitudinal gradients at different latitudes. Janzen compared the amount of overlap in temperature ranges between different altitudinal sites in Costa Rica and North America

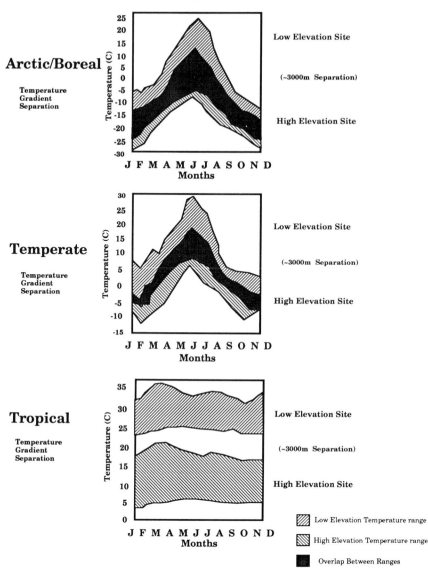

Figure 11.7 General latitudinal difference in seasonal overlap in temperature ranges (modified after Janzen, 1967).

in his analysis. The general finding of this investigation was that tropical sites often demonstrate a higher degree of separation between temperature ranges at low and high elevation sites throughout annual cycles. In other words, at no time of the year do the climates of a high temperature tropical foothill site and a cooler tropical montane site overlap. On the other hand, sites in temperate and boreal climatic zones exhibit dramatic seasonal variations in temperature ranges. This seasonal variation at higher latitude sites produces a greater frequency of climatic overlap between sites separated by large altitudinal gradients at higher latitudes. Figure 11.7 illustrates the general concept of this latitudinal difference in mountain climates.

The central implication of this finding is that similar topographic features may be viewed as significantly different obstacles to species migration at different latitudes because of the amount of climatic overlap exhibited between sites. A high latitude site may present a higher frequency of opportunities for plant species to establish and pass through critical life history phases at widely separated altitudinal sites. Therefore a seed transported over a high latitude ridge may establish itself and reproduce during the growing season at a site normally beyond the expected core range of the species. But a seed transported over a similar mountain in the tropics may not be able to establish and complete its life cycle because specific climatic requirements are not met during any time of the year outside of a narrow climatic belt. This type of analysis has been based exclusively on temperature ranges, and needs to be extended to seasonal evapotranspiration and soil moisture limits along altitudinal gradients to be a more effective approach to estimating latitudinal differences in mountain ecosystem sensitivity to climatic change.

A third general hypothesis concerning differential species responses to mountain climates at different latitudes concerns the altitudinal ranges of species at different latitudes. Stevens (1992) also uses sites in Costa Rica and North America to assess differences in the elevational ranges of various types of species. This analysis supports the hypothesis that the general climatic ranges of species are more extensive in high latitude sites than tropical sites. Stevens presents observations that suggest that the altitudinal ranges of montane tree species at different latitudes are significantly different. Tropical trees in Costa Rica exhibit narrow altitudinal ranges (approximately 500 m) with little variation between species while mid and high latitude tree species from sites in Tennessee (USA) and Alaska (USA) exhibit significantly wider elevation ranges (800–1,800 m) which increase with both latitude and altitude (Stevens, 1992).

This observed difference in elevation ranges of tree species at different latitudinal sites may have significant consequences for the response of vegetation to possible future climatic changes. The narrow elevation ranges of tropical species may be controlled by either increased competition in equatorial latitudes or more narrowly defined physiological response functions of the tropical tree species than their higher latitude counterparts. Because the empirical basis of this analysis does not identify the mechanism of range limitation, it is difficult to extrapolate particular responses for future climates. However, the current realized ranges of tropical tree species in Costa Rica suggest that mountain ecosystems at this latitude could be significantly more sensitive than previously expected to climatic disruption, either through changes in competitive advantages of species or direct physiological tolerance than higher latitude sites which contain wider ranging species.

CONCLUSIONS

The ability for persistent tree species to maintain viable populations along altitudinal gradients under changing climatic regimes will be affected by numerous interactions between existing populations and micro-site features. Any actual impacts of changing climatic conditions on species contained in mountain nature reserves will result from a highly complex cascade of environmental and ecological feedbacks which must be modelled using more physiologically mechanistic and temporally dynamic simulation models (Halpin, 1993; Shugart *et al.*, 1992). While significantly more difficult to develop, future models of mountain ecosystem dynamics will have to derive environmental site features and changing vegetation states interactively with spatial data bases in order to assess ecosystem features which can be validated through field observation or remote sensing. Simple conceptual models of the potential responses of vegetation to changing climatic conditions must be replaced with more detailed and mechanistic hypotheses of the complex interactions which govern mountain environments at different latitudes.

REFERENCES

Arno, S.T. (1984) 'Timberline: mountain and arctic frontiers', *The Mountaineers*, University of Washington, Seattle.

Barry, R.G. (1979) 'High altitude climates', in P.J. Webber (ed.), *High Altitude Geoecology*, Westview, Boulder.

Box, E.O. (1981) *Macroclimate and Plant Forms: An Introduction to Predictive Modeling in Phytogeography*, Junk, Haguer.

Budyko, M.I. (1974) *Climate and Life, International Geophysics Series*, 18, Academic Press, New York.

Davis, M.B. (1989) 'Lags in vegetation response to greenhouse warming', *Climatic Change*, 15: 75–82.

Davis, O.K. (1989) 'Ancient analogs for greenhouse warming of central California', in J.B. Smith and D.A. Tirpak (eds), *The Potential Impacts of Climate Change on the United States, Appendix D: Forests*, USEPA, Washington, DC.

Emmanuel, W.R., Shugart, H.H. and Stevenson, M.P. (1985) 'Climatic change and the broad-scale distribution of terrestrial ecosystem complexes', *Climatic Change*, 7: 29–43.

Flenley, J.R. (1979) *The Equatorial Rain Forest*, Butterworths, London.

Gerrard, A.J. (1990) *Mountain Environments: An Examination of the Physical Geography of Mountains*, MIT Press, Cambridge, Massachusetts.

Graham, R.W. (1988) 'The role of climate change in the design of biological reserves: the paleoecological perspective for conservation biology', *Conservation Biology*, 2 (4): 391–4.

Halpin, P.N. (1992) 'Potential impacts of climate change on protected areas: global assessments and regional analysis', in IUCN, *Proceedings of the IV World Parks Congress*, Caracas, Venezuela.

—— (1994) 'A GIS analysis of potential impacts of climate change on mountain ecosystems and protected areas', in M.F. Price and D.I. Heywood (eds), *Mountain Environments and Geographic Information Systems*, (in press), Taylor & Francis, London.

Halpin, P.N., Kelly, P.M., Secrett, C.M. and Smith, T.M. (1991) *Climate Change and Central American Forest Systems: Costa Rica Pilot Project* (symposium report).

Halpin, P.N. and Secrett, C.M. (in press) 'Potential impacts of climate change on forest protection in the humid tropics: a case study of Costa Rica', in *Impacts of Climate Change on Ecosystems and Species*, vol. 2, *Terrestrial Ecosystems*, IUCN, Gland, Switzerland.

Hansen, J., Fung, I., Lacis, A., Lebedef, S., Rind, D., Ruedy, R., Russel, G. and Stone, P. (1988) 'Global climate changes as forecast by the Goddard Institute for Space Studies three dimensional model', *Journal of Geophysical Research*, 93: 9341–64.

Holdridge, L.R. (1967) *Life Zone Ecology*, Tropical Science Center, San José, Costa Rica.

Holdridge, L.R., Grenke, W.C., Hatheway, W.H., Liang, T. and Tosi, J.A. Jr (1971) *Forest Environments in Tropical Zones: A Pilot Study*, Pergamon Press, Oxford.

Hudson, W.E. (1991) *Landscape Linkages and Biodiversity*, Island Press, Washington, DC.

Hunter, M.L., Jacobson, G.L. Jr and Webb, T. III (1988) 'Paleoecology and the coarse filter approach to maintaining biological diversity', *Conservation Biology*, 2 (4): 375–85.

IPCC (1989) 'Unmanaged ecosystems – biological diversity, adaptive responses to climate change', Resource Use and Management Subgroup of IPCC Working Group III, Working Paper.

Janzen, D.H. (1967) 'Why are mountain passes higher in the tropics?', *The American Naturalist* 101 (919): 233–49.

Kelly, P.M. (1991) 'Regional climate change scenarios for Costa Rica', Climatic Research Unit, University of East Anglia (unpublished project report).

Leemans, R. and Cramer, W.P. (1990) *The IIASA Database for Mean Monthly Values of Temperature, Precipitation and Cloudiness on a Global Terrestrial Grid*, WP–90–41, Austria International Institute for Applied Systems Analysis, Laxenburg.

Leemans, R. and Halpin, P.N. (1992) 'Biodiversity and global change', in B. Groombridge (ed.), *Biodiversity Status of the Earth's Living Resources*, World Conservation Monitoring Centre, Chapman Hall, London.

Lugo, A.E. (1988) 'Estimating reductions in the diversity of tropical forest species', in E.O. Wilson and Frances M. Peter (eds), *Biodiversity*, National Academy Press, Washinton, pp. 58–70.

MacArthur, R.H. (1972) *Geographical Ecology*, Harper & Row, New York.

Mackintosh, G., Fitzgerald, J. and Kloepfer, D. (1989) *Preserving Communities and Corridors*, Defenders of Wildlife, G.W. Press, Washington, DC.

McNeely, J.A. (1990) 'Climate change and biological diversity: policy implications', in M.M. Boer and R.S. de Groot (eds), *Landscape–Ecological Impact of Climatic Change*, IOS Press, Amsterdam.

Manabe, S. and Wetherald, R.T. (1987) 'Large scale changes in soil wetness induced by an increase in carbon dioxide', *J. Atm. Sci.*, 44: 1211–35.

Mitchell, J.F.B. (1983) 'The seasonal response of a general circulation model to changes in CO_2 and sea temperatures', *Q. J. Roy. Met. Soc.*, 109: 113–52.

201

Myers, N. (1980) 'The present status and future prospects of tropical moist forests', *Environmental Conservation*, 7: 101–14.

Noss, R. (1992) 'The wildlands project: land conservation strategy', in *Wild Earth*, Cenozoic Society, New York.

Peteet, D. (1987) 'Late Quaternary vegetation and climatic history of the montane and lowland tropics', in C. Rosenzweig and R. Dickinson (eds), *Climate–Vegetation Interactions*, University Corporation for Atmospheric Studies, Boulder, CO.

Peters, R.L. and Darling, J.D. (1985) 'The greenhouse effect and nature reserves', *Bioscience*, 35 (11): 707–17.

Prentice, I.C. *et al.* (1992) *Developing a Global Vegetation Dynamics Model: Results of the IIASA Summer Workshop*, IIASA, Laxenburg, Austria.

Rapaport, E.H. (1979) *Areography: Geographical Strategies of Species, Vol. 1*, 1st English edn, trans. B. Drausal, Pergamon, New York.

Sader, S.A. and Joyce, A.T. (1989) 'Deforestation rates and trends in Costa Rica: 1940 to 1983', *Biotropica* 20 (1): 11–19.

Sawyer, J. and Lindsey, A. (1971) *Vegetation of the Life Zones in Costa Rica*, Indiana Academy of Science Monograph no. 2.

Schlesinger, M. and Zhao, Z. (1988) 'Seasonal climatic changes induced by doubled CO_2 as simulated by the OSU atmospheric GCM/mixed layer ocean model', Climate Research Institute, Oregon State University, Cornvallis, OR.

Schwartz, M.W. (1992) 'Modelling effects of habitat fragmentation on the ability of trees to respond to climatic warming', *Biodiversity and Conservation*, 2: 51–61.

Scott, J.M., Csuti, B., Jacobi, J.D. and Estes, J.B. (1987) 'Species richness: a geographic approach to protecting future biodiversity', *Bioscience*, 37: 782–8.

Secrett, C.M. (1992) 'Adapting to climate change; a strategy for the tropical forest sector', International Institute for Environment and Development Project Paper, London.

Sedjo, R.A. and Solomon, A.M. (1989) 'Climate and forests', in N.J. Rosenberg, W.E. Easterling, P.R. Crosson and J. Darmstadter (eds), *Greenhouse Warming: Abatement and Adaptation*, Resources for the Future, Washington, DC.

Shugart, H.H., Smith, T.M. and Post, W.M. (1992) 'The potential for application of individual-based simulation models for assessing the effects of global change', *Annu. Rev. of Ecol. and Syst.*, 23: 15–38.

Slayter, R.O. and Noble, I.R. (1992) 'Dynamics of montane treelines', in A.J. Hansen and F. di Castri (eds), *Landscape Boundaries*, Springer Verlag, New York.

Smith, J. and Tirpak, D. (1989) *The Potential Effects of Climate Change on the United States*, Environmental Protection Agency Report to Congress, 2 vols, Washington, DC.

Smith, T.M., Shugart, H.H. and Halpin, P.N. (1990) 'Global forests, in progress reports on international studies of climate change impacts', USEPA, Washington, DC.

Smith, T.M., Shugart, H.H., Bonan, G.B. and Smith, J.B. (1992) 'Modeling the potential response of vegetation to global climate change', *Advances in Ecological Research*, 22: 93–113.

Solomon, A.M. (1986) 'Transient responses of forests to CO_2-induced climate change: simulation modeling experiments in eastern North America', *Oecologia*, 68: 567–9.

Stephenson, N.L. (1990) 'Climatic control of vegetation distribution: the role of the water balance', *The American Naturalist*, 135 (5): 649–70.

Stevens, G.C. (1992) 'The elevational gradient in altitudinal range: an extension of Rapaport's altitudinal rule', *The American Naturalist*, 140 (6): 893–911.

Stevens, G.C. and Fox, J.F. (1991) 'The causes of treeline', *Annu. Rev. of Ecol. and Syst.*, 22: 177–91.

Tosi, J.A., Jr (1969) *Mapa Ecologico Republica de Costa Rica*, Tropical Science Center, San José, Costa Rica.

Tosi, J.A., Watson, V. and Echeverria, J. (1992) 'Potential impacts of climate change on the productive capacity of Costa Rican forests: a case study', Tropical Science Center, San José, Costa Rica.

Tranquillini, W. (1979) 'Physiological ecology of the alpine timberline. Tree existence at high altitudes with special reference to the European Alps', *Ecological Studies*, 31, Springer, Berlin, Heidelberg.

Vuilleumier, F. and Monasterio, M. (1986) *High Altitude Tropical Biogeography*, Oxford University Press, New York.

Wardle, P. (1974) 'Alpine timberlines', in J.D. Ives and R.G. Barry (eds), *Arctic and Alpine Environments*, Harper & Row, New York.

Whittaker, R.H. (1975) *Communities and Ecosystems*, 2nd edn, Macmillan, New York.

Woodward, F.I. (1987) *Climate and Plant Distribution*, Cambridge University Press, Cambridge.

12

COMPARING THE BEHAVIOUR OF MOUNTAINOUS FOREST SUCCESSION MODELS IN A CHANGING CLIMATE

H. Bugmann and A. Fischlin

INTRODUCTION

In mountainous regions forests fulfil a multitude of functions. They protect settlements from avalanches or landslides; they regulate runoff, thereby helping to prevent erosion; forests and meadows make a varied mountain landscape and provide the environment necessary for various recreational activities; they hold a large fraction of the world's terrestrial carbon, and are also important carbon sequestering systems; finally, and not least, forests are exploited for fuel, pulpwood, and timber. Climatic changes may impact on all these functions (e.g., Bolin *et al.*, 1986; Davis, 1990). However, the complex topography in mountains leads to a large spatial variability of climate, soil, and other site factors, which makes it difficult to assess their influence on forest dynamics. Moreover, via the tree species composition, edaphic factors such as soil organic matter and nutrient availability may have large effects on above functions (Pastor and Post, 1985; Shugart *et al.*, 1986; Davis, 1990; Shugart, 1990). Therefore, in mountains it is more important to study the processes involved explicitly and in more details than in flat terrain.

Models of forest succession which are based on the gap dynamics hypothesis (Botkin *et al.*, 1972a, b; Shugart, 1984) operate on similar temporal and spatial scales. These gap models simulate the establishment, growth, and death of individual trees partly as a deterministic, partly as a stochastic process confined within small, often $\frac{1}{12}$ ha, plots. The actual forest succession on the ecosystem level is then averaged from the successional patterns simulated for many plots. These models offer the following advantages for studying the impact of climatic change. First, they are based on a well documented ecological theory of tree growth and plant competition (Watt, 1947; Bray, 1956; Curtis, 1959; Forman and Godron, 1981). Second, it has been shown that these models incorporate many essential mechanisms and exhibit realistic features of species succession in natural and semi-natural forest ecosystems, both in terms of their transient behaviour and of the equilibrium species composition reached after several centuries (Botkin *et al.*, 1972a, b; Shugart, 1984). Moreover, since forest gap models operate on small spatial scales from 10 m to 1 km, it appears particularly feasible to apply such models to mountainous forests in a complex topography.

Many authors have constructed forest gap models for a wide range of test sites, but all have assumed a constant climate (Botkin *et al.*, 1972a, b; Shugart and West, 1977; Doyle, 1981; Pastor and Post, 1985; Kienast, 1987; Leemans and Prentice, 1989; Bonan, 1992). From the realistic behaviour of forest gap models under current climates some authors have inferred that the models can be used to simulate the impacts of future climatic changes on species composition (e.g., Pastor and Post, 1988; Kienast, 1991). Other authors have tried to enhance the trustworthiness of these models by applying them for past constant climates (e. g., Solomon *et al.*, 1981; Lotter and Kienast, 1992) or for scenarios of past climatic change (e.g., Solomon *et al.*, 1980, 1981; Solomon and Tharp, 1985). However, there still remains considerable uncertainty concerning the appropriateness of the 'facts and concepts' incorporated in gap models, a view shared by some authors such as Solomon (1986: 568): 'the errors become amplified ..., generating flaws that are large enough to preclude direct application of the model ...'. Moreover, applying these models to a changing climate we found evidence pointing to a considerable input and structural sensitivity of forest gap models in terms of their temporal behaviour (Fischlin *et al.*, 1994). Yet we are not aware of any previous study that explicitly explores the applicability of these models for assessing the impact of climatic change.

Comparing the consistency and robustness of the results produced by several models applied to the same climate might be a means to explore the strengths and limits of forest gap models. This approach requires us to compare closely related members of the same family of models; from a viewpoint of systems theory, each of these models has the same base model and the same experimental frame (Zeigler, 1976; Fischlin, 1991). Such a study is especially interesting if we do not have unanimously accepted reasons for favouring *a priori* one of the models over the others. These prerequisites are met exactly by the family of models used in this study. The three forest gap models FORECE (Kienast, 1987), ForClim 1.1, and ForClim 1.3 (Fischlin *et al.*, 1994) can all simulate forest succession of unmanaged European forests. The models are built for the same experimental frame, that is, the same degree of resolution and the same temporal and spatial scales, and differ from each other only in the formulation of climatic influences and with respect to the modelling of some ecological processes.

Models of this type usually require climatic input parameters, such as monthly temperature plus precipitation means and variances. For exploring the applicability of such a model to climatic change, the models should be scrutinized under current and future climates. For the current climate within the Swiss Alps it is easy to derive these parameters from long-term measurements of weather. However, future climates must be estimated by additional means, for instance by the new methods of downscaling (Gyalistras *et al.*, 1994). They allow us to scale down global climates as projected by General Circulation Models (GCMs) to a particular weather station. Not only does this downscaling allow for 'best estimates' of a future changed climate, for example, based on the scenarios for greenhouse gas emissions (Houghton *et al.*, 1990), but also to quantify its variability.

Such comparisons have to concentrate on the most interesting impacts of climatic change, that is, those on which the listed functions of mountainous forests obviously depend the most. Included among those is the temporal behaviour of the species-specific biomasses, which must not differ beyond certain ranges from model to model if the models are to be considered reliable and applicable to climate change. In case the projected forests should differ substantially from model to model, it may at least be necessary to understand the reasons. Are the differences due to the location, the number of factors incorporated in the model, or the climate parameterization?

In this paper we compare and evaluate the behaviour of the chosen family of models with respect to the following questions: how similar – or how different – are the species compositions of the potential natural vegetation simulated by the three models (1) under present climates, (2) under future 'best estimate' climates downscaled from GCM results, and (3) under the variability of the downscaled best estimates? Since the downscaling yields site-specific data, we chose several representative test sites along an altitudinal gradient within the European Alps. We found that some models yielded similar and consistent results, in particular for current climates, but that they can disagree considerably in other situations.

MATERIAL AND METHODS

The following three forest gap models were used:

The first model, FORECE (Kienast, 1987), is a conventional gap model derived from LINKAGES (Pastor and Post, 1985) to accommodate European conditions and species.

The second model, ForClim 1.1, is a simplified descendant of FORECE and comprises fewer, that is, only the most fundamental, ecological processes (Bugmann, 1991; Bugmann and Fischlin, 1992). Based on a structural sensitivity analysis under the current climate the following processes were dropped: the modification of the rates of sapling establishment by (1) the annual mean and annual amplitude of monthly temperatures (temperature indicators after Ellenberg, 1986); (2) degree-days; (3) the influence of frost; (4) sprouting from tree stumps, a factor often of

Table 12.1 Characteristics and major current climate parameters of the test sites used to simulate the three forest succession models

Site	Location	Elevation (m above sea level)	Annual mean temperature (°C)	Annual precipitation sum (cm)	Potential natural vegetation (Ellenberg and Klötzli, 1972)
Berne	Swiss Plateau	540	8.4	100.1	Mixed deciduous forests dominated by beech (*Fagus silvatica* L.) and silver fir (*Abies alba* Miller)
Davos	Northern Alps	1560	3.0	101.1	Coniferous forests dominated by larch (*Larix decidua* Miller) and spruce (*Picea abies* L.)
Bever	Central Alps	1708	1.5	83.8	Coniferous forests dominated by larch (*Larix decidua* Miller) and Swiss Stone pine (*Pinus cembra* L.)

little importance in unmanaged forests; (5) the positive feedback of the presence of adult trees on seed availability ('scoring system' after Kienast, 1987). Moreover, instead of tracking individual trees ForClim 1.1 simulates only size cohorts. The parameterization of the climate is done in the same way as in FORECE.

The third model, ForClim 1.3, was developed from ForClim 1.1 by altering the mathematical formulations of the climatic factors (Fischlin *et al.*, 1994). The only difference from ForClim 1.1 is that it adopts a more reliable parameterization of climate by avoiding any implicit temperature and precipitation dependencies: (1) the calculation of the annual sum of degree-days is corrected for site-specific bias by linear regression; (2) sapling establishment is limited by the minimum of the actual mean temperatures of December, January, and February instead of the long-term mean January temperature, which avoids unrealistic threshold effects when climate changes; (3) the carrying capacity for above-ground biomass (parameter SOILQ in conventional gap models) is not assumed to be constant but is calculated based on long-term temperature and precipitation data to allow for simulations of climatic change (O'Neill and DeAngelis, 1981); (4) drought stress is calculated according to the outlines by Prentice and Helmisaari (1991) instead of using the 'dry days' approach (Pastor and Post, 1985), again avoiding threshold effects. Fischlin *et al.*, (1994) have described the exact mathematical formulations fully.

Table 12.2 Site-specific winter and summer temperature (T) and precipitation (P) changes projected for the year 2100 relative to current climatic conditions (1901–90 for Berne and Davos, 1901–80 for Bever)

Site		Winter (Dec.–Feb.)		Summer (Jun.–Aug.)	
		T (°C)	P (cm/month)	T (°C)	P (cm/month)
Berne	Trend(μ)	3.27	2.73	2.30	3.46
	Mean(s)	0.37	0.57	0.30	1.06
	Trend(s)	0.46	0.44	0.48	1.42
	T_oP_o	3.76	3.13	2.64	3.98
	T+P+	5.09	4.85	3.85	7.94
	T+P–	5.09	1.42	3.85	0.02
	T–P+	2.43	4.85	1.43	7.94
	T–P–	2.43	1.42	1.43	0.02
Davos	Trend(μ)	2.61	1.86	2.85	0.79
	Mean(s)	0.35	0.96	0.20	0.58
	Trend(s)	0.43	0.96	0.27	0.60
	T_oP_o	3.00	2.14	3.28	0.91
	T+P+	4.26	5.32	4.02	2.86
	T+P–	4.26	–1.03	4.02	–1.05
	T–P+	1.74	5.32	2.53	2.86
	T–P–	1.74	–1.03	2.53	–1.05
Bever	Trend(μ)	1.28	2.21	3.62	3.32
	Mean(s)	0.27	0.35	0.36	0.64
	Trend(s)	0.32	0.38	0.67	0.86
	T_oP_o	1.48	2.54	4.16	3.82
	T+P+	2.42	3.74	5.75	6.21
	T+P–	2.42	1.35	5.75	1.43
	T–P+	0.53	3.74	2.57	6.21
	T–P–	0.53	1.35	2.57	1.43

Source: These climatic scenarios are based on downscaled trends from a 'Business as Usual' transient ECHAM GCM run (Cubasch *et al.*, 1992; Gyalistras *et al.*, 1994).
Note: Trend(μ): linear trend of the mean (1986–2085); Mean(s): standard deviation in 2036 (= average of period 1986–2085); Trend(s): linear trend of the standard deviation (1986–2085). T_oP_o: 'best estimate' of changes for year 2100 extrapolated from Trend(μ). X±: lower/upper end of confidence interval for variable X, X± = X_o±2·s_{2100}, where s_{2100} is the standard deviation for 2100 extrapolated from Trend(s) and Mean(s).

The following reasons lead to the selection of the test sites Berne, Davos, and Bever (Table 12.1). These three sites represent three dominant belts of vegetation determined by altitude (Table 12.1; plant nomenclature according to Hess *et al.*, 1980), and long-term climate records have been compiled by the Swiss Meteorological Agency (Bantle, 1989; SMA, 1901–90), which allow us to calculate reliable long-term means and standard deviations of monthly temperatures and precipitation sums (Fischlin *et al.*, 1994).

Scenarios for future climates at the test sites (Table 12.2) were obtained by statistical downscaling which relates large-scale temperature and pressure anomalies (North Atlantic, Europe) to local weather anomalies by means of principal component analysis and canonical correlation

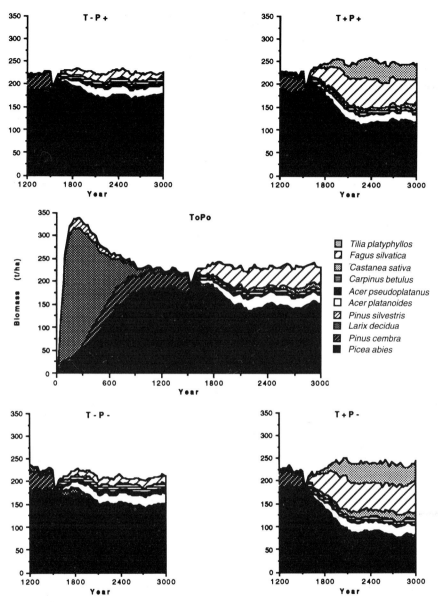

Figure 12.1 Species compositions simulated by the ForClim model version 1.3 at Davos (Table 12.1): first 1,500 years of primary forest succession in the current climate, second 1,500 years of secondary forest succession in response to a downscaled best estimate ($T_o P_o$) step change in the global climate (centre). The panels in the corners show the secondary succession or the step responses in function of the uncertainties inherent in the climatic change scenarios (Table 12.2). Species biomasses are shown from top to bottom in same sequence as listed in legend.

analysis (Gyalistras *et al.*, 1994). The data for this downscaling were provided by a 100-year (1986–2085) uncorrected transient run of the ECHAM General Circulation Model (Cubasch *et al.*, 1992) for the IPCC 'Business as Usual' Scenario A (Houghton *et al.*, 1990). Using the downscaled trends (Table 12.2) we computed the anomalies of the mean winter (Dec.–Feb.) and summer (Jun.–Aug.) temperature (T) and precipitation (P). The values obtained were added to the site-specific current monthly means and applied during six months each (summer: Apr.–Sep.; winter: Oct.–Mar.). Since any scenario of climatic change is based on essentially unknown assumptions about the future (Houghton *et al.*, 1990), we compared the behaviour of each model within the range of c. 95 per cent (±2σ) of those 84 downscaling models which performed best in the validation period (Gyalistras *et al.*, 1994). Figure 12.1 (centre) gives an example of typical simulation results as obtained with the three models for the best estimate of climatic change, whereas the effect of varying this estimate by ±2σ is shown in the corners.

At each site, the equilibrium states of the gap models were calculated for the present climate during the first 1,500 simulation years. Second, starting from the equilibrium states, the step response of the models was explored by imposing an instant climatic change based on the downscaled projections described above, and the simulations ended after 3,000 years (Figure 12.1). Average species biomasses were calculated from 200 stochastic runs (Bugmann and Fischlin, 1992). The equilibrium states were estimated by averaging the results over the periods 1,000–1,500 and 2,500–3,000 simulation years for current and future climates, respectively. The ForClim models were simulated on an Apple Macintosh IIfx computer using the simulation software ModelWorks and RAMSES (Fischlin *et al.*, 1990; Fischlin, 1991). Simulations with the model FORECE are less efficient and were thus executed on a SUN SS630 workstation.

The differences between the equilibrium states of species biomasses produced by the various models and climate scenarios were quantified using a percentage similarity coefficient (PS) (e.g., Prentice and Helmisaari, 1991), which relates any two sets of data $X = \{x_1, x_2, \ldots, x_n\}$ and $Y = \{y_1, y_2, \ldots, y_n\}$ as follows:

$$PS = 1 - \frac{\sum\limits_{i=1}^{n} |x_i - y_i|}{\sum\limits_{i=1}^{n} (x_i + y_i)} \tag{1}$$

where $0 < PS < 1$. This coefficient can be interpreted as the fraction of values common to both sets of data. It offers the following advantage: not only does it track differences in the relative distributions of the x_i and y_i values (e.g., species-specific biomasses), but it also declines the larger the difference between the sums Σx_i and Σy_i (e.g., total biomass) becomes.

RESULTS

In this study we focused on the comparison among steady states of biomasses computed as t/ha per tree species. All three models can reach a singular steady state. Simulation experiments showed that the equilibrium biomasses calculated from 200 simulation runs have a standard error smaller than 10 per cent of their mean. These steady states are reached after a rather long transient behaviour lasting between 400 (Figure 12.2 left) and more than 700 years (Figure 12.1, Figure 12.2 right).

The estimated steady states appear not to depend on the initial states. This allows us to compare the results obtained with species compositions observed in real forests, although there exist no precise field data on the true initial states at the three test sites.

For the current climate the steady states simulated by the three models show realistic species compositions at all three sites. The mixed deciduous forest at Berne is dominated by common beech (*Fagus silvatica* L.) and silver fir (*Abies alba* Miller); subalpine coniferous forests at Davos are composed mainly of Norway spruce (*Picea abies* L.) and European larch (*Larix decidua* Miller) (Figure 12.1 centre); simulated forests at Bever are dominated by European larch and Swiss stone pine (*Pinus cembra* L.) (Figure 12.2).

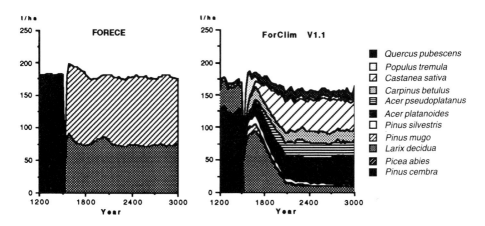

Figure 12.2 Species compositions simulated by the forest succession model FORECE (left) and ForClim model version 1.1 (right) at Bever (Table 12.1). Both simulations represent step responses to the downscaled best estimate ($T_o P_o$) of the changed climate as projected by the ECHAM GCM for the end of the next century (Table 12.2). Species biomasses are shown from top to bottom in same sequence as listed in legend.

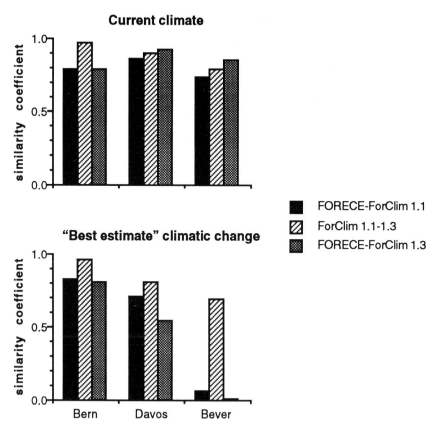

Figure 12.3 Similarity coefficients computed for all possible pairs of the three model versions at the three test sites: (top) under current climate, (bottom) under the downscaled best estimate (T_oP_o) of the changed climate projected by the ECHAM GCM for the end of the next century (Table 12.2).

Moreover, the simulated species compositions compare well with phytosociological descriptions of natural and semi-natural communities by Ellenberg and Klötzli (1972) and Ellenberg (1986). For the FORECE model, this has been discussed in detail by Kienast and Kuhn (1989). The large percentage similarity coefficients (Figure 12.3 top) show the good agreement among all three models for the current climate.

Depending on the test site, we observed diverging similarity coefficients among projected steady states under the best estimate scenario for the climate at the end of the next century: The smallest differences between the models were found at Berne (540 m a.s.l.), medium ones at Davos (1,560 m a.s.l.), and marked differences were found at Bever (1,708 m a.s.l.) (Figure 12.3 bottom). Thus, the degree of divergence among the three models correlates positively

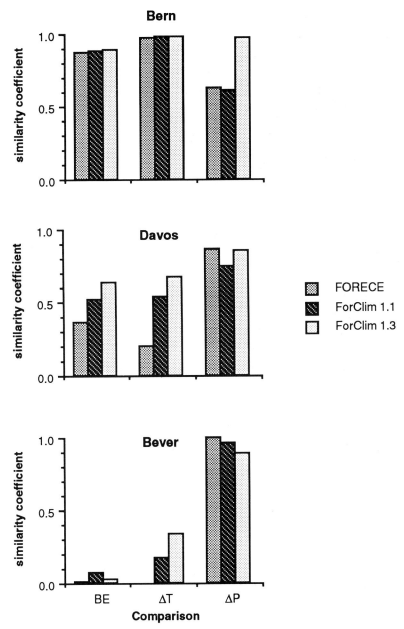

Figure 12.4 Similarity coefficients computed between pairs of species steady states simulated by the same model while modifying climatic conditions: BE – similarity between current climate and downscaled best estimate (T_oP_o); ΔT – similarity between T–P+ and T+P+ scenarios (temperature gradient); ΔP – similarity between T+P+ and T+P– scenarios (precipitation gradient) (Table 12.2).

with elevation. Moreover, the similarity coefficients between the species compositions in the current climate and those in the future best estimate climate also decrease with increasing altitudes (Figure 12.4, column BE), that is, the higher the site the more different the communities might become due to the climatic change.

Not only is the downscaled variability in the input data of the same order of magnitude as the estimates of the uncertainty inherent in GCM simulations, but the response of the forest models remains within a similar range (Fischlin *et al.*, 1994). The simulations with each model performed at each test site along the borders of the scenario ranges showed the following: the higher the site, the less similar are a model's steady states, which were generated while the climatic scenario was modified (Figure 12.1, Figure 12.4).

Precipitation changes appear to influence species composition mainly at lower elevations (Figure 12.4, column ΔP). Only at Berne do the similarity coefficients diverge. This behaviour can be explained by a threshold effect associated with the way drought effects are modelled (Fischlin *et al.*, 1994). Moreover, since we found in other studies that low dry sites are sensitive to changes of precipitation, we expect that the response of the models at Basel or Sion to changing precipitation sums could be even stronger than that found at Berne (Figure 12.4, ΔP top).

Temperature changes have strong influences on species composition at higher elevations (Figure 12.4, column ΔT). The similarity coefficients are large and differ little at low altitudes (Figure 12.4, top). They become smaller and diverge more with increasing elevation of the test sites. This augmented temperature sensitivity was found in all models, and it corroborates the expected temperature dependency of the alpine treeline. At these elevations the different formulations of degree-days and winter temperature are responsible for model divergence, whereas they are of little significance at lower elevations.

DISCUSSION

Not only do the three models produce in eight out of nine comparisons for current climate (89 per cent) consistent results with large inter-model similarity coefficients (PS > 0.75; Figure 12.3 top), but their behaviour appears also to be in good accordance with field data wherever they are available and have been produced by model equations which conform with the current ecological theory of the processes governing the species composition of a forest stand. These findings corroborate the results and expectations of many other authors (Botkin *et al.*, 1972a, b; Shugart and West, 1977; Doyle, 1981; Shugart, 1984; Pastor and Post, 1985; Kienast, 1987; Leemans and Prentice, 1989; Bonan, 1992). On this basis alone it would not be possible to favour one model over another, especially since they all have been built for similar purposes and are all applicable to central European forests.

On the other hand, the comparisons among the three models did reveal that their response to climatic change differed markedly in twenty-three out of forty-five cases (51 per cent). Thus, their application to assessing the impact of climatic change might not be as easy as some authors have thought.

What were the causes for the different behaviours? Under a changed climate the models respond more strongly to the complete elimination of factors that depend on the ecology or the climate (ForClim 1.1) than to slight modifications of model equations (ForClim 1.3). In particular, the elimination of the temperature indicators after Ellenberg (1986) in the model ForClim 1.1 has strong effects at the sites Davos and Bever. The use of these indicators in FORECE may be questioned. First, many species are excluded arbitrarily because of the indicators' discrete nature. Second, it is the current weather which really influences the establishment of saplings, not the long-term mean difference of the temperature between the warmest and the coldest month. ForClim 1.1 does not require us to assume such statistical relations, and is based to a larger extent on plausible, causal mechanisms.

Moreover, the steady state species compositions for current climate as simulated by the ForClim models are often more plausible (Ellenberg and Klötzli, 1972; Ellenberg, 1986), e.g., for *Acer platanoides* L. at Berne and *Larix decidua* L. at Bever. At sites closer to the precipitation limited treeline ForClim 1.3 simulates the climatic influences more plausibly. It avoids threshold effects due to discrete functions (Fischlin *et al.*, 1994), which are responsible for some of the observed strong dissimilarities (Figure 12.4) and appear rather to be artefacts. Based on these findings we favour ForClim 1.3 over the other two models.

Although some results are contradictory and the exact species compositions might not always be predictable, it is possible to draw several conclusions from the results. As a consequence of assessed climatic changes strong responses in species composition cannot be ruled out, and in certain environmental conditions they are even likely. This conforms again with the findings from earlier studies, although they have not been able to use GCM downscaled climate scenarios (Solomon, 1986; Pastor and Post, 1988; Overpeck *et al.*, 1990; Kienast, 1991; Fischlin *et al.*, 1994).

In particular, as expressed by the sequence of the Bever, Davos, and Berne sites (Figure 12.4 left), the high forests appear to be more susceptible than the lower ones. This pattern supports findings by IPCC that subalpine forests might be especially susceptible (Izrael *et al.*, 1990) and also corroborates the climatological interpretation of tree rings from subalpine zones (e.g., Kienast and Schweingruber, 1986).

CONCLUSIONS

Within a large range of altitudes the models of forest succession studied respond to a climatic change by adapting their species compositions. The response of the models to climates downscaled from transient GCM simulations (Cubasch *et al.*, 1992; Gyalistras *et al.*, 1994) for the end of the next century based on the IPCC 'Business as Usual' scenario contrasts in some cases sharply with the steady states for current climatic conditions. Since the steady states of all the models appear to be globally stable, this statement can be made independently of the exact course of the primary succession in a constant climate (Figure 12.1) as well as of the secondary succession following climatic changes. The successional transient response to a step in the climate abates only after about 400 up to a maximum of 700 years (Figure 12.1, Figure 12.2 right). This is of the same order of magnitude as that at which abyssal oceans respond (Flohn and Fantechi, 1984; Cubasch *et al.*, 1992).

For forests within the European Alps the results suggest that those at lower altitudes might be most susceptible to a drier climate, whereas at higher altitudes, especially in the subalpine zone, species compositions might be more susceptible to a warmer climate (Figure 12.4). In general we tentatively conclude that the forests least susceptible to climatic change are at mid-altitudes, that is, in the montane zone. However, these findings may not generally hold in other climates, since the three test sites represent only a small fraction of possible climates in the parameter space of temperature and precipitation. More thorough and systematic sensitivity analysis is necessary before general conclusions can be drawn (Fischlin *et al.*, 1994), in particular, we have to compare not only the steady states, but also transient behaviour.

The members of our family of forest gap models show similar results under present climate (Figure 12.2), but may produce markedly diverging steady state species compositions under a changed future climate. The bigger the divergences, the more crucial becomes the question of whether one model can be favoured over the others and for which reasons. The fact that a gap model performs well in the present climate is only a necessary precondition, but not a sufficient one to furnish a model applicable for a detailed study of the impact of climatic change on forests; additional criteria are needed to pick a model from a family. Among all studied models we favour ForClim 1.3 for the following reasons: first it exhibits realistic behaviour at more sites, that is, it satisfies the necessary precondition better; second, its mathematical formulation is more rigorous and contains fewer parameterizations, and third, its equations are based on causal relationships to a larger extent (Fischlin *et al.*, 1994).

From the observed discrepancies among the models and from their explanations, we surmise that our findings are likely to be generally valid: Many conventional gap models are similar to the models we analysed, hence, without major modifications many of them are likely not to be robust and versatile enough to be used for detailed assessments of the impact of climatic change. Yet, it appears promising to improve and revise them, first, by scrutinizing their relationship between climate and ecoprocesses, second, by searching for potential inconsistencies among alternative formulations, third, by reformulating their equations to enhance their robustness under varying climatic regimes, and fourth, by validating them against various, for example, past, changing climates. The resulting forest models are likely to yield more reliable projections in studies of future impacts of climatic change on our forests.

REFERENCES

Bantle, H. (1989) *Programmdokumentation Klima-Datenbank am RZ-ETH Zürich*, Swiss Meteorological Agency, Zürich.

Bolin, B., Döös, B.R., Jaeger, J. and Warrick, R.A. (1986) *The Greenhouse Effect, Climatic Change and Ecosystems*, John Wiley & Sons, Chichester.

Bonan, G.B. (1992) 'A simulation analysis of environmental factors and ecological processes in North American boreal forests', in H.H. Shugart, R. Leemans and G.B. Bonan (eds), *A Systems Analysis of the Global Boreal Forest*, Cambridge University Press, Cambridge, pp. 404–27.

Botkin, D.B., Janak, J.F. and Wallis, J.R. (1972a) 'Some ecological consequences of a computer model of forest growth', *J. Ecol.*, 60: 849–72.

——, —— and —— (1972b) 'Rationale, limitations and assumptions of a north-eastern forest growth simulator', *IBM J. Res. Develop.*, 16: 101–16.

Bray, J.R. (1956) 'Gap-phase replacement in a maple-basswood forest', *Ecology*, 37: 598–600.

Bugmann, H. (1991) 'Development of ForClim-P, a simplified forest gap model for workstations and personal computers', Zürich, internal paper, Systems Ecology Group, ETHZ.

Bugmann, H. and Fischlin, A. (1992) 'Ecological processes in forest gap models – analysis and improvement', in A. Teller, P. Mathy and J.N.R. Jeffers (eds), *Responses of Forest Ecosystems to Environmental Changes*, Elsevier Applied Science, London and New York, pp. 953–4.

Cubasch, U., Hasselmann, K., Höck, H., Maier-Reimer, E., Mikolajewicz, U., Santer, B.D. and Sausen, R. (1992) 'Time-dependent greenhouse warming computations with a coupled ocean-atmosphere model', *Climate Dynamics*, 8: 55–69.

Curtis, J.T. (1959) *The Vegetation of Wisconsin*, University of Wisconsin Press, Madison.

Davis, M.B. (1990) 'Biology and palaeobiology of global climate change: Introduction', *Trends in Ecol. and Evol.*, 5: 269–70.

Doyle, T.W. (1981) 'The role of disturbance in the gap dynamics of a montane rain forest: an application of a tropical forest succession model', in D.C. West, H.H. Shugart and D.B. Botkin (eds), *Forest Succession: Concepts and Application*, Springer, New York, pp. 56–73.

Ellenberg, H. (1986) *Vegetation Mitteleuropas mit den Alpen in ökologischer Sicht*, 4th edn, Verlag Eugen Ulmer, Stuttgart.

Ellenberg, H. and Klötzli, F. (1972) 'Waldgesellschaften und Waldstandorte der Schweiz', *Eidg. Anst. Forstl. Versuchswes., Mitt.*, 48: 587–930.

Fischlin, A. (1991) 'Interactive modeling and simulation of environmental systems on workstations', in D.P.F. Möller (ed.), *Proc. of the 4th Ebernburger Working Conference on the Analysis of Dynamic Systems in Medicine, Biology, and Ecology, 5–7 April 1990*, Ebernburg, Bad Münster, BRD, Informatik-Fachberichte 275, Springer, Berlin, pp. 131–45.

Fischlin, A., Bugmann, H. and Gyalistras, D. (1994) 'Sensitivity of a forest ecosystem model to climate parameterization schemes', *Env. Pollution*.

Fischlin, A., Roth, O., Gyalistras, D., Ulrich, M. and Nemecek, T. (1990) 'Model-Works: an interactive simulation environment for work stations and personal computers', Internal Report no. 8, Systems Ecology Group, ETH Zürich.

Flohn, H. and Fantechi, R. (eds) (1984) *The Climate of Europe: Past, Present and Future*, Reidel, Dordrecht.

Forman, R.T.T. and Godron, M. (1981) 'Patches and structural components for a landscape ecology', *Bioscience*, 31: 733–40.

Gyalistras, D., Storch, H. von, Fischlin, A. and Beniston, M. (1994) 'Linking GCM generated climate scenarios to ecosystems: case studies of statistical downscaling in the Alps', *Clim. Res.*

Hess, H.E., Landolt, E. and Hirzel, R. (1980) *Flora der Schweiz*, 4 Bde., 2. ed., Birkhäuser, Basel and Stuttgart.

Houghton, J.T., Jenkins, G.J. and Ephraums, J.J. (eds) (1990) *Climate Change – the IPCC Scientific Assessment*, Report prepared for IPCC by Working Group 1, Cambridge University Press, Cambridge.

Izrael, Y.A., Hashimoto, M. and Tegart, W.J.M. (eds) (1990) *Climate Change – the IPCC Impact Assessment*, Report prepared for IPCC by Working Group 2, Cambridge University Press, Cambridge.

Kienast, F. (1987) *FORECE – A Forest Succession Model for Southern Central Europe*, Oak Ridge National Laboratory, Oak Ridge, Tennessee, ORNL/TM–10575.

—— (1991) 'Simulated effects of increasing atmospheric CO_2 and changing climate on the successional characteristics of Alpine forest ecosystems', *Landscape Ecology*, 5: 225–38.

Kienast, F. and Kuhn, N. (1989) 'Simulating forest succession along ecological gradients in southern central Europe', *Vegetatio*, 79: 7–20.

Kienast, F. and Schweingruber, F.H. (1986) 'Dendroecological studies in the Front Range, Colorado, U.S.A.', *Arctic and Alpine Research*, 18: 277–88.

Leemans, R. and Prentice, I.C. (1989) *FORSKA, A General Forest Succession Model*, Techn. Rep. of the Institute of Ecological Botany, University of Uppsala, Sweden.

Lotter, A. and Kienast, F. (1992) 'Validation of a forest succession model by means of annually laminated sediments', in M. Saarnisto and A. Kahra (eds), *Proceedings of the INQUA Workshop on Laminated Sediments*, 4–6 June 1990, Lammi, Finland. Geological Survey of Finland, Special paper series 14: 25–31.

O'Neill, R.V. and DeAngelis, D.L. (1981) 'Comparative productivity and biomass relations of forest ecosystems', in D.E. Reichle (ed.), *Dynamic Properties of Forest Ecosystems*, IBP publication no. 23, Cambridge University Press, Cambridge, pp. 411–49.

Overpeck, J.T., Rind, D. and Goldberg, R. (1990) 'Climate-induced changes in forest disturbance and vegetation', *Nature*, 343: 51–3.

Pastor, J. and Post, W.M. (1985) *Development of a Linked Forest Productivity-soil Process Model*, US Dept of energy, ORNL/TM–9519.

—— and —— (1988) 'Response of northern forests to CO_2-induced climate change', *Nature*, 334: 55–8.

Prentice, I.C. and Helmisaari, H. (1991) 'Silvics of north European trees: compilation, comparisons and implications for forest succession modelling', *For. Ecol. Manage.*, 42: 79–93.

Shugart, H.H. (1984) *A Theory of Forest Dynamics. The Ecological Implications of Forest Succession Models*, Springer, New York.

—— (1990) 'Using ecosystem models to assess potential consequences of global climatic change', *Trends in Ecol. and Evol.*, 5: 303–7.

Shugart, H.H. and West, D.C. (1977) 'Development of an Appalachian deciduous forest succession model and its application to assessment of the impact of the chestnut blight', *J. Env. Mgmt.*, 5: 161–79.

Shugart, H.H., Antonovsky, M. Ya., Jarvis, P.G. and Sandford, A.P. (1986) 'CO$_2$, climatic change, and forest ecosystems', in B. Bolin, B.R. Döös, J. Jaeger and R.A. Warrick (eds), *The Greenhouse Effect, Climatic Change and Ecosystems*, John Wiley & Sons, Chichester.

SMA (Swiss Meteorological Agency) (1901–90) *Annalen der Schweizerischen Meteorologischen Anstalt*, Swiss Meteorological Agency, Zürich.

Solomon, A.M. (1986) 'Transient response of forests to CO$_2$-induced climate change: simulation modeling experiments in eastern North America', *Oecologia*, 68: 567–79.

Solomon, A.M. and Tharp, M.L. (1985) 'Simulation experiments with late quaternary carbon storage in mid-latitude forest communities', in E.T. Sundquist and W.S. Broecker (eds), *The Carbon Cycle and Atmospheric CO$_2$: Natural Variations Archean to Present*, American Geophysical Union, Washington DC. Geophysical Monograph, 32: 235–50.

Solomon, A.M., Delcourt, H.R., West, D.C. and Blasing, T.J. (1980) 'Testing a simulation model for reconstruction of prehistoric forest-stand dynamics', *Quat. Res.*, 14: 275–93.

Solomon, A.M., West, D.C. and Solomon, J.A. (1981) 'Simulating the role of climate change and species immigration in forest succession', in D.C. West, H.H. Shugart, and D.B. Botkin (eds), *Forest Succession: Concepts and Application*, Springer, New York, pp. 154–77.

Watt, A.S. (1947) 'Pattern and process in the plant community', *J. Ecol.*, 35: 1–22.

Zeigler, B.P. (1976) *Theory of Modelling and Simulation*, John Wiley, New York.

13

ECOLOGICAL ASPECTS OF CLIMATICALLY-CAUSED TIMBERLINE FLUCTUATIONS

Review and outlook

F.-K. Holtmeier

INTRODUCTION

Expected global changes of the natural environment are often attributed to man-caused increases of CO_2 and other trace gases which might enhance the greenhouse effect in the atmosphere. Considering the fact that all living beings on earth ultimately depend on warmth, not much imagination is needed to suppose that vegetation has to respond to the warming trend in one way or the other. Change of species composition and horizontal and altitudinal shift of vegetation zones and belts might be expected. There are already impressive computed maps and other graphs showing the future position of vegetation zones and altitudinal belts, which indicate the space of potential change of ecological conditions (Kauppi and Posch, 1988; Environment Canada, 1989; Ozenda and Borel, 1991).

However, these general predictions are necessarily based on very simple assumptions. As to polar timberline, for example, the extent of expected northward shift is estimated by extrapolating the rough coincidence of the present timberline and the position of the 10°C-isotherm (July), or by summer heat accumulation (growing-degree days; Kauppi and Posch, 1988; Solomon, 1989). The predictions concerning the altitudinal belts and the upper timberline follow the same pattern (cf. Ozenda and Borel, 1991).

From the ecological point of view these schemata may disguise rather than clarify very complex phenomena, because its great local and regional variety is not sufficiently considered. Tree growth at timberline is never controlled by mean temperatures, which do not really exist, but rather by the climate character, that is, annual temperature regime and extreme events, such as late or early frosts, long-lasting or missing snow cover, drought, and other factors. Moreover, tree growth and position of timberline do not correspond

220

linearly to changes of thermal conditions (e.g. Fritts, 1969; Graumlich and Brubaker, 1986). How rapidly timberlines advance does not depend primarily on growth conditions but rather on reproduction. Reproduction success usually varies spatially and temporally, as can be seen from many studies at the present timberline. Thus, continued warming does not necessarily mean consistently good reproduction.

Whenever timberline dynamics are considered, timescale is also of great importance. In the long term (centuries to millenia), species adaptation to changing environment or replacement of non-adapted species must also be taken into consideration. As to frost hardiness, the most hardy ecotypes of mountain birch (*Betula pubescens ssp. tortuosa*) in Lapland, for example, developed near the upper timberline (Kallio *et al.*, 1983). To be able to predict future development, we have to understand the present ecological situation and the postglacial history of climate and timberline.

The following paragraphs illustrate some aspects of timberline dynamics that have to be considered when predicting timberline response to climatic change.

NATURAL TIMBERLINE AND CLIMATIC CHANGE

Climatically caused timberlines that have not been altered by man may appear as lines in many places. In most cases, however, timberline is a more or less wide ecotone including the transitional zone between closed forest and treeline which is usually formed by deformed trees. This ecotone is characterized by ecological conditions different from closed forest as well as from polar and alpine tundra (see Holtmeier, 1985; Holtmeier and Broll, 1992).

In general, the upper timberline is caused by heat deficiency, which is a very complex factor. It implies short growing seasons, long-lasting snow cover and thereby low soil temperatures, delay of root growth and nutrient uptake, late flushing of buds, late and early frost, frost-drought, etc. Since photosynthetic efficiency, net carbon gain, and thus tree growth decrease with altitude, it has been argued that the upper limit of tree growth might be caused by zero production of organic dry matter (Boysen-Jensen, 1932, 1949; Sarvas, 1970). However, as shown by Schultze *et al.*, (1967) for bristlecone pine in the White Mountains (California), carbon balance may even become negative during unfavourable years. However, these results do not necessarily mean that in the long term tree growth at its upper and polar limits is controlled by zero net carbon balance. From field observations in many other high-mountain areas, and experimental studies, it became obvious that if one is to explain the upper or polar limit of tree growth, one must take ripening of current shoots and needles into consideration. If tissues cannot completely mature during the growing season because of unavailable organic matter, low temperatures, and late frosts, shoots and needles will easily

Table 13.1 Quality and germination capacity of seeds of *Picea engelmannii* on Niwot Ridge, Colorado Front Range, USA

Elevation (m)	3150 %	3350–3500 %
Endosperm and embryo missing	52.9	60.3
Endosperm present, embryo missing or less than 50%	18.3	32.7
Necrosises, embryo not available	4.8	6.6
Germination capacity	24.0	0.4

succumb to climatic injuries in winter (Tranquillini, 1967, 1979; Holtmeier, 1971, 1974, 1980; Wardle, 1971, 1974), usually before reaching a growth limit caused by zero dry matter production.

In addition, reproduction from seeds occurs only at long intervals and may even be missing for many decades. Although the highest located trees occasionally produce abundant cone crops, they usually do not produce viable seeds (Table 13.1). Usually the upper limit of production of viable seeds is located below the physiological limit of tree growth (Figure 13.1).

In any case, the average age of the high-altitude forests will increase if the climate deteriorates, because of insufficient reproduction. The trees already established, however, may survive for a long time if they are able to resist extreme climatic influences. An example of this are the bristlecone pines – many of them older than 2,000 years – at timberline in the White Mountains (California), in the Snake Range (Nevada), and in the Rocky Mountains of southern Colorado and northern New Mexico (LaMarche and Mooney, 1967, 1972; LaMarche, 1969, 1973; Krebs, 1972, 1973; Brunstein and Yamaguchi, 1992). Finally, however, the timberline will recede if the trees cannot reproduce themselves (Figure 13.1).

High-altitude forests formed by tree species that are able to rejuvenate by layering (formation of adventitious roots), such as spruce or fir, are characterized by greater persistence, because regeneration by layering still goes on at temperatures that would exclude any sexual reproduction. Thus, the present position of such timberlines may be out of balance with climate for hundreds or even thousands of years, and the most advanced, usually crippled trees reflect a more favourable previous climate (Larsen, 1965, 1980, 1989; Ives, 1973, 1978; Nichols, 1974, 1975a, 1975b, 1976; Elliott, 1979; Hansen-Bristow, 1981; Ives and Hansen-Bristow, 1983; Holtmeier, 1985, 1986a).

Under favourable climatic conditions the upper limit of production of viable seeds may become located at timberline or close to the limit of arborescent growth (Figure 13.1). Abundant natural regeneration may occur within the timberline ecotone, and, in the case of long-distance seed dispersal,

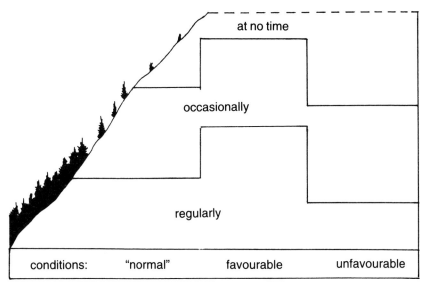

Figure 13.1 Vertical shift of the upper limit of production of viable seeds under the influence of climatic fluctuations.

even above the present tree limit. Thus, extensive regeneration could be observed within the timberline ecotone and locally far beyond the most altitudinally advanced trees in many high mountain areas of the northern hemisphere and at the polar timberline in Eurasia and North America during the middle of this century (Holtmeier, 1979; Heikkinen, 1984a, 1984b).

Hence one might expect the recent warming to cause a similar effect. However, up to the present no definite evidence has been reported that would suggest improved tree growth and regeneration at timberline which could be attributed to the most recent increase in temperature. On the contrary, many of the trees that became established during the favourable decades around the middle of this century have already died or become crippled by climatic influences (Holtmeier, 1974, 1979; Kullman, 1983, 1989, 1990).

We have only so far considered the influence of increased temperature on tree growth at timberline. Enriched CO_2, however, may also increase dry organic matter production (fertilization), as evidenced by many studies (e.g. Kramer, 1981; Kimball, 1983; Gates, 1985; Conroy *et al.*, 1986; Luxmoore *et al.*, 1986). At timberline natural CO_2 levels are lower than those at lower elevations (Graybill, 1987). Tranquillini (1979) estimated a decline of 10–20 per cent in photosynthetic performance of trees at timberline due to lower natural carbon dioxide. Thus, several different authors suggest that substantially enriched CO_2 enhances tree growth at timberline (LaMarche *et al.*, 1984). Bristlecone pines (*Pinus longaeva*) on the White Mountains (California) and limber pines (*Pinus flexilis*) on Mt Jefferson (Nevada), for example, showed an increase in radial growth that started in the middle of the last century and persisted through the early 1980s (LaMarche *et al.*, 1984). Until about 1960 it was more or less in line with rising summer temperatures. However, for the last 20 years no climatic warming trend is reflected by the long-term regional data set (Bradley *et al.*, 1982). Consequently the authors concluded that enhanced radial growth must be due to the increasing concentration of atmospheric CO_2. Stockton (1984) offers another hypothesis. He explains enhanced tree growth on these relatively dry sites by the recent increase of precipitation in the American Southwest.

One should in general be cautious in making too many hasty generalizations about these findings. Global CO_2 has increased, but that is not necessarily so at the very windy timberline environment. At wind-exposed sites – which usually is most of the ecotone – the concentration of CO_2 is usually still very low.

Moreover, taking into account the complexity of timberline ecosystems, it does not make sense to focus only on that popular factor, although we have to think about it. There are many other factors that control tree growth. Thus, subalpine trees in the Cascade Mountains, for example, were unaffected by enriched CO_2. Instead growth has declined since the favourable 1940s, following the regional trends of temperature (Graumlich and Brubaker, 1986; Graumlich *et al.*, 1989). In the southern Sierra Nevada as well, direct CO_2-fertilization is not important for subalpine conifers, but precipitation during the previous winter and temperature during the current summer interact in controlling tree growth (Graumlich, 1991). In addition, it was evidenced that the response of trees to these two variables can be nonlinear (see also Fritts, 1976; Graumlich and Brubaker, 1986) and that substantiated variations in species-to-species response do occur. Kienast and Luxmoore (1988), who studied tree cores from 34 sites in the northern hemisphere, found that increased growth in any of their tree-ring chronologies could not be explained by higher atmospheric CO_2 only. On the other hand, Hari and Arovaara (1988) found growth of Scots pine (*Pinus sylvestris*) at the subarctic timberline in northern Finland 15.5–43.3 per cent greater than expected when related to the predicted climatic data. These differences might be attributed to enriched CO_2. However, the authors themselves hesitate to consider their results to be unambiguous, because they are highly sensitive to an autocorrelation parameter that predicts current growth on the basis of past growth. Thus far, there is no evidence of a significant correlation of enriched CO_2 and an increase of tree growth at timberlines.

We found plant-available phosphorus, for example, to be the likely limiting factor at the upper timberline on Niwot Ridge (Colorado; Holtmeier and Broll, 1992) and in northernmost Finland (Schreiber, 1991). In this case, for example, any increase of other 'fertilizers' (e.g. nitrogen) could exacerbate the situation. Plant-available phosphorus could also be low because it had probably been immediately taken up by mycorrhiza (see also Haimi et al., 1992).

In addition, excessive input of nitrogen can cause too early bud break and prolonged annual shoot elongation which may increase frost damage (see literature review in Skre, 1988). It also changes the allocation pattern of carbon and nutrients within the trees (Hinrichsen, 1986). More carbon will be allocated in the root system and more nutrients might be taken up by the trees. Thus, it depends on the local decomposition rate whether the additional atmospheric nitrogen becomes a factor detrimental to tree growth at timberline. We also have to take the possible influence of acid deposition into consideration. Moreover, we do not know how mycorrhiza will respond to a climatically induced changing environment. Altogether, the effects of the different climatic factors, the 'fertilizer effect' of CO_2, and the effects of other nutrients need to be examined under timberline conditions before establishing any far-reaching hypotheses.

ANTHROPOGENIC TIMBERLINE AND CLIMATIC CHANGE

The ecological situation at timberline is particularly complex in areas where anthropogenic disturbances are interfering with natural factors, as is the case in the European Alps and many other high mountain ranges of Eurasia that have been settled since prehistoric times. In the Alps, for example, the timberline was lowered by 150–400 m with respect to its uppermost position during the postglacial thermal optimum. This was in response to alpine pasturing, local mining, and salt works during the Middle Ages (Holtmeier, 1974, 1986b).

As a consequence of the extensive deforestation of high elevation areas, avalanches, landslides, soil erosion, and torrential washes became more frequent and a permanent threat to the people living in the mountain valleys. Reafforestation up to the climatic limit of tree growth, combined with or encouraged by artificial constructions when necessary, has proved to be the best way to reduce such catastrophes.

At present, the climatic limit of tree growth is located above the actual forest limit as is clearly evidenced by the invasion of trees into abandoned or rarely used alpine pastures by trees. This invasion was probably facilitated by the warming during the middle of this century. In any event, tree growth above the closed forests is hampered more by unfavourable site conditions than one would expect considering the fact that the forest limit was lowered by man. When the former forest was removed, the windflow near the soil surface and the amount of solar radiation became strongly influenced by the local topography (Figures 13.2, 13.3). Sites relatively favourable to tree

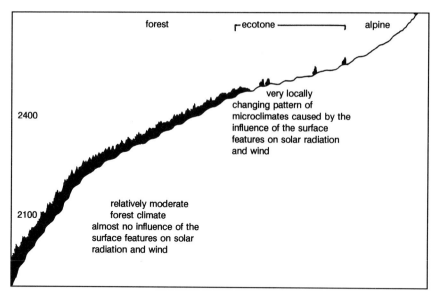

1. Situation during postglacial thermal optimum

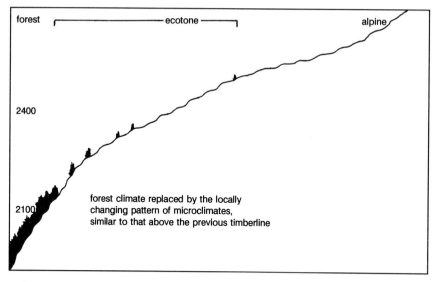

2. Situation at present

Figure 13.2 Change of ecoclimatic conditions following the lowering of the upper timberline as a result of anthropogenic disturbances.

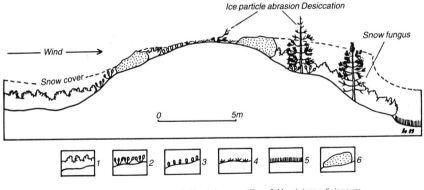

1 *Rhododendron ferrugineum*; 2 *Vaccinium myrtillus*; 3 *Vaccinium uliginosum*, *Empetrum nigrum*; 4 *Loiseleuria procumbens*, lichens (*Alectoria ochroleuca*, *Thamnolia vermicularis*, *Cladonia* spec.), *Juncus trifidus*; 5 *Trichophorum caespitosum*, *Eriopherum scheuchzeri*, *Carex* spec.; 6 Boulder

Figure 13.3 Influence of the topography on microclimate, vegetation, and tree growth in abandoned alpine pastures.

growth are close to sites where reafforestation is totally prevented by the present microclimatic conditions. The anthropogenic timberline has become as pronounced an ecological boundary as the natural timberline had been before (Holtmeier, 1974; 1986b).

At exposed sites which lack or are only occasionally covered with snow in winter, young trees suffer from desiccation, frost damage, and ice particle abrasion. On leeward slopes and in other places characterized by heavy snow accumulation, the growing season may be too short, and evergreen conifers are seriously damaged or even killed by parasitic snow fungi (e.g. *Phacidium infestans, Herpotrichia juniperi*).The rise of temperature during this century could obviously not compensate for the change of microclimates caused by the removal of the former high-altitude forests. Contrary to what is widely believed, the influence of the increasing size of trees and tree clusters on microclimates in particular and other site conditions (such as depth and duration of snow cover and resulting length of the growing season, soil moisture, and soil temperature, etc.), will not necessarily improve the growing conditions enough to make the forest move upwards in a more or less continuous front. Even in the long term, successful restocking will be confined to favourable locations.

CONCLUDING REMARKS

If global warming continues for a long time, timberline will also certainly be affected. However, such a general statement is commonplace and has no practical meaning as long as the potential change of regional and local climates is not known. Regionalized scenarios might help to improve predictions, but they cannot be better than the data they are based on – and these data are still insufficient in most timberline areas.

Moreover, timberline is a biological boundary that will not linearly respond to changing temperatures. Thus, it is time to get away from the misconception that timberline advance would run parallel to an altitudinal (upper timberline) or horizontal (polar timberline) shift of any isotherm considered essential to tree growth.

In addition, global warming cannot be expected to cause a synchronous adjustment of environmental conditions and timberline to a complex climatic change. More likely, the adjustment would be asynchronous, as can be concluded from climatically caused timberline fluctuations in the past (LaMarche and Mooney, 1967, 1972; LaMarche, 1973, 1977; Andrews et al., 1978; Elliott, 1979; Elliott and Short, 1979; Kullman, 1988, 1989; Kullman and Engelmark, 1991; Graumlich, 1991; MacDonald and Gajewski, 1992). Changes in the treeline in western, central, and eastern Canada appear to have been asynchronous. While total lack of seedlings is reported by Elliott (1979) on the northern treeline in the Ennadai Lake area (central Canada), sexual regeneration is abundant at the treeline on the Labrador Peninsula (Elliott and Short, 1979; Elliott-Fisk, 1983). Payette et al. (1989), who studied treeline vegetation in northern Quebec, could not substantiate a positive response to global warming at exposed treeline sites. They suppose that the amplitude and magnitude of recent warming is not large enough to compensate at such sites what is left of the negative effects of the Little Ice Age. Scott et al. (1987) reported from the Churchill region (Hudson Bay, Manitoba) that climatic warming resulted in an increase in tree population but the treeline did not change.

Timberline response to climatic change will be different in arid or semi-arid regions when compared to humid environments. As evidenced by studies in the Mt Washington Area in Nevada (LaMarche and Mooney, 1972), timberlines under arid conditions depend at least as much on sufficient moisture supply as upon thermal conditions during the growing season (see also Höllermann, 1978). In addition, global change will not affect different tree species at timberline in the same way. It can be concluded from timberline history in the Scandes Mountains, for example, that growth conditions for pine (*Pinus sylvestris*) deteriorated during the past 600–5,300 years, while conditions for mountain birch (*Betula pubescens ssp. tortuosa*) improved (Kullman, 1983, 1987, 1988). Moreover, it can be concluded from many field studies on treeline in different regions that negative influences such as forest fires (Kuramoto and Bliss, 1970; Douglas and Ballard, 1971; Henderson, 1973; Nichols, 1975) or mass infestations by insects (Nuorteva, 1963; Holtmeier, 1974, 1985) that occur during long-term site history may be of greater importance to the present site conditions than the current climate (Holtmeier, 1993). Thus, if possible, predictions on possible effects of environmental change should always be backed up by information on site history.

In discussing timberline advance, we must also define the period of time considered. It may be obvious from the available data that local or regional fluctuations may run counter to the long-term global warming trend. This is the case in the northern hemisphere in the period between 1940 and 1970. This fluctuation resulted in a decline of many trees that had become established during the favourable decades before. This is contrary to the optimistic assumption – made by Blüthgen (1942) in view of abundant young growth during the 1920–1940 period – that timberlines were progressing to their postglacial maximum positions. Thus, for a possible timberline advance, at least some 100 years of thermal conditions more favourable than at present must be postulated.

Not least important, since tree growth is never controlled by average temperatures, future studies on timberline or treeline response to climatic change should focus on change of range, frequency, seasonal occurrence and variability of extremes rather than on any average value.

We also have to keep in mind that tree growth and reproduction are primarily controlled by strongly contrasting microclimatic and soil-ecological conditions within the ecotone itself rather than by the altitudinal gradients of temperature. Thorough regional studies are required, especially on reproduction (production of viable seeds, germination success, survival rate of seedlings, spatial pattern of regeneration, etc.). Without such data, hypotheses on response of timberline to changing climate will remain pure speculation.

REFERENCES

Andrews, J.T., Carrara, P.E., Bartos, F. and Stuckenrat, R. (1978) 'Holocene stratigraphy and geochronology of four bogs (3,700 a.s.l.), San Juan Mountains, SW Colorado, and implications to the neoglacial record', *Geological Society of America, Abstracts with program*, 5 (6): 460–1.

Blüthgen, J. (1942) 'Die polare Baumgrenze', *Veröffentlichungen des Deutschen Wissenschaftlichen Institutes zu Kopenhagen*, Reihe I, Arktis 10.

Boysen-Jensen, P. (1932) *Die Stoffproduktion der Pflanze*, Fischer Verlag, Jena.

—— (1949) 'Causal plant geography', *Dansk. Vidensk. Selsk. Biol. Medd.*, 21 (3).

Bradley, R.S. (1974) 'Secular changes of precipitation in the Rocky Mountains and adjacent western states', Ph.D. thesis, University of Colorado.

Bradley, R.S., Barry, R.G. and Kiladis, G. (1982) 'Climatic fluctuations of the western United States during the period of instrumental records', *Contribution*, 42, Department of Geology and Geography, University of Massachusetts, Amherst.

Brunstein, F.C. and Yamaguchi, D.K. (1992) 'The oldest known Rocky Mountain bristlecone pines (Pinus aristata Engelm.)', *Arctic and Alpine Research*, 24 (3): 253–6.

Conroy, J.P., Smillie, R.M., Küppers, M., Bevege, D.I. and Barlow, E. W. (1986) 'Chlorophyll A fluorescence and photosynthetic and growth responses of Pinus radiata to phosphorus deficiency, drought stress, and high CO_2', *Plant Physiology*, 81: 423–9.

Dahms, A. (1984) 'Die natürliche Vermehrung verschiedener Baumarten im oberen

Waldgrenzbereich der Colorado Front Range in ökologischer Sicht', unpublished master's degree thesis, Department of Geography, University of Münster.

Douglas, G.W. and Ballard, T.M. (1971) 'Effects of fire on alpine plant communities in the North Cascades, Washington', *Ecology*, 52: 1058–64.

Elliott, D.L. (1979) 'The current regenerative capacity of the northern Canadian trees, Keewatin, N.W.T., Canada: some preliminary observations', *Arctic and Alpine Research*, 11 (2): 243–51.

Elliott, D.L. and Short, S.K. (1979) 'The northern limit of trees in Labrador: A discussion', *Arctic*, 32 (3): 201–6.

Elliott-Fisk, D.L. (1983) 'The stability of the northern Canadian tree-limit', *Annals of the American Geographers*, 73; 560–76.

Environment Canada (1989) Atmospheric Environment Service, Changing Atmosphere fact sheet, 'The greenhouse effect: impacts on the Arctic'.

Fritts, H.C. (1969) 'Bristlecone pine in the White Mountains of California: growth and ring-width characteristics', *Papers of the Laboratory of Tree-Ring Research* 4, University of Arizona Press, Tucson, Arizona.

—— (1976) *Tree-rings and Climate*, Academic Press, New York.

Gates, H. (1985) 'Global biospheric response to increasing atmospheric carbon dioxide concentration', in B.R. Strain and J.D. Cure (eds), *Direct Effects of Increasing Carbon Dioxide on Vegetation*, US Department of Energy Report DOE/ER–0238, Washington, pp. 171–84.

Graumlich, L.J. (1991) 'Subalpine tree growth, climate, and increasing CO_2: an assessment of recent growth trends', *Ecology*, 72 (1): 1–11.

Graumlich, L.J. and Brubaker, L.B. (1986) 'Reconstruction of annual temperature (1590–1979) for Longmire, Washington, derived from tree rings', *Quaternary Research*, 25: 223–34.

Graumlich, L.J., Brubaker L.B. and Grier, Ch.C. (1989) 'Long-term trends in forest net primary productivity: Cascade Mountains, Washington', *Ecology*, 70 (2): 405–10.

Graybill, D.A. (1987) 'A network of high elevation conifers in the western US for detection of tree-ring growth response to increasing atmospheric carbon dioxide', in G.C. Jacoby and J.W. Hornbeck (eds), *Proceedings of the International Symposium on Ecological Aspects of Tree-Ring Analysis*, US Department of Energy Conference Report DOE/CONF–8608144: 463–74.

Haimi, J., Huhta, V. and Boucleham, M. (1992) 'Growth increase of birch seedlings under the influence of earthworms – a laboratory study', *Soil Biology and Biochemistry* 24 (12): 1525–8.

Hansen-Bristow, K.J. (1981) 'Environmental controls influencing the altitude and form of the forest-alpine tundra ecotone, Colorado Front Range', Ph.D. thesis, University of Colorado, Boulder.

Hari, P. and Arovaara, H. (1988) 'Detection of CO_2 induced enhancement in the radial increment of trees: evidence from northern timberline', *Scandinavian Journal of Forest Research* 3: 64 –74.

Heikkinen, O. (1984a) 'Climatic changes during recent centuries as indicated by dendrochronological studies, Mount Baker, Washington, U.S.A.', in N.-A. Mörner and W. Karlén (eds), *Climatic Changes on a Yearly to Millenial Basis*, Springer, Munich, pp. 353–61.

—— (1984b) 'Forest expansion in the subalpine zone during the past hundred years, Mount Baker, Washington, U.S.A.', *Erdkunde*, 38: 194–202.

Henderson, J.A. (1973) 'Composition, distribution and succession of subalpine meadows in Mount Rainier National Park, Washington', unpublished Ph.D. thesis, Oregon State University, Corvallis.

Hinrichsen, D. (1986) 'Multiple pollutants and forest decline', *Ambio*, 15: 258–65.

Höllermann, P.W. (1978) 'Geoecological aspects of the upper timberline in Tenerife, Canary Islands', *Arctic and Alpine Research*, 10 (2): 365–82.

Holtmeier, F.-K. (1971) 'Waldgrenzstudien im nördlichen Finnisch-Lappland und angrenzenden Nordnorwegen', *Report Kevo Subarctic Research Station*, 8: 53–62.

—— (1974) 'Geoökologische Beobachtungen und Studien an der subarktischen und alpinen Waldgrenze in vergleichender Sicht (nördliches Fennoskandien/ Zentralalpen)', *Erdwissenschaftliche Forschung*, VIII.

—— (1979) 'Remarks on oscillations of the arctic and alpine timberline', *Acta Universitas Oulensis*, A.82, Geol. 3: 165–71.

—— (1980) 'Influence of wind on tree physiognomy at the upper timberline in the Colorado Front Range', *New Zealand Forest Service Technical Paper*, 70: 247–61.

—— (1985) 'Die klimatische Waldgrenze – Linie oder Übergangssaum (Ökoton)? – Ein Diskussionsbeitrag unter besonderer Berücksichtigung der Waldgrenzen in den mittleren und hohen Breiten der Nordhalbkugel', *Erdkunde*, 39 (4): 271–85.

—— (1986a) 'Über Bauminseln (Kollektive) an der klimatischen Waldgrenze – unter besonderer Berücksichtigung von Beobachtungen in verschiedenen Hochgebirgen Nordamerikas', *Wetter und Leben*, 38: 121–39.

—— (1986b) 'Die obere Waldgrenze unter dem Einfluß von Klima und Mensch', *Abhandlungen aus dem Westfälischen Museum für Naturkunde zu Münster*, 48 (2/3): 395–412.

—— (1993) 'Timberlines as indicators of climatic changes: problems and research needs', *Paläoklimaforschung/ Paleoclimate Research*, 9: 211–22.

Holtmeier, F.-K. and Broll, G. (1992) 'The influence of tree islands and micro-topography on pedoecological conditions in the forest–alpine tundra ecotone on Niwot Ridge, Colorado Front Range, U.S.A.', *Arctic and Alpine Research*, 24 (3): 216–28.

Ives, J.D. (1973) 'Studies in high altitude geoecology of the Colorado Front Range. A review of the research program of the Institute of Arctic and Alpine Research', *Arctic and Alpine Research*, 5 (3, pt. 2): 65–6.

—— (1978) 'Remarks on the stability of timberline', *Erdwissenschaftliche Forschung*, 11: 313–17.

Ives, J.D. and Hansen-Bristow, K.J. (1983) 'Stability and instability of natural and modified upper timberline landscapes in the Colorado Rocky Mountains, U.S.A.', *Mountain Research and Development*, 3 (2): 149–55.

Kallio, P., Niemi, S., Sulkinoja, M. and Valanne, T. (1983) 'The Fennoscandian birch and its evolution in the marginal forests zone', in P. Morisset and S. Payette (eds), *Tree-line Ecology, Proceedings of the Northern Quebec Tree-line Conference*, Centre d' Études Nordique: 101–20.

Kauppi, P. and Posch, M. (1988) 'A case study of the effect of CO_2-induced climatic warming on forest growth and the forest sector: A. Productivity reactions of northern boreal forest', in M.L. Parry, T.R. Carter and N.T. Konijin (eds), *The Impact of Climatic Variations on Agriculture*, vol. I, Kluwer Acad. Publ., Dordrecht, pp. 183–95.

Kienast, F. and Luxmoore, R.J. (1988) 'Tree-ring analysis and conifer growth responses to increased atmospheric CO_2 levels', *Oecologia*, 76: 487–95.

Kimball, B.A. (1983) 'Carbon dioxide and agricultural yield: an assemblage and analysis of 430 prior observations', *Agronomy Journal*, 75: 779–88.

Kramer, P.J. (1981) 'Carbon dioxide concentration, photosynthesis and dry matter production', *BioScience*, 31: 29–33.

Krebs, P.V. (1972) 'Dendrochronology and the distribution of bristlecone pine (Pinus aristata Engelm.) in Colorado', Ph. D. thesis, University of Colorado.

—— (1973) 'Dendrochronology of bristlecone pine (Pinus aristata Engelm.) in Colorado', *Arctic and Alpine Research*, 5: 149–50.

Kullman, L. (1983) 'Past and present treelines of different species in the Handölan Valley, Central Sweden', *Nordicana*, 47: 25–45.

—— (1987) 'Little Ice Age decline of a cold marginal Pinus sylvestris forest in the Swedish Scandes', *New Phytology*, 196: 567–84.

—— (1988) 'Holocene history of the forest–alpine tundra ecotone in the Scandes Mountains (Central Sweden)', *New Physiology*, 108: 101–10.

—— (1989) 'Rapid decline of upper montane forests in the Swedish Scandes' *Arctic*, 42 (3): 217–26.

—— (1990) 'Dynamics of altitudinal tree-limits in Sweden: a review', *Norsk Geografisk Tidsskrift*, 44: 103–16.

Kullman, L. and Engelmark, O. (1991) 'Historical biogeography of Picea abies (L.) Karst. at its subarctic limit in northern Sweden', *Journal of Biogeography*, 18: 63–70.

Kuramoto, R.T. and Bliss, L.C. (1970) 'Ecology of subalpine meadows in the Olympic Mountains, Washington', unpublished Ph.D. thesis, University of Illinois, Urbana.

LaMarche, V.C. (1967) 'Altithermal timberline advance in western United States', *Nature*, 213: 980–2.

—— (1969) 'Environment in relation to age of bristlecone pines', *Ecology*, 50: 53–9.

—— (1973) 'Holocene climatic variations, inferred from treeline fluctuations in the White Mountains, California', *Quaternary Research*, 3: 632–60.

—— (1977) 'Dendrochronological and paleoecological evidence for holocene climatic fluctuations in the White Mountains', *Erdwissenschaftliche Forschung*, 13: 151–5.

LaMarche, V.C. and Mooney, H.A. (1967) 'Altithermal timberline advance in western United States', *Nature*, 213: 980–2.

—— (1972) 'Recent climatic change and development of the bristlecone pine (Pinus longaeva Bailey) krummholz zone, Mt Washington, Nevada', *Arctic and Alpine Research 4*, (1): 61–72.

LaMarche, V.C., Graybill, D.A., Fritts, H.C. and Rose, M.R. (1984) 'Increasing atmospheric carbon dioxide: tree-ring evidence for growth enhancement in natural vegetation', *Science*, 225: 1019–21.

Larsen, J.A. (1965) 'The vegetation of the Ennedai Lake Area, NWT: studies in arctic and subarctic bioclimatology', *Ecological Monographs*, 35 (1): 37–59.

—— (1980) *The Boreal Ecosystem*, Academic Press, New York.

—— (1989) 'The northern forest border in Canada and Alaska', *Ecological Studies*, 70, Springer Verlag, New York, Berlin, Heidelberg, London, Paris, Tokyo.

Luxmoore, R.J., O'Neill, E.G., Ells, J.M. and Rogers, H.H. (1986) 'Nutrient uptake and growth response of Virginian pine to elevated atmospheric carbon dioxide', *Journal of Environment Quality*, 15: 244–51.

MacDonald, G.M. and Gajewski, K. (1992) 'The northern treeline of Canada', in D.G. Janelle (ed.), *Geographic Snapshots of North America Commemorating the 24th Congress of the International Geographical Union and Assembly*, Guildford Press, London, pp. 34–7.

Nichols, H. (1974) 'Arctic North America palaeoecology: the recent history of vegetation and climate deduced from pollen analysis', in J.D. Ives and R.G. Barry (eds), *Arctic and Alpine Environments*, Methuen, London.

—— (1975a) *Palynological and Paleoclimatic Study of the Late Quaternary Displacement of the Boreal Forest-Tundra Ecotone in Keewatin and Mackenzie N.W.T.*,

Canada, Institute of Arctic and Alpine Research, University of Colorado, Occasional Paper 15.

—— (1975b) 'The time perspective in northern ecology: palynology and the history of the Candian forest', *Proc. Circumpolar Conference on Northern Ecology*, 15–18 Sept., Ottawa, National Research Council of Canada, pp. 157–64.

—— (1976) 'Historical aspects of the northern Canadian treeline', *Arctic*, 29 (1): 8–47.

Nuorteva, P. (1963) 'The influence of Oporinia autumnata (Bkh.) (Lep. Geometridae) on the timber-line in subarctic conditions', *Annales Entomologicae Fennicae*, 29: 270–7.

Ozenda, P. and Borel, J.-L. (1991) 'Les conséquences possible des changements climatiques dans l'arc alpine', Rapport FUTURALP 1.

Payette, S., Filion, L., Delwaide, A. and Bégin, C. (1989) 'Reconstruction of the tree-line vegetation response to long-term climate change', *Nature*, 341: 429–32.

Sarvas, R. (1970) 'Temperature sum as a restricting factor in the development of forest in the subarctic', *Ecology and Conservation*, (Ecology of the Subarctic Regions, UNESCO-symposium Helsinki 1966), UNESCO, Paris, pp. 79–82.

Schreiber, H. (1991) 'Untersuchungen zur Standortdifferenzierung im Waldgrenzbereich am Koahppeloaivi (Utsjoki, Finnisch Lappland)', unpublished diploma thesis, Department of Geography, University of Münster.

Schultze, E.D., Mooney, H.A. and Dunn, E.L. (1967) 'Wintertime photosynthesis of bristlecone pine (Pinus aristata) in the White Mountains of California', *Ecology*, 48: 1044–7.

Scott, P.A., Hansell, R.I.C. and Fayle, D.C.F. (1987) 'Establishment of white spruce populations and responses to climatic change at the treeline, Churchill, Manitoba, Canada', *Arctic and Alpine Research*, 19 (1): 45–51.

Skre, O. (1988) 'Frost resistance in forest trees: a literature survey', *Meddelser fra Norsk Institutt for Skogforskning* 40 (9): 1–35.

Solomon, A.M. (1989) 'Measures for the protection of forests in temperate zones', *Enquêtekommission, Vorsorge zum Schutz der Erdatmosphäre, Hearing on 'Schutz der Wälder in mittleren und nördlichen Breiten'*, 8 June 1989, pp. 141–7.

Stockton, C.W. (1984) 'An alternative hypothesis to direct CO_2 fertilization as a cause of increased tree growth during 1850–1980 in Central Nevada', Preliminary Report, Laboratory of Tree Ring Research, University of Arizona, Tucson.

Tranquillini, W. (1967) 'Über die physiologischen Ursachen der Wald- und Baumgrenze', *Mitteilungen der Forstlichen Bundesversuchsanstalt Mariabrunn*, 75: 457–87.

—— (1979) 'Physiological ecology of the alpine timberline. Tree existence at high altitudes with special reference to the European Alps', *Ecological Studies*, 31, Springer, Berlin, Heidelberg, New York.

Wardle, P. (1971) 'An explanation for alpine timberline', *New Zealand Journal of Botany*, 9 (3): 371–402.

—— (1974) 'Alpine timberline', in J.D. Ives and R.G. Barry (eds), *Arctic and Alpine Environments*, Methuen, London, pp. 371–402.

14

RECENT CHANGES IN THE GROWTH AND ESTABLISHMENT OF SUBALPINE CONIFERS IN WESTERN NORTH AMERICA

D.L. Peterson

INTRODUCTION

Several studies suggest that recent changes in the condition of forest resources in western North America may be related to changes in global atmospheric conditions. For example, growth rates of *Pinus longaeva*, *P. aristata*, and *P. flexilis* in the southwestern United States have increased significantly since the mid-nineteenth century. Recent growth increases have also been reported for *P. contorta* and *P. albicaulis* in the state of California, and for *Tsuga mertensiana*, *Larix lyallii*, and *Abies amabilis* in the state of Washington. Encroachment of subalpine tree species on meadows during the twentieth century in the Olympic Mountains, Cascade Mountains, and other areas of western North America suggests that recent climate patterns have affected treeline community composition and structure, as well as growth of established trees. Continued increases in tree establishment in these areas may ultimately lead to changes in the location and structure of the subalpine life zone. In addition, the combination of elevated CO_2, climate change, and changes in seed source availability may create new community types.

Subalpine forests and meadows comprise a large portion of high elevation ecosystems in mountain-protected areas of western North America. The dynamics of subalpine forests are extremely complex, encompassing a wide range of environmental controls. Temperature, precipitation, snowmelt, and storm frequency affect the growth and productivity of these communities, and any changes could alter the location of the subalpine/alpine and montane/subalpine ecotones. Timing, quantity, and distribution of precipitation (primarily snowfall) are particularly important at high elevation. This paper summarizes recent changes in growth and distribution of subalpine

234

tree species in western North America, and discusses the relationship of climate and other environmental factors to these changes. The effects of changes in critical environmental factors will no doubt vary among species and regions.

SENSITIVITY OF THE SUBALPINE ZONE TO CLIMATE CHANGE

Subalpine forest and meadow ecosystems comprise an important, climatically sensitive component of mountainous regions of western North America (Peterson, 1991). Changes in temperature, precipitation, snowpack, storm frequency, and fire all could affect the growth and productivity of these systems, resulting in substantial shifts in the location of ecotones between subalpine/alpine zones and montane/subalpine zones (Canaday and Fonda, 1974).

Subalpine forests of western North America provide an excellent opportunity to examine responses to past climate variability. Trees in the subalpine zone are frequently over 500 years old, and respond to climatic variations over annual to century-long timescales. The magnitude of climatic variation they have experienced in the past may be comparable to projections of future climate as a result of increased concentration of greenhouse gases. The population dynamics of subalpine tree species can be used to interpret climatic conditions under which trees have regenerated, and can indicate how subalpine forest/meadow ecotones changed in the past. Preserved pollen and plant fossils can be used to examine subalpine vegetation distribution during different climatic periods of the Holocene.

Much of the recent literature on the effects of potential climate change have focused on changes in the growth and distribution of low-elevation forests (e.g., Woodman, 1987; Davis, 1989; Franklin et al., 1991). In western North America, most low-elevation forests are sensitive to soil moisture deficits during relatively dry summers (Peterson et al., 1991; Graybill et al., 1992). Although subalpine forests have been the subject of considerably less study, it appears that snowpack is an important limiting factor to growth, with respect to length of growing season (Graumlich, 1991; Peterson, 1993). Seedling establishment in subalpine meadows is limited by duration of snowpack (Fonda, 1976) and disturbed by fire (Little, 1992). Summer temperature also affects the growth (positively) of mature subalpine conifers (Graumlich, 1991; Peterson, 1993) and the survival (negatively) of seedlings (Little, 1992).

There are several reports of recent increases in the growth of subalpine conifer species in western North America (Innes, 1991). Recent increases in the abundance of subalpine conifer populations have also been documented at several locations. There is a variety of possible explanations for these phenomena, including the impact of changes in the atmospheric environment and climate. This paper reviews recent reports of changes in the growth and distribution of subalpine conifers in western North America, and discusses some of the possible causes of these changes.

SUBALPINE TREE GROWTH

The first prominent report of a recent increase in growth of subalpine coniferous species was published by LaMarche *et al.* (1984). They reported dramatic increases in the growth rate of *Pinus longaeva/P. aristata* (bristle-cone pine) and *P. flexilis* (limber pine) in California and Nevada. The extreme age of these trees, combined with the fact that radial growth has increased since 1850, makes this a particularly interesting result. The authors suggested that elevated levels of CO_2 associated with combustion fossil fuels may be enhancing their growth and productivity, perhaps through increased water-use efficiency. A more recent examination of these data corroborates the growth increase phenomenon and restates that CO_2 fertilization is the hypothesized cause of the increase (Graybill and Idso, 1993). However, there is some disagreement about whether the increases in growth found in these studies (which included sampling of strip-bark trees) are representative of the populations as a whole, and about the factors causing the growth increase (Cooper and Gale, 1986).

We subsequently studied basal area growth trends of *P. contorta* (lodge-pole pine) and *P. albicaulis* (whitebark pine) at sites above 3,000 m elevation in the east-central Sierra Nevada of California, and were surprised to find that a high proportion of these trees had recent growth increases as well (Peterson *et al.*, 1990). The onset of the increase was normally between 1850 and 1900, as found by LaMarche *et al.* (1984), and growth has been particularly rapid during the past 30 years or so. No strip-bark trees were sampled in this study, so arguments directed toward this issue in the studies *P. longaeva*, *P. aristata*, and *P. flexilis* are not applicable to *P. contorta* and *P. albicaulis*.

There are other reports of recent growth increases in subalpine conifers of western North America (Innes, 1991) (Table 14.1). Jacoby (1986) found radial growth increases in *P. contorta* in the San Jacinto Mountains of southern California, but did not identify a strong causal factor despite detailed climatic analysis. Graumlich *et al.* (1989) found increases in the growth and productivity of *Abies amabilis* (Pacific silver fir) and *Tsuga mertensiana* (mountain hemlock) in the Cascade Mountains of Washington State, and suggested that these trends were related to increased temperature.

There are also reports of recent growth increases in European conifers (Innes, 1991), such as *Picea abies* (Norway spruce) (Kienast and Luxmoore, 1988; Briffa, 1992) and *Abies alba* (silver fir) (Becker, 1989), although these species are generally found at locations below the subalpine zone. Both increased CO_2 (Kienast and Luxmoore, 1988) and temperature (Becker,

Table 14.1 Summary of studies that found recent increases in the growth of
subalpine conifers in western North America

Species	Reference
Abies amabilis	Graumlich *et al.* (1989)
Pinus albicaulis	Peterson *et al.* (1991)
	LaMarche *et al.* (1984)
Pinus aristata	Graybill and Idso (1993)
Pinus balfouriana	Graybill and Idso (1993)
	Jacoby (1986)
Pinus contorta	Peterson *et al.* (1991)
Pinus flexilis	LaMarche *et al.* (1984)
	LaMarche *et al.* (1984)
Pinus longaeva	Graybill and Idso (1993)
Tsuga mertensiana	Graumlich *et al.* (1989)

1989; Briffa, 1992) have been suggested as potential causal factors for the growth increases.

It should be noted that not all studies of subalpine conifers have found recent growth increases. For example, Graumlich (1991) did not find increased radial growth in *P. balfouriana* (foxtail pine), *P. flexilis*, and *Juniperus occidentalis* (western juniper) in the Sierra Nevada. It is difficult to compare the various studies of tree growth discussed here because the studies employed a diversity of sampling and analytical techniques to evaluate growth patterns. It is possible that an analysis of time series of growth index values (based on ring-widths) could result in different interpretations from an analysis of the same dataset based on time series of basal area growth.

As noted above, there is a range of potential explanations for recent growth increases in subalpine conifers. The possibility of CO_2 fertilization has been supported by experimental studies (Graybill and Idso, 1993), but is extremely difficult to demonstrate for mature trees in the field. Increased temperature is another potential cause, but its relationship with growth is correlative and is also difficult to demonstrate for mature trees. An additional climatic factor that may be a more likely cause of growth increases is changes in the annual duration of snowpack, which affects length of growing season. Unfortunately, the long-term relationship of snowpack to tree growth has not been adequately investigated (snowpack data are often difficult to obtain).

Fertilization through nitrogen deposition is another proposed cause of growth increases. Although nitrogen deposition is relatively low in western North America, it is probably somewhat higher now than in the past because of the combustion of fossil fuels. Many subalpine forests are located in sites with shallow soils and relatively low fertility, so even a small increase in nitrogen input could have some impact over several decades. Finally, the growth increases may simply be the result of normal stand dynamics.

Relatively little is known about the growth and ecological characteristics of subalpine forest ecosystems. Although the observed increases appear abnormal compared to lower elevation species, they may in fact be a normal phenomenon that reflects the natural range of variance in growth of subalpine species.

SPATIAL PATTERNS OF SUBALPINE FOREST ESTABLISHMENT

Recent increases in tree establishment in subalpine meadows have been documented in mountainous regions throughout western North America (Rochefort et al., 1994). Most locations show an expansion of the forest-margin after 1890, with establishment peaks between 1920 and 1950. Additional establishment peaks have been identified on a local basis. Most investigators have concluded that increases in tree establishment are the result of a warmer climate following the Little Ice Age (Franklin et al., 1971; Kearney, 1982; Heikkinen, 1984; Butler, 1986). It is unclear if establishment patterns signify a long-term directional change or short-term variation in relatively stable ecotones, regardless of the potential cause(s).

The majority of studies on subalpine tree establishment have been conducted in the Pacific Northwest region (British Columbia [Canada]; Washington, Oregon [USA]) (Woodward et al., 1991; Rochefort et al., 1994). Eight separate studies have documented large increases in populations of *Abies lasiocarpa* (subalpine fir), *A. amabilis*, *T. mertensiana*, *Larix lyallii* (subalpine larch), and *Chamaecyparis nootkatensis* (Alaska yellow-cedar) (Table 14.2). All of these species experienced increases in establishment between 1920 and 1950. This was generally a period of lower snowpacks, which probably allowed seedlings to become established during a longer growing season. Winter precipitation limits subalpine tree growth and establishment in the Pacific Northwest, which has a maritime climate with wet winters and dry summers; high summer temperature can also limit tree establishment, because shallow-rooted seedlings are subject to soil moisture stress (Little, 1993).

There are three species in the Sierra Nevada and White Mountains of California for which increases in establishment have been documented: *P. balfouriana*, *P. contorta*, and *P. longaeva* (Table 14.2). Soil moisture stress is clearly a limiting factor in this area, which is dominated by a Mediterranean climate with very dry summers. There is no consistency among the different locations in this region with respect to temporal patterns of establishment. There is no documentation of establishment during the past 20 years.

Studies conducted in the Rocky Mountains have documented increases in subalpine tree establishment for the following species: *A. lasiocarpa*, *P. contorta*, and *Picea engelmannii* (Engelmann spruce) (Table 14.2). This region is dominated by a continental climate, with low precipitation and cold

Table 14.2 Summary of studies which found recent increases in the invasion of meadows by subalpine conifers

Species	Study location	Period of invasion	Reference
Abies lasiocarpa *Tsuga mertensiana*	British Columbia (Canada)	1919–1939	Brink (1959)
Abies lasiocarpa *Larix lyallii* *Tsuga mertensiana*	Washington	1919–1937	Arno (1970)
Abies amabilis *Abies lasiocarpa* *Chamaecaypris nootkatensis* *Tsuga mertensiana*	Washington	1920–1950	Lowery (1972)
Abies amabilis *Tsuga mertensiana*	Washington	1923–1943	Douglas (1972)
Abies amabilis *Abies lasiocarpa* *Tsuga mertensiana*	Washington	1925–1934, 1940–1944	Heikkinen (1984)
Abies lasiocarpa *Larix lyallii* *Tsuga mertensiana*	Oregon, Washington	1894–1920, 1925–1950	Franklin *et al.* (1971)
Tsuga mertensiana	Washington	1920–1950	Agee and Smith (1984)
Abies lasiocarpa *Tsuga mertensiana*	Washington	1923–1933, 1943–1948, 1953–1960	Fonda and Bliss (1969)
Pinus balfouriana	California	1890–1895, 1897–1987	Vankat and Major (1978) Scuderi (1987)
Pinus contorta	California	1905–1936, 1948–1973	*Helms* (1987)
Pinus longaeva	California	1850–1940	LaMarche (1973)
Abies lasiocarpa *Picea engelmannii*	Alberta (Canada)	1940–1960, 1965–1973	Kearney (1982)
Abies lasiocarpa *Picea engelmannii* *Pinus contorta*	Idaho	1895–1915, 1925–1950	Butler (1986)

Source: Rochefort *et al.* (1994)
Note: Only those studies that attributed the invasion to climatic factors are listed. Locations are in the United States except where indicated.

winters. There was some consistency in temporal patterns of establishment, especially during 1940–1950, which was a period with a warmer, wetter climate.

It is unclear whether observations of subalpine tree invasions are isolated examples, or part of a broad pattern in western North America. There are insufficient data from locations other than the Pacific Northwest to speculate about the geographic extent of this phenomenon. Tree invasion in subalpine meadows is widespread in the Pacific Northwest, and trees are currently establishing at a rapid rate in many locations (Rochefort and Peterson, 1991; Woodward et al., 1991), especially in meadows dominated by ericaceous species. Much of the establishment is occurring in concavities and other landscape positions where snow would normally accumulate and inhibit germination and survival (personal observation). As trees become established, tree clumps act as black bodies to increase the absorption of radiation, snowmelt occurs progressively earlier, and tree survival adjacent to the tree clump is further enhanced. This progression of events has been termed 'contagious dispersion' (Payette and Filion, 1985).

FUTURE CHANGES IN TREE GROWTH AND ESTABLISHMENT

Current data on subalpine tree growth for western North America is too sparse to infer that growth increases are a broad regional phenomenon. Additional data from other sites are needed to quantify growth trends in subalpine species. Furthermore, consistent sampling and analytical methods should be applied, so that different data-sets can be compared.

However, there is sufficient information on long-term growth trends and shorter-term response of growth to climate to make some general predictions about potential growth under future climate scenarios. If the climate becomes warmer and drier, as predicted by general circulation models, growth rates of subalpine conifers will probably increase. This growth increase would depend on the seasonality of precipitation. A decrease in snowfall would be particularly beneficial to species such as A. lasiocarpa and P. engelmannii (Ettl and Peterson, 1991; Peterson, 1993). However, warmer summer temperatures could cause summer soil moisture deficits which would be detrimental to growth. It is unknown how future growth patterns will be influenced by increased concentrations of CO_2. Any potential growth changes must, of course, be considered with respect to the effects of climate change on interspecific competition and disturbance.

It is difficult to develop regional predictions for changes in subalpine forest dynamics in western North America under different future climate scenarios. The key to subalpine tree establishment is reduction in the magnitude of limiting factors. Depth and duration of snowpack is probably an important factor in all regions of western North America, especially in the Pacific

Northwest. However, the relative importance of snowpack and other limiting factors (such as summer temperature) varies by region and perhaps by species. The impacts of climate, competition, and disturbance (such as fire) are especially critical to subalpine forest dynamics, because the conditions present at the time of stand establishment can determine species composition and stand structure for hundreds of years in the future.

ACKNOWLEDGEMENTS

I thank members of the Cooperative Park Studies Unit at the University of Washington, whose ideas and research have contributed to many of the concepts in this paper: Gregory Ettl, Ronda Little, David W. Peterson, Regina Rochefort, David Silsbee, and Andrea Woodward. June Rugh assisted with editing. Research was supported by the National Park Service Global Change Program and Pacific Northwest Region of the National Park Service.

REFERENCES

Agee, J.K. and Smith, L. (1984) 'Subalpine tree establishment after fire in the Olympic Mountains, Washington', *Ecology*, 65: 810–19.

Arno, S.T. (1970) 'Ecology of alpine larch (*Larix lyallii* Parl.) in the Pacific Northwest', Ph.D. Thesis, University of Montana, Missoula, Montana.

Becker, M. (1989) 'The role of present and past vitality of silver fir forests in the Vosges mountains of northeastern France', *Canadian Journal of Forest Research*, 19: 1110–17.

Briffa, K.R. (1992) 'Increasing productivity of "natural growth" conifers in Europe over the last century', in T.S. Bartholin, B.E. Berglund, D. Eckstein and F.H. Schweingruber (eds), *Tree Rings and the Environment: Proceedings of the International Dendrochronological Symposium*, Lundqua Report 34, Lund University, Lund, Sweden, pp. 64–71.

Brink, V.C. (1959) 'A directional change in the subalpine forest-heath ecotone in Garibaldi Park, British Columbia', *Ecology*, 40: 10–16.

Butler, D.R. (1986) 'Conifer invasion of subalpine meadows, Central Lemhi Mountains, Idaho', *Northwest Science*, 60: 166–73.

Canaday, B.B. and Fonda, R.W. (1974) 'The influence of subalpine snowbanks on vegetation pattern, reproduction, and phenology', *Bulletin of the Torrey Botanical Club*, 101: 340–50.

Cooper, C. and Gale, J. (1986) 'Carbon dioxide enhancement of tree growth at high elevation: commentary', *Science*, 231: 859–60.

Davis, M. B. (1989) 'Insights from palaeoecology on global change', *Bulletin of the Ecological Society of America*, 70: 222–8.

Douglas, G.W. (1972) 'Subalpine plant communities of the Western North Cascades, Washington', *Arctic and Alpine Research*, 4: 147–66.

Ettl, G.J. and Peterson, D.L. (1991) 'Growth and genetic response of subalpine fir (*Abies lasiocarpa*) in a changing environment', *The Northwest Environmental Journal*, 7: 357–9.

Fonda, R.W. (1976) 'Ecology of alpine timberline in Olympic National Park', *Proceedings, Conference on Scientific Research in National Parks*, 1: 209–12.

Fonda, R.W. and Bliss, L.C. (1969) 'Forest vegetation of the montane and subalpine zones, Olympic Mountains, Washington', *Ecological Monographs*, 39: 271–96.

Franklin, J.F., Moir, W.H., Douglas, G.W. and Wiberg, C. (1971) 'Invasion of subalpine meadows by trees in the Cascade Range, Washington and Oregon', *Arctic and Alpine Research*, 3: 215–24.

Franklin, J.F., Swanson, F.J., Marmon, M.E., Perry, D.A., Spies, T.A., Dale, V.H., McKee, A., Ferrell, W.K., Means, J.E., Gregory, S.V., Lattin, J.D., Schowalter, T.D. and Larsen, D. (1991) 'Effects of global climate change on forests in northwestern North America', *The Northwest Environmental Journal*, 7: 233–54.

Graumlich, L.J. (1991) 'Subalpine tree growth, climate, and increasing CO_2: an assessment of recent growth trends', *Ecology*, 72 (1): 1–11.

Graumlich, L.J., Brubaker, L.B. and Grier, C.C. (1989) 'Long-term trends in forest net primary productivity: Cascade Mountains, Washington', *Ecology*, 70: 405–10.

Graybill, D.A. and Idso, S.B. (1993) 'Detecting the aerial fertilization effect of atmospheric CO_2 enrichment in tree-ring chronologies', *Global Biogeochemical Cycles*, 7: 81–95.

Graybill, D.A., Peterson, D.L. and Arbaugh, M.J. (1992) 'Coniferous forests of the Colorado Front Range', in R.K. Olson, D. Binkley and M. Böhm (eds), *Response of Western Forests to Air Pollution*, Springer-Verlag, New York, pp. 365–401.

Heikkinen, O. (1984) 'Forest expansion in the subalpine zone during the past hundred years, Mount Baker, Washington, U.S.A.', *Erdkunde*, 38: 194–202.

Helms, J.A. (1987) 'Invasion of *Pinus contorta* var. *murrayana* (Pinaceae) into mountain meadows at Yosemite National Park, California', *Madroño*, 34: 91–7.

Innes, J.L. (1991) 'High-altitude and high-latitude tree growth in relation to past, present and future global climate change', *The Holocene*, 1: 174–80.

Jacoby, G.C. (1986) 'Long-term temperature trends and a positive departure from the climate-growth response since the 1950s in high elevation lodgepole pine from California', in *Proceedings of the NASA Conference on Climate–Vegetation Interactions*, Report OIES–2, University Corporation Atmospheric Research, Boulder, Colorado, pp. 81–3.

Kearney, M.S. (1982) 'Recent seedling establishment at timberline in Jasper National Park, Alberta', *Canadian Journal of Botany*, 60: 2283–7.

Kienast, F. and Luxmoore, R.J. (1988) 'Tree-ring analysis and conifer growth responses to increased atmospheric CO_2 levels', *Oecologia*, 76: 487–95.

LaMarche, V.C. (1973) 'Holocene climatic variations inferred from treeline fluctuations in the White Mountains, California', *Quaternary Research*, 3: 632–60.

LaMarche, V.C., Graybill, D.A., Fritts, H.C. and Rose, M.R. (1984) 'Increasing atmospheric carbon dioxide: tree ring evidence for growth enhancement in natural vegetation', *Science*, 225: 1019–21.

Little, R.L. (1993) 'Subalpine tree regeneration following fire: effects of climate and other factors', Master's Thesis, University of Washington, Seattle, Washington.

Lowery, R.F. (1972) 'Ecology of subalpine zone tree clumps in the North Cascade Mountains of Washington', Ph.D. Thesis, University of Washington, Seattle, Washington.

Payette, S. and Filion, L. (1985) 'White spruce expansion at the tree line and recent climatic change', *Canadian Journal of Forest Research*, 15: 241–51.

Peterson, D.L. (1991) 'Sensitivity of subalpine forests in the Pacific Northwest to global climate change', *The Northwest Environmental Journal*, 7: 349–50.

Peterson, D.L., Arbaugh, M.J., Robinson, L.J. and Derderian, B.R. (1991) 'Growth trends of whitebark pine and lodgepole pine in a subalpine Sierra Nevada forest, California, U.S.A', *Arctic and Alpine Research*, 22: 233–43.

Peterson, D.W. (1993) 'Dendroecological study of subalpine conifer growth in the

North Cascade Mountains', Master's Thesis, University of Washington, Seattle, Washington.

Rochefort, R.M. and Peterson, D.L. (1991) 'Tree establishment in subalpine meadows of Mount Rainier National Park', *The Northwest Environmental Journal*, 7: 354–5.

Rochefort, R.M., Little, R.L., Woodward, A. and Peterson, D.L. (1994) 'Changes in subalpine tree distribution in western North America: a review of climate and other factors', *The Holocene* (in press).

Scuderi, L.A. (1987) 'Late-Holocene upper timberline variation in the southern Sierra Nevada', *Nature*, 325: 242–4.

Vankat, J.L. and Major, J. (1978) 'Vegetation changes in Sequoia National Park, California', *Journal of Biogeography*, 5: 377–402.

Woodman, J.N. (1987) 'Potential impact of carbon dioxide-induced climate changes on management of Douglas-fir and western hemlock', in W.E. Shands and J.S. Hoffman (eds), *The Greenhouse Effect, Climate Change, and U.S. Forests*, The Conservation Foundation, Washington, DC, pp. 277–83.

Woodward, A., Gracz, M.B. and Schreiner, E.S. (1991) 'Climatic effects on establishment of subalpine fir (*Abies lasiocarpa*) in meadows of the Olympic Mountains', *The Northwest Environmental Journal*, 7: 353–4.

15

DYNAMICS AND FUNCTIONING OF *RHODODENDRON FERRUGINEUM* SUBALPINE HEATHLANDS (NORTHERN ALPS, FRANCE)

Mineral nitrogen availability in the context of global climate change

A. Pornon and B. Doche

INTRODUCTION

It is now known and accepted that global climatic change could modify the dynamics and the functioning of mountain ecosystems. Indeed, at high altitudes, climate is considered the major limiting factor for many plants. The pathways by which this change in climate could act upon plant communities, both directly or indirectly, are numerous. Example of this are the effects of increased levels of atmospheric CO_2 on photosynthesis, and the effects of temperature increase on the metabolism or the absorption of nutrients by the plants, etc. The manner in which climatic change could act by indirect pathways is not as yet very well studied. However it has been shown that temperature and humidity may affect biological cycles (Blackman, 1936; Dadykin, 1958; Bonneau, 1980; Tavant, 1986; Roze, 1986) – mainly the nitrogen cycle. The quantity of available nitrogen acts on plant populations by means of their biomass, productivity, reproduction, and intra-specific competition (Aarssen and Burton, 1990; Bonneau, 1980; Le Tacon and Millier, 1970; Le Tacon, 1972; Vermeer and Berendse, 1983; Roze, 1986; Hull and Hooney, 1990). In turn, soil temperature and humidity can affect nitrogen availibility by either of two pathways.

(1) Direct pathway – Blackman (1936) and Dadykin (1958) have found nitrogen uptake by plants to be inhibited by low spring soil temperature. Conversely, high spring soil temperature can allow a better uptake.

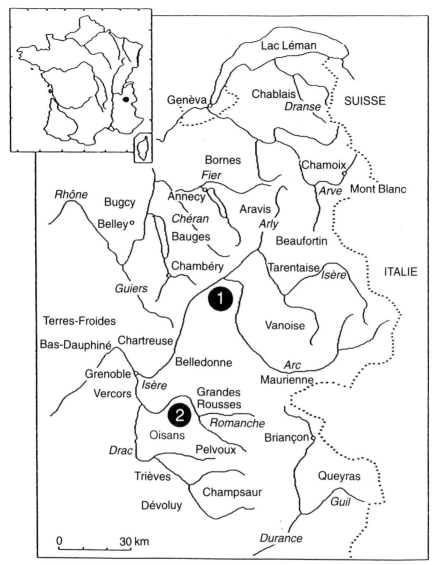

Figure 15.1 The study sites in the northern Alps (France): 1 – Belledonne;
2 – Taillefer.

(2) Indirect pathway – in the moist, cold and acid soils (that is to say high-altitude soils) such processes as ammonification and above all nitrification are reduced. As a result, the quantity of mineral nitrogen (which is the form mostly used by plants) is reduced. In addition, this mineral nitrogen occurs principally as $N-NH_4^+$ ions with only negligible

245

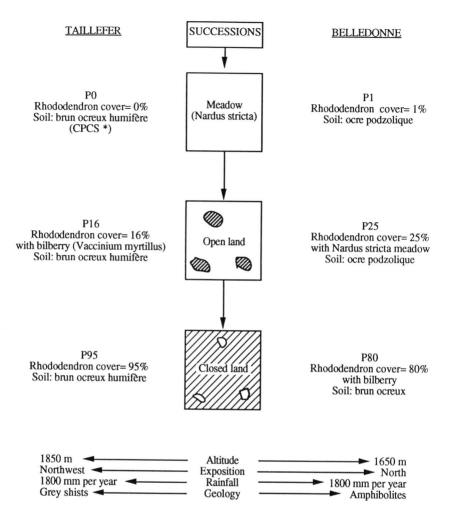

TAILLEFER   SUCCESSIONS   BELLEDONNE

P0
Rhododendron cover= 0%
Soil: brun ocreux humifère
(CPCS *)

Meadow
(Nardus stricta)

P1
Rhododendron cover= 1%
Soil: ocre podzolique

P16
Rhododendron cover= 16%
with bilberry (Vaccinium myrtillus)
Soil: brun ocreux humifère

Open land

P25
Rhododendron cover= 25%
with Nardus stricta meadow
Soil: ocre podzolique

P95
Rhododendron cover= 95%
Soil: brun ocreux humifère

Closed land

P80
Rhododendron cover= 80%
with bilberry
Soil: brun ocreux

1850 m	Altitude	1650 m
Northwest	Exposition	North
1800 mm per year	Rainfall	1800 mm per year
Grey shists	Geology	Amphibolites

Figure 15.2 Description of sites and the successions of vegetation.

amounts of nitrates being present (Bonneau, 1980; Tavant, 1986).

As a result of global climatic change, the soil temperature will increase the amount of mineral nitrogen, and therefore the balance of the $N-NH_4^+$ and $N-NO_3^-$ ions will be modified. Thus, interactions between particular plant communities could be affected. Indeed, the ability of plants to absorb $N-NH_4^+$ or $N-NO_3^-$ varies. It has been observed that woody plants, such as spruce, are able to use nitrogen in both $N-NH_4^+$ and $N-NO_3^-$ forms (Hoffmann, 1966; Hoffmann and Fiedler, 1966), while grasses are only able to use $N-NO_3^-$ ions. More recently though, it has been observed that if $N-NO_3^-$ ions are used more by woody plants, it is only in the case where its soil concentration is greater than the concentration of the $N-NH_4^+$ (Boxman and Roelofs, 1988; Scheromm and Plassard, 1988). All these investigations show that mineral nitrogen forms clearly influence the developmental processes of high-altitude plant communities.

In this paper, we show how soil temperature acts on the amounts of both $N-NH_4^+$ and $N-NO_3^-$ present in the soil and how that in turn could influence the productivity and the dynamics of the *Rhododendron ferrugineum* populations of two similar successions; from meadow to closed heathland in the subalpine level.

SPECIES STUDIED AND STUDY SITE

Rhododendron ferrugineum is an evergreen shrub (Ericaceous) with a height of 70 cm, which grows at the subalpine level from about 1,600 to 2,200 m. In the northwestern Alps, it forms very large heathlands on the north and northwest facing slopes.

The study sites are 'Collet d'Allevard' (45° 25'N, 6° 10'E) in the Belledonne massif, and 'Côte des Salières' (45° 4'N, 5° 58'E) in the Taillefer massif (Figure 15.1), in France.

At each study site, we studied the succession from meadow to closed heathland, the environmental features of which are summarized in Figure 15.2.

METHODS

The demography of the *Rhododendron* populations was carried out by studying the age of all the stands within a 100 m² area in six different plots both at Belledonne and at Taillefer. The age of stands is given by counting the annual rings on the widest stem of each individual.

For the productivity of shoots (1992), the results were obtained from collecting material in twenty 6.25 dm² squares along three separate transects in each plot of the two sites. The space between two squares was 2 m.

In order to calculate the productivity of the shoots, the material from each square was dried at 105°C and then weighed. The data shown in Table 15.1 are the averages of twenty samples. For all the averages, the confidence interval was calculated with an error risk of 5 per cent.

The soil mineral nitrogen content was studied as follows: the amounts of $N-NO_3^-$ and $N-NH_4^+$ were measured both in undisturbed humus and in humus samples which had been incubated *in situ* for one month. These measures were repeated each month for three months during the plant's growing season. The samples were incubated in order to prevent leaching by rainfall and uptake by the plants of the mineral nitrogen. The data concerning

Table 15.1 Current-year shoot productivity (g of dry weight /m^2) for different *Rhododendron* populations (p = 0.05)

Plots	P 25	P 80	P 16	P 95
Shoots (g DW/m^2)	329.1 ± 28.83	417.21 ± 46.52	229.56 ± 34.99	254.09 ± 34.53
Fruit (%)	4.6	2.75	6	2
Leaves and Twigs	95.5	97.25	94	98
		Taillefer		Belledonne
Meadow Productivity (g DW/m^2)		580		100

the undisturbed humus was obtained from ten samples of the upper soil level (5 cm), while for incubated humus the data was obtained from four incubations. For the incubation of the humus, we used a method similar to one proposed by Roze (1986). $N-NH_4^+$ and $N-NO_3^-$ were extracted according to the method of Wheatley *et al.* (1989). The amount of $N-NH_4^+$ was measured by the blue indophenol method (Nelson and Dorich, 1983) and the amount of $N-NO_3^-$ was measured by High Performance Liquid Chromatography (HPLC, Dionex 4500 i).

Soil temperature was measured every week during the growing season with hygrometic probes Wescor POT–55 connectable to a Wescor HR–33T microvoltmeter buried in the soil at depths of 10 and 25 cm.

RESULTS

The dynamics of the *Rhododendron* populations

A *Rhododendron ferrugineum* heathland, which is not in competition with other offensive woody species, needs 150 years to extend throughout a meadow area and to occupy 80–90 per cent of it (Pornon, 1993).

In order to compare the colonization efficiency of *Rhododendron*, we calculated the population growth rate (Figure 15.3).

In P 80, the growth rate is constant with time and the gradient reaches 1.4. In almost closed heathlands, the installation of new individuals is possible and the colonization is continued by vegetative means.

Early in the colonization in P 25, the growth rate is smaller than that in P 80 because cattle probably have been a controlling influence on the youngest stands.

At Taillefer, in P 16, the growth rate is smaller than in P 80 and P 25 because the environmental factors are not so well suited to the *Rhododendron*.

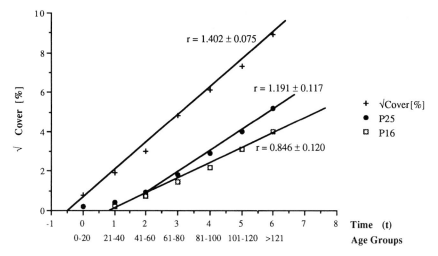

Figure 15.3 Population growth rate for Belledonne (P 80, P 25) and for Taillefer (P 16).

Current-year shoot productivity

Table 15.1 shows shoot productivity for 1992. It may be observed that the productivity shows the same pattern as the dynamics of *Rhododendron*; the plots in which the dynamics are more efficient are the ones in which the productivity is the greatest. In P 80, the shoot productivity was 417.21 g DW/m^2 for the year 1992 and for P 25 and P 16, the shoot productivity was 329 g DW/m^2 and 229 g DW/m^2 respectively.

To summarize: the three following factors influence the development and functioning of the heathlands at the different study sites:

- The degree of population cover: in closed heathlands the productivity is higher than in open heathlands.
- The grazing pressure: in P 80 the slope is steeper. Consequently it was not easily accessible to animals. In the past, therefore, the conditions for the development of the *Rhododendron* were more favourable.
- The pedological characteristics: at Belledonne the pedological characteristics are much more suited to the *Rhododendron* than at Taillefer .

The soil mineral nitrogen

The highest mineral nitrogen content of the incubated samples was found at Belledonne (Figure 15.4) during the period June–July. It can also be observed that nitrogen is mostly in the form of $N\text{-}NH_4^+$. Open heathlands produce the greatest amounts of this element (30 mg/kg DW of soil) whereas closed

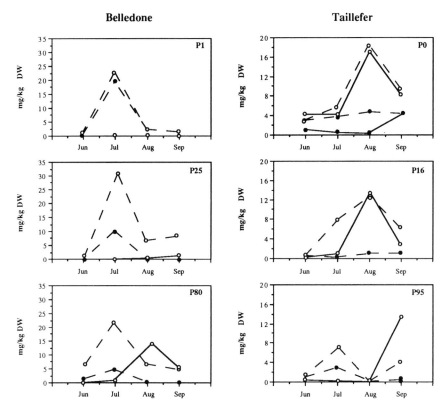

Figure 15.4 Nitrogen concentration (dashed line – as N-NH$_4^+$; continuous line – as N-NO$_3^-$) in humus incubation 'in situ' during one month (open circles); in undisturbed humus (closed circles). All samples taken on the 12th of each month.

heathlands produce the smallest (20 mg/kg DW of soil) conversely to its higher biomass production.

In the undisturbed humus layer of the meadow, the concentration of N-NH$_4^+$ (20 mg/kg DW) was found to be similar to that of the incubated humus (24 mg/kg DW). These results indicate that for this period at Belledonne, the meadow plants either cannot at all, or can but only in very small quantities, use N-NH$_4^+$ ions as a nutrient. On the contrary, in the open and closed heathlands, the results show that *Rhododendron* can readily use this nutrient element. The concentration of N-NH$_4^+$ in the undisturbed humus of the open and closed heathlands (9 and 4 mg/kg DW of soil respectively) were found to be very different to that of the incubated humus (31 and 21 mg/kg DW of soil). There is a loss of N-NH$_4^+$ in the meadow soils during July–August which is probably caused by the element being taken up by bacterial biomass during this period. Indeed, in acid soils

N-NH$_4^+$ is not transformed into NH$_3$ gas (Roze, 1986) and, as is well known, it is in the N-NH$_4^+$ form that the nitrogen is absorbed by the bacterial biomass in the case where both N-NH$_4^+$ and N-NO$_3^-$ coexist. This 'reorganization' of N-NH$_4^+$ is likely to occur in the soils of *Rhododendron* heathlands but its effects are not evident. However, there is probably strong competition between the bacteria and Ericaceous plants for this nitrogen nutrient.

The results also show that nitrates remained at a constantly low level (except for the July–August period in closed heathland) in both the undisturbed humus and the incubated humus (< 4 mg/kg DW). Soil nitrate concentration is dependent on spring nitrification which is in turn dependent on the soil spring temperatures. Therefore, if the temperatures are too low, they prevent nitrification and this results in a too low soil nitrate concentration.

The total amounts of mineral nitrogen measured for Taillefer and Belledonne (Figure 15.4) were similar (30 mg/kg DW of soil), and at both sites the closed heathlands had the lowest available nitrogen.

At Taillefer however, there were differences in the functioning of the soil: (i) the peak of mineralization (July–August period) was late and occurred at the end of the *Rhododendron* vegetative season; and (ii) in the meadow and open heathland, the nitrification was greater by a factor of 50 than for meadow and open heathland at Belledonne as soil N-NO$_3^-$ concentration was clearly higher.

Soil temperature and nitrogen mineralization

Figure 15.5 shows the soil temperature during the growing season of 1991 and 1992 for the plots at Belledonne.

It can be seen that the underground temperature for both 1991 and 1992 is lower for closed heathlands than for meadows and such is the trend for nitrogen mineralization. In all the plots, during the second part of plant growth (August), the temperature of the soil was lower by 3 or 4°C in 1992 than in 1991.

It can be seen (Figure 15.6) that in 1992 high N-NH$_4^+$ concentration corresponds to a low nitrate concentration. On the contrary, during 1991, due to an increase in soil temperature, there was a low N-NH$_4^+$ concentration and a high nitrate concentration.

These results indicate that at high altitudes, the soil temperature strongly acts on nitrification.

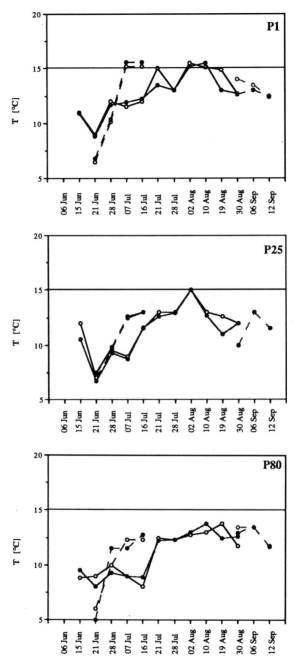

Figure 15.5 Soil temperature for Belledonne. Dashed line – year 1991; continuous line – year 1992; open circles – 10 cm depth; closed circles – 25 cm depth.

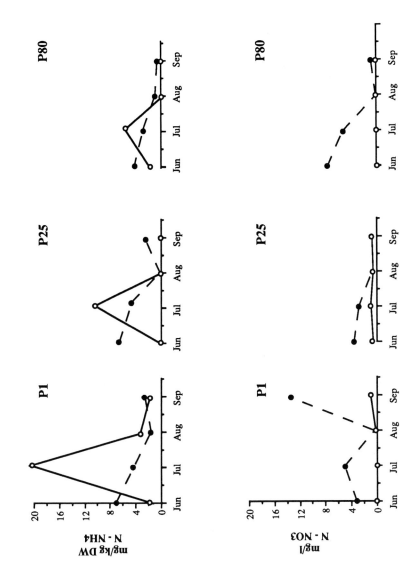

Figure 15.6 Nitrogen concentration as N-NO$_3^-$ and as N-NH$_4^+$ in soils of the Belledonne. All samples taken on the 12th of each month. Closed circles – year 1991; open circles – year 1992 (p = 0.05).

DISCUSSION

Mineral nitrogen and the biology of the *Rhododendron* population

Much research has demonstrated the interactions between the availability of nitrogen and the development of plant populations. The ecosystems at high altitudes, when compared with ecosystems at low altitudes (Brittany heathlands, Roze, 1986) have a much lower soil nitrogen concentration by a factor of between two and six according to the plant communities (Bonneau, 1980).

Some researchers (Bonneau, 1980; Tavant, 1986) have indicated that in mountain milieus, nitrogen is mainly in the form of $N-NH_4^+$ ions. Our own results for 1992 showed that in reality, however, the situation is more complicated.

At Taillefer, mineral nitrogen exists both as $N-NH_4^+$ and $N-NO_3^-$ in meadows and in closed heathlands, while it is only present as $N-NH_4^+$ at Belledonne. During 1992, the soil humidity and temperature were not significantly different between the two sites and therefore do not explain the above differences between Taillefer and Belledonne. On the contrary, the soil's pH (H_2O) is clearly higher at Taillefer (from 4.7 to 5.5 according to the soil level and the plots) than at Belledonne (from 3.8 to 4.2). pH is often noted as being an influential factor acting on mineralization and above all on nitrification (Le Tacon, 1972; Roze, 1986; Salsac *et al.*, 1987).

Soil temperature is an another influencing parameter. In the case when spring soil temperatures are higher (as for 1991 compared to 1992), even in acid soils (Belledonne), nitrification occurs, so nitrate concentration therefore increases.

These variations of different mineral nitrogen forms are very important because:

(1) The ability of the plant to absorb $N-NH_4^+$ and $N-NO_3^-$ is different. For Gramineous plants, $N-NO_3^-$ is generally the main nutrient (Salsac *et al.*, 1987) and the environmental factor most limiting to growth of annual grasses is the availability of nitrate nitrogen (Hull and Muller, 1977; Jenny *et al.*, 1980). On the contrary, woody plants are able to absorb both $N-NH_4^+$ and $N-NO_3^-$ as nutrients (Boxman and Roelofs, 1988; Scheromm and Plassard, 1988; Hoffmann, 1966; Hoffmann and Fiedler, 1966). On the other hand, Le Tacon (1972) noted that the growth of Ericaceous plants is better on soils with $N-NH_4^+$ compared to soils with $N-NO_3^-$. Our own results illustrate in detail these two different pathways of plant nutrition. In comparison with Belledonne, where mineral nitrogen was essentially as $N-NH_4^+$, the productivity of the Taillefer's meadow was greater as mineral nitrogen was equally distributed as $N-NO_3^-$ and as $N-NH_4^+$. Taillefer's heathlands had smaller productivities than those at Belledonne, which is likely as $N-NH_4^+$, being the favoured nutrient for *Rhododendron*, was present in a smaller

254

concentration. The lower productivity of Taillefer's heathlands could not be explained only by the low levels of $N\text{-}NH_4^+$ but also by the possible toxicity of $N\text{-}NO_3^-$.

(2) The stability of $N\text{-}NH_4^+$ and $N\text{-}NO_3^-$ in the soil is different. $N\text{-}NH_4^+$ constitutes a relatively stable pool and indeed in acid soil its leaching by rainfull, 'reversion' in clay, and disappearance as NH_3^+ gas are very limited, while $N\text{-}NO_3^-$ is able to disappear by leaching and the 'denitrification' pathway.

Global climatic change and ecosystem functioning

It is evident that with global climatic change, as the soil temperature increases, the balance between the two mineral nitrogen ions will be

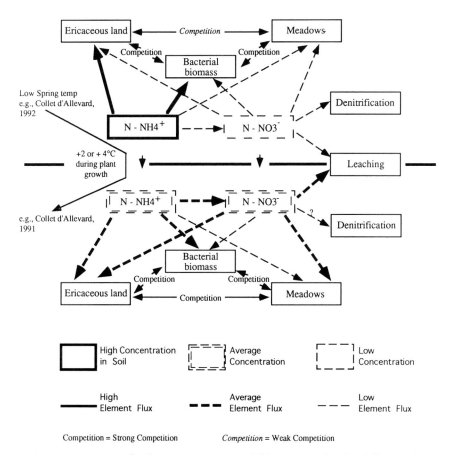

Figure 15.7 Action of soil temperature on available nitrogen flux for different plant populations.

modified. A pattern may be proposed such that soil temperature could act on the nitrogen flux and interact between particular mixed plant communities (Figure 15.7). With low spring temperatures, mineral nitrogen is essentially as $N-NH_4^+$ and consequently there is competition between the Ericaceous and the soil bacteria and there is little nitrate available for the meadow grasses. If soil temperature increases about 2–4°C during the growing season, $N-NH_4^+$ concentration decreases and $N-NO_3^-$ concentration increases. $N-NO_3^-$ is mainly used by grasses, but there exists a strong competition between Ericaceous, grasses and soil bacteria. Where this occurs a lower productivity of the heathlands also takes place.

The division of nutrients between different species is a probable explanation for why the closed heathlands, in which *Rhododendron* is almost the single species, are more productive than open heathlands in which *Rhododendron* coexists with many other species.

Another effect of the soil temperature increase is that mineralization occurs earlier during the growing season and therefore so does the increase in the availability of mineral nitrogen. This effect particularly concerns sites such as Taillefer where the peak of mineralization occurs at the end of the shoot growth period.

As the $N-NO_3^-$ pool is unstable, the increase in temperature could cause a considerable loss of this element by leaching and denitrification.

CONCLUSION

In a high-altitude environment, climatic conditions are critical to plant communities. Slight fluctuations in temperature and humidity over a long-term period can largely affect the functioning of mountain ecosystems. Our results for the years 1991 and 1992 show the effects of temperature on the nitrogen cycle, nitrogen being a major plant nutrient. With high spring temperatures, the $N-NO_3^-$ soil concentration increases while the concentration of $N-NH_4^+$ decreases. It is expected that this situation would lead to a strong competition for nitrogen between the Ericaceous and other plant species, particularly because there is also considerable loss by leaching and denitrification. This is an example of how climatic change could affect the productivity and dynamics of a heathland.

This suggested course of events does not apply to N-flux as it is thought that, with the help of mycorrhizae, *Rhododendron* is able to use organic nitrogen as is seen with *Calluna vulgaris* (Pearson, 1971; Pearson and Read, 1973a and b, 1975; Stribley and Read, 1974a and b). In this case, there is no competition.

The difficulty involved in deducing the long-term effects of climatic change on mountain ecosystems is having only short-term analysis methods at our disposal. In order fully to understand the consequences of global climatic change, long-term studies would be necessary.

ACKNOWLEDGEMENTS

This research was subsidized by EGPN Contract (n°92 217: 'Impacts des changements climatiques sur les lisières d'altitude'). We wish to thank C. Denardo for her participation in the field and for the English translation, and also G. Girard, A. and J. P. Guichard for realizing the technical aspects of this article.

REFERENCES

Aarsen, L.W. and Burton, S.M. (1990) 'Maternal effects at four levels in Senecio vulgaris (Asteraceae) grown on a soil nutrient gradient', *Americ. J. Bot.*, 77 (9): 1231–40.

Blackman, G.E. (1936) 'The influence of temperature and available nitrogen supply on the growth of pasture in the spring', *J. Agric. Sci.*, 26: 620–47.

Bonneau, M. (1980) 'Production d'azote mineral dans divers types de landes du Massif Central', *Ann. Sci. Forest.*, 37 (3): 173–88.

Boudot, J.P. (1982) 'Relation entre l'altération minérale et le cycle de l'azote sur matériel chloriteux du massif schisto-grauwackeux vosgien', Thèse d'Etat Sci. Nat., Université Nancy I.

Boxman, A.W. and Roelofs, J.G.M. (1988) 'Some effects of nitrate versus ammonium nutrition on the nutrient fluxes in Pinus sylvestris seedlings. Effects of mycorrhizal infection', *Can. J. Bot.*, 66: 1091–7.

C.P.C.S. (1976) *Communication de Pédologie et de Cartographie du sol. Classification des sols*, ENSA, Grignon.

Dadykin, V.P. (1958) 'Plant physiological research problems of the far North', *Probl. North* (National Research Council of Canada), 1: 205–16.

Hoffmann, F. (1966) 'Untersuchungen zür stickstoffernährung junger Koniferen. II – Die aufnahme von ammonium und nitratstickstof durch fichtensamlinge unter verschiedemen', *Bedingungen, Arch. F., Forstwesen*, 15 (10): 1093–103.

Hoffmann, F. and Fiedler, H. (1966) 'Die stickstoffernährung junger Koniferen', *Biologische Rundschau*, 4 (3): 138–55.

Hull, J.C. and Hooney, H.A. (1990) 'Effects of nitrogen on photosynthesis and growth rates of four California annual grasses', *Acta Oecologica*, 11 (4): 453–68.

Hull, J.C. and Muller, C.H. (1977) 'The potential for dominance by Stipa pulchra in a California grassland', *Am. Midl. Naturalist*, 97: 147–75.

Jenny, H., Ulamrs, J. and Martin, W.E. (1980) 'Greenhouse assay of fertility of California soils', *Hilgardia*, 20: 1–18.

Le Tacon, F. (1972) 'Disponibilité de l'azote nitrique et ammoniacal dans certains sols de l'Est de la France. Influence sur la nutrition de l'épicéa commun', *Ann. Sci. Forest.*, 30 (2): 183–203.

Le Tacon, F. and Millier, C. (1970) 'La nutrition minérale de l'épicéa commun en sols carbonatés et en sols décarbonatés; essai sur le comportement du calcaire et du manganèse', *Ann. Sci. Forest.*, 27 (1): 63–88.

Nelson, D.W. and Dorich, R.A. (1983) 'Dosage de l'ammonium', *Soil Sci. Soc. Am. J.*, 47: 833–6.

Pearson, V. (1971) 'The biology of the mycorrhiza in the Ericaceae'. Ph.D. Thesis, University of Sheffield.

Pearson, V. and Read, D.J. (1973a) 'The biology of mycorrhiza in the Ericaceae. I – The isolation of the endophyte and synthesis of mycorrhizas in aseptic culture', *New Phytol.*, 72: 371–9.

—— and —— (1973b) 'The biology of mycorrhiza in the Ericaceae. II – The transport of carbon and phosphorus by endophyte and the mycorrhiza', *New Phytol.*, 72: 1325–31.

—— and —— (1975) 'The physiology of the mycorrhizal endophyte of Calluna vulgaris', *Trans. Br. Mycol. Soc.*, 64 (1): 1–17.

Pornon, A. (1993) 'Dynamique des landes subalpines à Rhododendron ferrugineum; exemple du Collet d'Allevard (Massif de Belledonne, Alpes du Nord)', *116ème Congrès des Sociétés Savantes*, Chambéry, 1991, pp. 1–12.

Roze, F. (1986) 'Le cycle de l'azote dans les landes bretonnes', Thèse de Doctorat ès Sci. Nat., Université Rennes.

Salsac, L., Chaillou, S., Morot-Gaudry, J.F., Lesaint, C. and Jolivet, E. (1987) 'Nitrate and ammonium nutrition in plants', *Plant Physiol. Biochem.*, 25 (6): 805-12.

Scheromm, P. and Plassard, C. (1988) 'Nitrogen nutrition of no mycorrhized pine (Pinus pinaster) grown on nitrate or ammonium', *Plant Physiol. Biochem.*, 26: 261–9.

Stribley, D. and Read, D.J. (1974a) 'The biology of mycorrhiza in the Ericaceae. III – Movement of carbon-14 from host to fungus', *New Phytol.*, 73: 731–41.

—— and —— (1974b) 'The biology of mycorrhiza in the Ericaceae. IV – The effect of mycorrhizal infection on uptake of ^{15}N from labelled soil by Vaccinium macrocarpon', *New Phytol.*, 73: 1149–55.

Tavant, Y. (1986) 'Dynamique saisonnière des matières organiques et de l'azote des sols forestiers brunifiés et calcimagnésiques humifères des séquences bioclimatiques du Jura', Thèse d'Université, Sciences de la vie, University de Besançon.

Vermeer, J.G. and Berendse, F. (1983) 'The relationship between nutrient availability, shoot biomass and species richness in grassland communities', *Vegetatio*, 53: 121–6.

Wheatley, R.E., Macdonald, R. and Macsmith A. (1989) 'Extraction of nitrogen from soils', *Biol. Fertil. Soils*, 8: 189–90.

16

A DENDROCHEMICAL PERSPECTIVE ON THE EFFECTS OF CLIMATE CHANGE IN THE OZARK HIGHLANDS

R. Guyette

INTRODUCTION

Predicting the effects of global change on vegetation is an important aspect of climate change research. Here the response of soils and vegetation in the Ozark Mountains to atmospheric change is evaluated through the window of tree-ring chemistry. Three examples utilizing tree-ring chemistry will be used to predict the chemical sensitivity of soils and vegetation in mountain environments, that is:

- drought-induced calcium deficiency;
- the serial variance of element concentrations in xylem growth increments;
- and soil pH change.

The objectives of this paper are to examine the chemical sensitivity of soils and examine why soils in mountain areas may be particularly sensitive to atmospheric change. Element concentrations in radial growth increments of eastern red cedar (*Juniperus virginiana*) will be used to illustrate changes in the chemistry of soils and trees. The work presented here is a dendrochemical perspective based on methods and data from several related studies (Guyette and McGinnes, 1987; Guyette *et al.*, 1989, 1991, 1992a, 1992b; Guyette and Cutter, 1994; Cutter and Guyette, 1994).

Dendrochemistry

Research using element concentrations in tree rings to investigate environmental change was initiated within the last 30 years. The term dendrochemistry (Amato, 1988) has been coined to describe the combination of dendrochronology and elemental analyses of tree rings. Studies of the distribution of toxic metals such as Pb and Cd (Baes and Ragsdale, 1981; Guyette *et al.*, 1991; Hagemeyer and Breckle, 1986; Symeonides, 1979) in tree radii is used to monitor environmental pollution. Other trace metals, such as Fe, Al, and Cu and Mo in tree rings have been used as indicators of pollution (Baes and McLaughlin, 1984; Guyette *et al.*, 1987). Changes in Ca and Mg availability were examined in relation to acid deposition (Bondietti *et al.*, 1989, 1990). The effects of liming on radial element concentration was documented by McClenahen *et al.* (1989). Kairiukstis and Kocharov (1990) associated changes in climate with radial changes in potassium and ring-width.

Geography and climate

Mountainous areas of the Ozark Highlands are the St Francis Mountains, the eroded Ozark Plateau, and the Boston Mountains. The St Francis Mountains are composed of volcanic and igneous rocks. The Boston Mountains and Ozark Plateau are uplifted and dissected plateaux composed mainly of limestones, dolomites, and sandstones. A maximum elevation of nearly 700 m is reached in the Boston Mountains of Arkansas and a minimum elevation of 116 m is found near the drainages of the eastern border of Ozark Plateau. All Ozark soils are unglaciated.

The climate is continental-humid. Very little orographic lifting by mountains occurs. The Great Plains are located 700 km to the west – the direction of the prevailing winds and storm tracts. Annual precipitation in the Ozarks ranges from 91 cm in the north to 117 cm in the south. Spring is the wettest season, followed by summer, fall, and winter. Summer drought is an important limiting factor for plant growth. Temperatures range from 32 to $-30\,°C$.

The sensitivity of mountain soils

Many soils in the Ozark Mountains are particularly sensitive to atmospheric change for three reasons, namely:

- they lie on steep slopes,
- they are shallow to bedrock,
- they have a high land surface to volume ratio.

Steep slopes are greatly influenced by changes in radiation, precipitation, and temperature. The soils of slopes with south and west aspects are especially sensitive to changes in radiation and temperature because of their angle of incidence. Increases in precipitation can accelerate the rate of nutrient cycling on slopes by increasing the leaching of water-soluble ions and compounds.

Soils in the Ozark Highlands are often shallow to bedrock because of steep slopes and high rates of erosion. The low soil volume limits the nutrient pool and water-holding capacity. The most important aspect of the shallow soils that concerns atmospheric change is the high land surface to volume ratio of mountain soils. Atmospheric inputs will be proportionally greater per mass unit to thin soils than deep soils. Surface area is a critical factor for the time of completion of any chemical reaction involving solids. The high exposure, per unit volume, of shallow mountain soils to the atmosphere increases their sensitivity to changes in temperature, precipitation volume and chemistry, and dry chemical inputs. Temperature, a major factor controlling chemical reactions and biological activity, can have a much greater effect on thin soils near the surface.

MATERIALS AND METHODS

Eastern red cedar is a coniferous evergreen which grows throughout eastern North America. This species has many characteristics that make it suitable for dendrochemistry. Old growth red cedar (300 to 900 years) can be found on the steep slopes of several different bedrock substrates.

Translocation of elements between annual growth increments is species and element dependent (Cutter and Guyette, 1994). Eastern red cedar has many characteristics that minimize radial translocation of elements. Xylem characteristics that minimize radial translocation of elements within the heartwood include:

- a dry (< 25 per cent moisture content) heartwood,
- low radial heartwood permeability,
- aspirated pits,
- a distinct heartwood–sapwood boundary (McGinnes and Dingeldein, 1969).

Samples (one increment core per tree) were taken with a 17 mm self-cleaning electric hole saw from a location on the tree which was defect-free above and below the boring site. All cores were cross-dated and dried before instrumental analyses. The sapwood was separated from the heartwood and analysed as a discrete unit. The very narrow sapwood–heartwood transition zone was included in the sapwood sample. The heartwood growth increments were divided into 20-year sections centred on the first year of odd decades for chemical analysis.

The soil under each tree was sampled to estimate the bulk soil pH. A soil auger was used to collect 50 g samples of soil. Composite samples were collected from each horizon and stored in labelled plastic bags in a freezer. After removing the litter layer, each of three soil horizons was sampled: an upper horizon with significant organic matter accumulation (typically an A horizon) (0–13 cm), an eluvial horizon (typically an E horizon) (13–24 cm), and an illuvial horizon (typically an argillic B horizon) (24–45+ cm). Horizons were distinguished by colour, texture, and structure. When distinct horizons were not visually identifiable, composite samples were taken at the surface (0–10 cm), midway to bedrock, and at the bedrock surface. Soil was thawed and was sifted through a #10 mesh screen before pH and Mn determinations were made.

Soil acidity was measured as pH with a glass Ag/AgCl combination electrode with a KCl reference electrode. Ten-gramme samples of soil were mixed with 10 ml of 0.01 M $CaCl_2$. The solution was then stirred at ten-minute intervals, and the pH was measured after thirty minutes. Individual (unweighted) horizon pH values were averaged together for each site to establish a mean pH.

Xylem analyses were done by splitting out 0.5–1.0 g samples from cores. Wood was digested in nitric-perchloric acid for elemental analyses by inductively coupled plasma optical emissions spectroscopy (ICP) with a Jarrell-Ash Model 1100 Mark III, controlled by a Digital Equipment Company 11/23+ computer. Standards were run every 10–15 samples.

RESULTS AND DISCUSSION

Climate-induced Ca deficiency

Drought-induced Ca deficiency demonstrates how climate affects plant nutrition in the Ozarks. Calcium uptake by plants is primarily non-metabolic and controlled by the bulk flow of the soil solution to the plant roots, and thus Ca uptake is dependent on soil moisture and transpiration. Water uptake has been correlated with the uptake of Ca in agricultural species (Lazaroff and Pitman, 1966; Tibbitts, 1979) and trees (Vigouroux *et al.*, 1989). The non-metabolic uptake of these elements via bulk flow through the roots rests on sound physiological and empirical considerations (Barber, 1986).

Since Ca nutrition is linked to climate by soil moisture and transpiration, Ca availability and deficiency in low Ca soils is particularly sensitive to climate. The distribution of Ca concentrations in the sapwood of eastern red cedar growing over rhyolite substrate in the St Francis Mountains of the Ozark Highlands reflects the low availability of Ca in some soils (Figure 16.1). Dieback of apical meristematic tissue and cell wall deformation is a symptom of Ca deficiency (Salisbury and Ross, 1992). Xylem trachied cells and heartwood anomalies occur in trees with the apical dieback and low sapwood Ca. Recent cell and heartwood anomalies occurred during the drought years of 1980 and 1983. These xylem anomalies are characterized by malformed cell walls and a break in the normal formation of heartwood. These xylem anomalies link drought years and Ca deficiency.

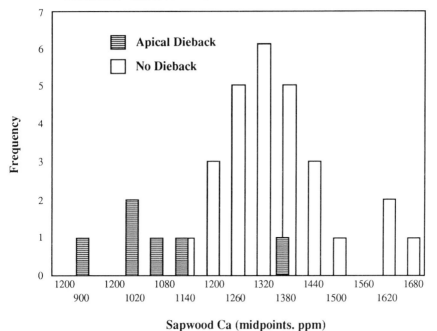

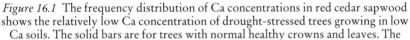

Figure 16.1 The frequency distribution of Ca concentrations in red cedar sapwood shows the relatively low Ca concentration of drought-stressed trees growing in low Ca soils. The solid bars are for trees with normal healthy crowns and leaves. The white bars are for trees that exhibit recent meristematic dieback of the branches and leaves as well as anomalous heartwood formation.

This example of drought-induced nutrient deficiency has important implications for the prediction of the response of vegetation to climate change. Climate change, whether characterized by increases or decreases in precipitation, will affect the availability of soil nutrients which have a strong non-metabolic uptake component such as Ca, Mg, and B. Soils and plants in which these essential nutrients are scarce will be affected by climate change first. Increased precipitation may allow the entry of new species into natural plant communities which previously had been excluded by a limiting mineral nutrient. Decreases in precipitation will decrease the availability of Ca, Mg, and B. This may limit growth and in extreme drought years cause mortality of species, such as eastern red cedar. Climate-induced Ca deficiency may be magnified by increases in tree size and competition. As the quantity of meristematic tissue increases, Ca requirements will be greater.

Heartwood element variability and bedrock

Climate, bedrock, and time interact in governing variability in soil characteristics. The rate of chemical weathering of bedrock is one of the critical factors in determining the sensitivity of shallow soils and associated vegetation to atmospheric change. When bedrock, such as fractured limestone, weathers rapidly it dominates the chemical characteristics of the soil. When bedrock, such as monolithic fine-grained granite, weathers slowly (that is, provides less chemical input per unit of time), atmospheric inputs such as dry deposition and precipitation can have a much greater influence on soil and tree chemistry. Bedrock that is readily subject to chemical weathering, should provide more temporal continuity to the soil environment of natural communities.

In the Ozarks, the temporal variability of element concentrations in tree rings over the last 300 years indicates a complex relationship to bedrock and climate (Table 16.1). The standard deviations of time series of strontium (Sr), barium (Ba), lead (Pb), and cadmium (Cd) from trees growing over rhyolite bedrock are greater than those for trees growing in dolomite-derived soils. The larger temporal variability in element concentrations in tree radii reflects a greater degree of atmospheric influence over the last 300 years on the chemistry of soils derived from rhyolite. The serial variability of Zinc (Zn) and magnesium (Mg) shows little, if any, difference between trees growing in soils over rhyolite and dolomite bedrock. Zinc is equally available (Guyette *et al.*, 1992a) in dolomite- and rhyolite-derived soils and tree means reflect this. Magnesium, on the other hand, is more than ten times as available in dolomitic soils as in rhyolite soils, yet the means and standard deviations are similar. This indicates a strong biological control of Mg uptake at these levels.

The temporal variability of xylem Ca concentrations by bedrock type is perhaps the most interesting in its relation to climate. Calcium concentrations from trees in soils derived from dolomite have greater temporal variability than those from trees in soils derived from rhyolite (Table 16.1). Thus, despite the more rapid chemical weathering of dolomite, which provides a stable, even excess supply of soil Ca, the temporal variability of Ca concentrations in xylem radii is greater for trees growing over dolomite than for trees growing over rhyolite. A possible explanation of the high Ca variability of growth increments lies in the non-metabolic uptake of Ca by plants (Barber, 1986) and the low availability of phosphorus in shallow dolomitic soils. Calcium uptake has been related to climate via mass flow uptake and transpiration (Kirkby, 1979; Vigouroux, *et al.*, 1989; Guyette *et al.*, 1991). Ca concentrations are tied to plant water uptake but only secondarily to plant growth. Because Ca uptake is not tied to growth (metabolic uptake), soil infertility limits the growth of trees on dolomite sites which are relatively low in phosphorus (Guyette *et al.*, 1992a) but does not

Table 16.1 Bedrock type, the number of trees, average number of heartwood increments analysed from each tree radius (N), mean element concentration in ppm (MEAN), and the average standard deviation (STD) of the time series of concentrations for eastern red cedar (*Juniperus virginiana*)

Element	Bedrock	Number of trees	Variable	Average
Zn	Rhyolite	20	N	9.30
			MEAN	0.48
			STD	0.28
Zn	Dolomite	29	N	12.75
			MEAN	0.41
			STD	0.26
Sr	Rhyolite	23	N	8.65
			MEAN	5.97
			STD	0.74
Sr	Dolomite	26	N	13.00
			MEAN	1.34
			STD	0.12
Mg	Rhyolite	23	N	8.65
			MEAN	74.84
			STD	12.11
Mg	Dolomite	27	N	11.44
			MEAN	81.97
			STD	12.97
Ca	Rhyolite	24	N	8.50
			MEAN	669.11
			STD	62.39
Ca	Dolomite	34	N	13.20
			MEAN	743.66
			STD	107.89
Cd	Rhyolite	23	N	9.04
			MEAN	0.02
			STD	0.03
Cd	Dolomite	27	N	12.44
			MEAN	0.00
			STD	0.00
Pb	Rhyolite	22	N	8.95
			MEAN	1.05
			STD	0.68
Pb	Dolomite	27	N	12.44
			MEAN	0.00
			STD	0.00

limit transpiration and Ca uptake. Thus, the carbon–Ca ratio is highly variable on these sites compared to the more fertile rhyolite sites.

The standard deviation of time series of element concentrations from tree rings is also statistically related to the mean radial element concentration, somewhat like the heteroscedastic nature of ring-width series (Cook *et al.*, 1990) but between trees. Table 16.2 gives the correlations among the means

Table 16.2 Correlation coefficients among the means and standard deviations of element time series from serial xylem increments of eastern red cedar. The number of time series correlated, one from each tree, is given

	Ca	Cd	Mg	Sr	Pb	Zn
Element correlation	0.44	0.90	0.51	0.71	0.85	0.72
No of trees	58	50	50	49	49	49

and standard deviations of time series of several elements from xylem growth increments.

Soil pH and atmospheric change

Changes in climate have been predicted to influence soil pH (Scharpenseel *et al.*, 1990). The acidity of some soils in the Ozark Mountains is particularly sensitive to changes in precipitation quantity and chemistry, temperature, and biological activity. Sensitivity varies greatly by substrate and the ratio of acidic to base cations on the exchange complex of soil particles.

Evidence for the pH sensitivity of some Ozark soils to atmospheric change comes from manganese concentrations in tree rings. Manganese concentrations in tree rings have been used to monitor long-term changes in soil pH (Guyette *et al.*, 1992b). Results have indicated that change in soil pH, for whatever reason, will be largest for soils in the transition between Ca and Al buffering. In the Ozarks, pH change over time is reflected in the increased variance of Mn time series from trees in soils that range in pH between 4.55 and 5.85. This change in soil pH was predicted by Reuss and Johnson (1986), in discussing the buffering capacity of a soil with respect to pH change, suggest that maximum pH depression per unit of H^+ input occurs at base saturation when transition from cation depletion to Al mobilization effects occur.

The implication for climate change in mountain environments is that pH will change first and foremost in soils between Ca and Al buffering with a low cation exchange capacity. Increases in precipitation and temperature, especially on soils with steep slopes, will cause acidification. On the other hand, decreases in precipitation may result in an increase in soil pH.

An important consequence of a reduction in soil pH illustrated by xylem element concentrations is the release from the soil of metals into the biosystem. The solubility, and consequently the uptake, of many toxic metals, such as Pb, is highly dependent on soil pH (Lindsay, 1986). Xylem concentrations of Pb in eastern red cedar are correlated ($r = -0.72$) with soil pH (Guyette, *et al.*, 1991). Uptake of many other elements (Table 16.3) into xylem is also highly correlated with soil pH. Climate-induced pH change

Table 16.3 Correlation coefficients among the pH of soil horizons and element concentrations (ln = natural log transformation) in eastern red cedar sapwood[1]

Variable	O–pH	A–pH	B–pH	M–pH
Trees on all substrates				
ln barium	−.70**	−.81**	−.75**	−.82**
	62	58	36	64
ln strontium	−.66**	−.74**	−.78**	−.76**
	65	61	37	67
ln manganese	−.64**	−.67**	−.75**	−.79**
	65	61	37	67
ln zinc	−.42**	−.25*	−.35*	−.35**
	65	61	37	67
calcium	−.03	−.06	−.01	−.02
	65	61	37	67
ln aluminium	−.47**	−.47**	−.59**	−.54**
	61	57	36	63
boron	−.38**	−.37**	−.41**	−.41**
	61	57	36	63
ln (Ca/Ba)	+.72**	+.81**	+.79**	+.83**
	61	58	35	63
Trees on dolomite				
ln magnesium	−.44*	−.25	−.44*	−.40*
	32	29	21	33
ln phosphorus	+.33	+.50**	+.41	+.39*
	32	29	21	33
iron	−.34	−.35*	−.54*	−.51**
	29	27	16	31

Notes: Elements that did not correlate with soil pH were Cu and K. O-pH is the pH of the upper, organic horizon (typically an A horizon) (0–13 cm), A-pH is the pH of the eluvial horizon (typically an E horizon) (13–24 cm), B-pH is the pH of the illuvial horizon (typically an argillic B horizon) (24–45+ cm). M-pH is the average pH of the three horizons. All substrates include rhyolite, dolomite, limestone, chert, and sandstone bedrock.
[1] The number (n) of trees is given below the correlation coefficients
* is the P ≤ 0.05 level; ** is the P ≤ 0.01 level

may have an important influence on the entry of toxic metals into the terrestrial food chain as well as changing the availability of some essential plant nutrients.

CONCLUSIONS

Climate change will affect the availability of plant nutrients, particularly those such as Ca, Si, and B which have a strong non-metabolic uptake component. The effects of climate and atmospheric change will be strongest for plants and natural communities whose growth and species are limited by essential nutrients with a non-metabolic, passive, and transpiration dependent uptake patterns.

Vegetation in mountain areas may be the first to undergo modifications because of the sensitivity of mountain soils to climate and atmospheric change. Many mountain soils are particularly sensitive to changes in climate and atmospheric deposition because of their high land surface to volume ratio, low volume, steep slope and high nutrient cycling rates. Climate-induced change in soil pH will occur first in the shallow soils of mountain regions that are on steep slopes and in soils that are in transition between Ca and Al buffering.

REFERENCES

Amato, I. (1988) 'Tapping tree-rings for the environmental tales they tell', *Analyt. Chem.*, 60 (19): 1103A–7A.

Baes, C.F., III and McLaughlin, S.B. (1984) 'Trace elements in tree rings: evidence of recent and historical air pollution', *Science*, (Washington, DC) 224: 494–7.

Baes, C.F., III and Ragsdale, H.L. (1981) 'Age-specific lead distribution in xylem rings of three tree genera in Atlanta', *Georgia. Environ. Pollut.*, Ser. B. 2: 21–35.

Barber, S.A. (1986) *Soil Nutrient Bioavailability*, John Wiley & Sons, Chichester.

Bondietti, E.A., Baes, C.F., III and McLaughlin, S.B. (1989) 'Radial trends in cation ratios in tree rings as indicators of the impact of atmospheric deposition of forests', *Can. J. For. Res.*, 19: 586–94.

Bondietti, E.A., Momoshima, N., Shortle, W.C. and Smith, K.T. (1990) 'A historical perspective on divalent cation trends in red spruce stemwood and the hypothetical relationship to acidic deposition', *Can. J. For. Res.*, 20: 1950–8.

Cook, E.R., Briffa, K., Shiyatov, S. and Mazepa, V. (1990) 'Tree-ring standardization and growth-trend estimation', in E.R. Cook and L.A. Kairiukstis (eds), *Methods of Dendrochronology*, Kluwer Academic Pub., Dordrecht.

Cutter, B.E. and Guyette, R.P. (1994) 'Species choice in dendrochemistry', *J. of Environmental Quality* (in press).

Guyette, R.P. and Cutter, B.E. (1993) 'Barium and manganese trends in tree-rings as monitors of sulphur deposition', *Water, Air, and Soil Pollution* (in press).

Guyette, R.P. and McGinnes, E.A., Jr (1987) 'Potential in using elemental concentrations in radial increments of old growth eastern red cedar to examine the chemical history of the environment', *Proc. International Symposium of Ecological Aspects of Tree-Ring Analysis*, compiled by G.C. Jacoby and J.W. Hornbeck, US Dept of Energy CONF–8608144, National Tech. Info. Serv., Springfield, VA, pp. 671–80.

Guyette, R.P., Cutter, B.E. and Henderson, G.S. (1989) 'Long-term relationships between molybdenum and sulfur concentrations in red cedar tree-rings', *J. Environ. Qual.*, 18: 385–9.

——, —— and —— (1991) 'Long-term correlations between mining activity and levels of Pb and Cd in tree-rings of eastern red cedar', *J. Environ. Qual.*, 20: 146–50.

——, —— and —— (1992a) 'Inorganic concentrations in the wood of eastern red cedar grown on different sites', *Wood Fiber Sci.* 24: 133–40.

——, —— and —— (1992b) 'Reconstructing soil pH from manganese concentrations in tree-rings', *For. Sci.*, 38 (4): 727–37.

Hagemeyer, J. and Breckle, S.-W. (1986) 'Cadmium in den Jahrringen von Eichen: Untersuchungen zur Aufstellung einer Chronologie der Immissionen', 'Cadmium in oak tree-rings: investigations on a chronology of the emissions', *Agnew. Botanik.*, 60: 161–74 (in German).

Kairiukstis, L.A. and Kocharov, G.E. (1990) 'Measuring chemical ingredients in tree rings', in E.R. Cook and L.A. Kairiukstis (eds), *Methods of Dendrochronology*, Kluwer Academic Pub., Dordrecht.

Kirkby, E.A. (1979) 'Maximizing calcium uptake by plants', *Commununications of the Soil Sciences and Plant Analysis*, 10: 89–113.

Kramer, P.J. and Kozlowski, T.T. (1979) *Physiology of Woody Plants*, Academic Press, New York.

Lazaroff, N. and Pitman, M.G. (1966) 'Calcium and magnesium uptake by barley seedlings', *Aust. J. Biol. Sci.*, 19: 991–1005.

Lindsay, W.L. (1979) *Chemical Equilibria in Soils*, John Wiley & Sons, New York.

McClenahen, J.R., Vimmerstedt, J.P. and Scherzer, A.J. (1989) 'Elemental concentrations in tree rings by PIXE: statistical variability, mobility, and effects of altered soil chemistry', *Can. J. For. Res.*, 19: 880–8.

McGinnes, E.A., Jr and Dingeldein, T.W. (1969) *Selected Wood Properties of Eastern Redcedar (*Juniperus virginia L.*) Grown in Missouri*, University of Missouri-Columbia Agric. Exper. Sta. Research Bulletin 960.

Reuss, J.O., and Johnson, D.W. (1986) 'Acid deposition and the acidification of soils and waters', in *Ecological Studies*, vol. 59, Springer-Verlag, New York.

Salisbury, F.B. and Ross, C.W. (1992) *Plant Physiology*, Wadsworth Pub. Co., Belmont, California.

Scharpenseel, H.W., Schomaker, M. and Ayoub, A. (1990) *Soils on a Warmer Earth*, Elsevier, Amsterdam.

Symeonides, C. (1979) 'Tree ring analyses for tracing the history of pollution: applications to a study in Northern Sweden', *J. Environ. Qual.*, 8: 482–6.

Tibbitts, T.W. (1979) 'Humidity and plants', *BioScience*, 29 (6): 358–63.

Vigouroux, A., Bussi, C. and Berger, J.F. (1989) 'Importance of water consumption for calcium nutrition of trees', *Ann. Sci. For.*, 46: 369s–371s.

17

PHENOLOGY AS A TOOL IN TOPOCLIMATOLOGY

A cross-section through the Swiss Jura Mountains

F. Bucher and F. Jeanneret

INTRODUCTION: PHENOLOGY AS A TOPOCLIMATIC TOOL

Phenology is a rather old, but not very well known branch of climatology. Phenology deals with the dates of events in the life of plants or animals. Phenological observations are an excellent tool to model topoclimatic conditions. The development of plants does not correlate only with climatic elements and conditions for there are other factors which influence phenology, such as soil, water, vegetation type, agricultural techniques, urbanization, etc. Some of the factors which influence phenology also reflect meteorological conditions. These include site parameters such as altitude, latitude, continentality, slope, exposition, topographical situation (bottom of the valley, basin, slope, terrace, pass, summit) and hydrographic situation (proximity of surface or ground water). A complete analysis should include all these factors, regardless of their significance. Phenological observations, therefore, can offer very interesting elements to a comparative climatic characterization of sites (Jeanneret, 1971; Primault, 1984). Phenology is a simple method, which needs no instruments. Hence the observations can be made nearly everywhere, thereby allowing a great spatial density. Handling of data is, however, rather complex. The advantages of phenology become particularly evident in mountain areas, where observations can be made at numerous sites and hence offer a degree of spatial differentiation which is usually not possible with climatic stations. In addition, phenology could become an important tool for assessment of potential impacts of climate changes in mountain areas (Jeanneret, 1972).

270

OBSERVATIONS IN SWITZERLAND

In earlier centuries, only isolated phenological observations were carried out. In 1950 the Swiss Meteorological Institute founded a national phenological network, consisting of about 120 stations (Defila, 1986, 1991). This network provides representative information on general conditions on a national scale. But phenological observations can also show very local variations of atmospheric conditions. This is particularly important in Jura mountains, where the topoclimatic differences are great even in rather small areas. These differences are due to variations in altitude, exposure, local winds, relative topography and other factors. The very low costs for data collection are balanced by the costs for analysis and surveys. In 1970, the Geography Institute of the University of Berne set up a mesoclimatological network in the Canton of Berne (5,000 km² with an approximate density of one station per 30 km². This density was later reduced to one station per 150 km².) The network complements that of the Swiss Meteorological Institute with a density of approximately one phenological station per 300 km². The observations were designed to meet topoclimatic needs, which includes planning purposes, agroclimatology, tourism, and other applications. Observations of fog and snow in winter and of plant phenology in summer are performed by observers recording topoclimatic differences within their area. Climatic variations can also be traced with the reaction of plants to changing climatic conditions. The results of the first period and the full development of the meso- and topoclimatic network has been treated in different publications (Jeanneret, 1971; Messerli, 1978; Volz et al., 1978). The number of stations was later greatly reduced, particularly after the end of the main project in 1977. However, some 40 stations remain in activity, allowing the analysis of phenological observations for over twenty years. Every observer surveys several sites in his area with different exposures. It is possible to assess the local conditions for every landscape type, as well as their influence on different phenological and climatic conditions. The number of observations (phenological stadiums) is however much larger for the national network (72 stadiums, of which 66 stadiums are phytophenological). In the Bernese network, only ten stadiums were in the program from the beginning, and six since 1973.

PHENOLOGICAL CONDITIONS OF THE FOLDED JURA MOUNTAINS

The folded Jura Mountains are part of a secondary mountain chain in the northwest of Switzerland. It is a test area within the observation area of the network of the Geography Institute of the University of Berne. The first attempt to analyse phenological observations over a period of twenty years is conducted on a cross-section through the Swiss Jura Mountains (with

271

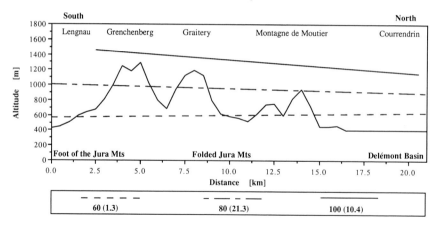

Figure 17.1 Cross-section through the Jura Mountains with the dates of blooming of hazel (*Coryllus avellana*). Day number and date (day and month).

summits ranging from 1,000 to 1,600 m). This cross-section extends from the lake of Bienne (430 m) in the lower Middle Land to the high plateau of the Franches-Montagnes (900 to 1,200 m) and to the Bassin de Delémont (415 m). The goal is to create a topographical model which allows one to predict phenological and therefore topoclimatic conditions within a mountainous area.

The phenological data from this area, together with the topographical data, can be used to extrapolate phenology. The phenological dates can be predicted with a suitable model for any point within the area. The observations of the Bernese network have been carried out using the same standards as the national network (Primault, 1971). However, comparison of the dates of both networks requires specific attention. The first task is a check of the observations, with an assessment of their quality – in particular, errors of coordinates, altitudes and dates.

Correlation between phenological dates and altitudes through linear regression techniques allows one to distinguish regional altitudinal gradients amongst different landscape types. The following results are derived from the analysis:

In the case of the blossoming of hazel (*Coryllus avellana*, Figure 17.1) in the test area between the end of February and middle of April, altitudinal gradients are not evident. An elevation of 100 m corresponds to a delay of between 5 and 7 days in blossoming. However, thermal inversions are frequent in the valleys and basins, resulting in blossoming on higher slopes earlier than on the valley floor.

The blooming of dandelion (*Taraxacum officinale*, Figure 17.2) represents the beginning of spring. It starts during April at low altitudes on the southern slope of the Jura, followed by the Bassin de Delémont with a delay of

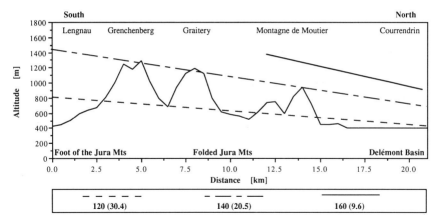

Figure 17.2 The blooming of dandelion (*Taraxacum officinale*).

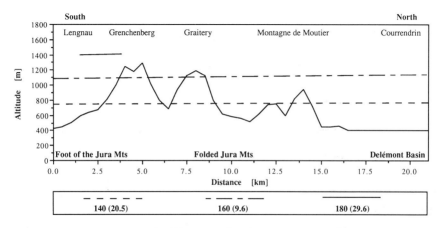

Figure 17.3 The blooming of apple trees (*Pyrus malus*).

approximately ten days at the same altitude. At an altitude of 1,000 m, however, the difference amounts to nearly one month.

For the blossoming of apple trees (*Pyrus malus*, Figure 17.3), the altitudinal gradient is rather flat, about 5 days per 100 m. The southern slope is no longer favoured, as in the case of the blossoming of dandelion at this time of the year (mid-April to the beginning of June). It must also be considered that apple trees are located in favoured spots (mainly in orchards close to villages).

The wheat harvest (*Triticum vulgare*, Figure 17.4) in July and August is a specific phenological event, because it is increasingly dependent more on advanced agricultural techniques than on climatic conditions alone. The patterns are similar to those of spring events. It is possible that clouds

273

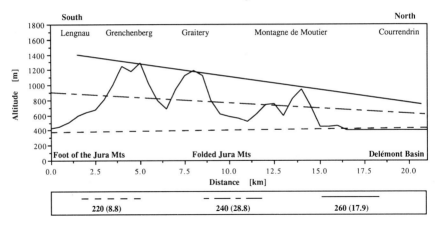

Figure 17.4 Wheat harvest (*Triticum vulgare*).

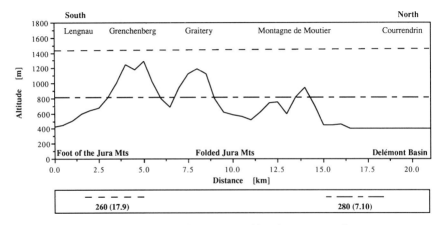

Figure 17.5 Colouring of the leaves of beeches (*Fagus sylvatica*).

and higher precipitation are responsible for later wheat harvests in the north.

Colouring of the leaves of beeches (*Fagus sylvatica*, Figure 17.5) in autumn is not easy to interpret. Colouring of leaves can be due to drought, cold or frost, and it is not possible to characterize regional differences. The altitudinal gradients are usually negative (minus 3 days per 100 mm), as the colouring of the leaves starts at higher altitudes.

This model reflects the phenological and consequently the topoclimatic conditions of the Jura Mountains. Blooming of hazel, dandelion, and apple trees shows that spring and summer arrive later in the basins and northern valleys than at the same altitude in the south. This means that the influence of altitude is less important in the northern part of the Jura Mountains where the climatic conditions are less favourable. From the observations, the influence of insolation and cloudiness is clearly visible. The southern slope of the Jura Mountains has specific climatic conditions, which are readily observable in spring (blooming of dandelion). Climatic conditions are less obvious in the summer and autumn because of thermal inversions and fog.

In summer, and particularly during the autumn, conditions are not as simple to analyse. The wheat harvest is a difficult phenological event, because of factors other than just climate which influence its mature phase. However, it shows a clear dependence on altitude. Colouring of the leaves of beeches does not show clear results and is further explained by Volz (1978).

AN EXAMPLE OF THE APPLICABILITY OF PHENOLOGICAL INFORMATION

Phenological observations can provide information about climatological conditions at a much higher spatial resolution than a nationwide meteorological network commonly does. Phenological observations do not correlate directly with individual climatic elements but give an idea about the environmental and more specifically the climatic conditions at a certain location (Volz, 1978). An interpolation of phenological point observations to a continuous surface, based on a transformation of phenological information into some simple climatological characteristics, seems to be quite promising. This interpolation will not be based on a geostatistical method such as kriging but with the help of a model, assuming that the spatial variations observed in phenological events are directly related to environmental differences (Klante, 1986; Branzi and Zanotti, 1989). A topoclimatological regionalization, that is, the aggregation of adjacently located samples showing similar climatic conditions, facilitates the description of climatic conditions at a specific site. The regionalization should therefore furnish thematically and spatially aggregated and continuous information about climatological and atmospheric conditions. This information can serve in the decision-making process for a wide range of planning applications.

A method for obtaining topoclimatological information at the regional scale

The procedure for moving from information on phenological events from randomly distributed point observations to a topoclimatological regionalization includes four steps (Figure 17.6):

- First, an empirical model, probably based on multiple linear regression, is set up. It relates the phenological events to so-called influencing factors. Model calibration is performed at phenological stations, where

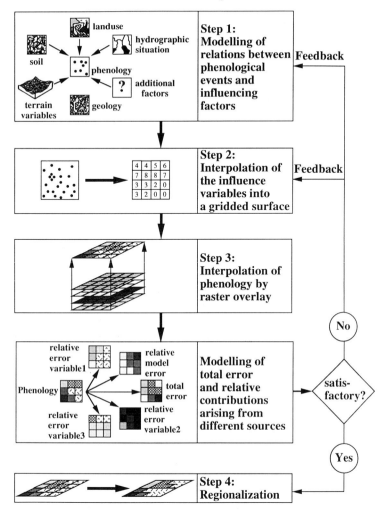

Figure 17.6 Schematic procedure to generate a meso- and topoclimatological regionalization.

additional information is available about the factors potentially influencing the variation of phenological events. The following factors, which must be carefully operationalized by means of suitable variables, are of prime concern:

(a) Elevation, slope, aspect and terrain position (that is, whether the site is located on a ridge, valley floor, mid-slope or a pass); these terrain parameters can be derived from a digital elevation model (DEM).
(b) Land-use; used to incorporate the influence of adjacent vegetation

types as well as built-up and paved areas.

(c) Hydrographic situation, that is, the proximity of a surface or groundwater body.

(d) Geology.

(e) Soil type.

(f) Additional factors which may need to be specified from case to case.

- In the second step, the variables which are finally included in the model equation have to be interpolated on continuous maps in a raster format. There are two reasons for this procedure:

  (a) Phenology, and more generally vegetation, is a feature which varies continuously over space, exhibiting gradual rather than abrupt changes. Raster maps are very well suited for representing continuous phenomena.

  (b) Polygon representations imply a level of accuracy which is not really included in the data.

- In step three, the phenological conditions which are originally only known at individual points are interpolated on to a continuous map. This interpolation is carried out by a raster overlay of the maps representing the various influencing factors and is based on equations established through the model developed in step 1.

- In the last step the regionalization is performed. This step involves the transformation of the phenological information to some simple climatological characteristics and the determination of the class membership for each grid cell. The procedure just outlined will be carried out in a GIS-environment in combination with a powerful statistical software package for the definition and calibration of the model. A well-defined interface between these two components will be established.

Error modelling for analysis optimization

In the environmental sciences, modelling tasks of this kind are increasingly performed with the help of GIS techniques. Quite often, however, the analysis does not include an assessment of the reliability of the results which makes them useless for real decision support in planning tasks. Unfortunately, results are often judged in terms of the quality of graphics used rather than on their intrinsic value (Burrough, 1991). Yet, it is only relatively recently that users are beginning to realize that information about the quality of modelling results is essential, should the output of quantitative models carry any weight in decision making (Goodchild and Dubuc, 1987). Assessment of results includes the modelling of errors or their propagation, arising from various sources such as data gathering by the observers, data manipulation within the GIS, or variations of the dependent variable not explainable by the influencing factors. Paying attention to this trend we will compute additional surfaces, which show the extent and distribution of uncertainty. Error modelling will be based on methods described by the Dutch researchers Heuvelink, Stein and Burrough (1989), which apply standard stochastic theory of error propagation to continuously differential arithmetical operations (like multiple regression) for manipulating gridded map data. These error models compute an estimate of total uncertainty as well as the relative contributions of errors that accrue from the regression model and the input data. Information about the relative contributions of errors contained in the model inputs (that is, model coefficients, and maps of individual variables) allows judgements to be made about subsequent analysis optimization. However, existing error models need to be even further refined. In particular, uncertainties arising from data capture and preprocessing (the so-called lineage of the data; NCGIA, 1988) must be separated from errors resulting from the integration of heterogeneous data (with respect to its spatial or temporal reference, level of accuracy) and represented in a more adequate way in future error propagation models. The first three steps of the procedure outlined previously have to be considered as a single iteration in the entire analysis. Each iteration produces two results: (i) a map with the continuous phenology and (ii) maps which show the extent and spatial distribution of uncertainties. Based on this information the decision can be made on whether the results are reliable enough to move on with the regionalization step, and to use the results for decision support in a planning task. Additionally, information about the relative contributions of typical error sources is yielded, giving an idea about the option which should be chosen to achieve better results. These options include:

- The use of a more suitable model to accomplish a better fit to the spatial variability of a phenological event (e.g., integration of additional factors, different model types, different variables to operationalize the factors).
- The partitioning of space into subregions and computation of individual parameters for these subregions to improve model calibration.
- The use of better methods for interpolating the raster maps as well as for the integration of data with different spatial or temporal reference, or level of accuracy.

CONCLUSIONS: APPLICATIONS OF A TOPOCLIMATOLOGICAL REGIONALIZATION

Our intention is to operationalize a method for topoclimatological regionalization for routine use based on the concepts presented here. The aim of topoclimatic regionalization is to provide thematically and spatially aggregated data for the daily work of environmental planners. These data offer a quick estimate of the climatic conditions in a study area and support decision making in planning applications. Potential applications are illustrated by the following examples:

Environmental monitoring:

- Impacts of changing environmental conditions on biological systems become particularly obvious in transitional areas (e.g., altitudinal boundaries). A topoclimatic regionalization enables the determination of the outlines of such border areas and makes it possible to recognize changes in environmental conditions.
- Evaluation and topoclimatological characterization of test sites for environmental modelling.

Agriculture:

- Land suitability assessment for specific crops.
- Feasibility estimation for an additional crop between the traditional cropping seasons in spring and autumn.
- Criterion for recommendations concerning suitable grass types.
- Criterion for estimating crop yield as a basis for planning and coordination of labour and harvesting equipment.

Civil engineering, housing development:

- Favourable phenological regions relate to preferable housing areas.

Some of these applications, particularly the agricultural ones, have already been in operation for years. However, they are based on phenological maps from the early seventies. The reason for this unsatisfactory situation is that the procedure for processing phenological observations into data suitable for decision making is currently far too time-consuming and has therefore never been operationalized so far. It is hoped that the proposed meso- and topoclimatological regionalization, the operationalization of the required method and its implementation in a GIS-environment will improve the conditions for environmental and agricultural planning. Beyond that, the development of a user-friendly interface within the GIS-environment will facilitate updating of the meso- and topoclimatic regionalization.

REFERENCES

Branzi, G.P. and Zanotti, A.L. (1989) 'Methods in phenological mapping', *Aerobiologia*, 5: 44–54.

Burrough, P.A. (1991) 'The development of intelligent geographical information systems', in *Proceedings of the EGIS '91 Conference*, Brussels, pp. 165–74.

Defila, C. (1986) 'Phänologische Karten in der Schweiz (gestern – heute – morgen)', *Arboreta Phaenologica*, 31: 61–9.

—— (1991) 'Pflanzenphänologie der Schweiz', *Veröffentlichungen der Schweizerischen Meteorologischen Anstalt*, 50.

Goodchild, M. F. and Dubuc, O. (1987) 'A model of error for choropleth maps, with applications to geographic information systems', in *Proceedings of the AutoCarto 8 Conference*, Falls Church (USA), pp. 165–74.

Heuvelink, G.B.M., Burrough, P.A. and Stein, A. (1989) 'Propagation of error in spatial modelling with GIS', *Int. Journal of Geographical Information Systems*, 3 (3): 303–22.

Jeanneret, F. (1971) 'Klimatologische Grundlagenforschung: Jura, Mittelland, Alpen', *Beiträge zur klimatologischen Grundlagenforschung*, 2. Geographisches Institut der Universität Berne.

—— (1972) 'Methods and problems of mesoclimatic surveys in a mountainous country: a research programme in the Canton of Berne, Switzerland', *Proceedings 7th Geography Conference*, New Zealand Geographical Society, Hamilton NZ, pp. 187–91.

—— (1974) 'Statistische und kartographische Bearbeitung phänologischer Beobachtungen – am Beispiel der Daten der Weizenernte 1970', *Informationen und Beiträge zur Klimaforschung*, 11, Geographisches Institut der Universität Berne.

—— (1991): 'Les mésoclimats du Jura central: une coupe phénologique', *Bulletin de la Société neuchateloise de geographie*, 35 = 'Die Mesoklimate des zentralen Juras: ein phänologischer Querschnitt', *Jahrbuch der Geographischen Gesellschaft Berne*, 57: 57–70.

Klante, B. (1986) 'Synthetische phänologische Karten', *Arboreta Phaenologica*, 31: 97–102.

Messerli, B. (1978) 'Klima und Planung-Ziele, Probleme und Ergebnisse eines klimatologischen Forschungsprogrammes im Kanton Berne', *Jahrbuch der Geographischen Gesellschaft von Berne*, Bd. 52/1975–76, pp. 11–22.

NCGIA (1988) 'The proposed standard for digital cartographic data', *The American Cartographer*, 15: 132–5.

Primault, B. (1971) (3rd edn) *Atlas phénologique = Phänologischer Atlas = Atlante fenologico*, Institut suisse de météorologie, Zürich.

—— (1984) 'Phänologie: Frühling, Frühsommer. Sommer, Herbst.' = 'Printemps, début de l'été. Eté, automne', in W. Kirchhofer *et al.*, *Klimaatlas der Schweiz = Atlas climatologique de la Suisse*, Bundesamt für Landestopographie = Office fédéral de topographie, Wabern-Berne, tab. 13.1 + 13.2.

Volz, R. (1978) 'Phänologische Karten von Frühling, Sommer und Herbst als Hilfsmittel für eine klimatische Gliederung des Kantons Berne', *Jahrbuch der Geographischen Gesellschaft von Berne*, vol. 52/1975–76, pp. 23–58.

Volz, R., Wanner, H. and Witmer, U. (1978) 'Zusammenfassung im Sinne einer regionalen Klimacharakterisierung', *Jahrbuch der Geographischen Gesellschaft von Berne*, vol. 52/1975–76, pp. 149–52 and 1 map.

18

DESIGN OF AN INTENSIVE MONITORING SYSTEM FOR SWISS FORESTS

J.L. Innes

INTRODUCTION

Many papers in this volume emphasize the potential importance of global change to mountain ecosystems. The possible effects of global change on Swiss forests have been examined using a modelling approach (Brzeziecki *et al.*, 1993; Bugmann and Fischlin, this volume). While models provide a valuable insight into the possible changes that might occur given changing environmental conditions, it is important that the predictions are backed up by observation. These are not only necessary for the initial model calibration, but enable the model to be tested and, if necessary, adjusted, through time. A major problem in relation to studies of changes in forests is the scarcity of basic data on forest ecosystem processes which can be used to calibrate the models (Wiersma *et al.*, 1991). In addition, there are few data available on changes in forest ecosystem processes in relation to environmental change. These provide important reasons for the field observation of processes acting in forest ecosystems.

The need for environmental monitoring in forests has been recognized for some time. As a result of observed changes in some mountain forest ecosystems in Germany and the former Czechoslovakia in the late 1970s, surveys of forest conditions were established in a number of countries and most European countries had implemented monitoring programmes by the mid-1980s. These programmes have started to yield much useful information, but it is generally recognized that the inventories (or at least the manner in which the data from them have been analysed) have not provided the answers about the dynamics of forest health that were initially being sought (Innes, 1993a). The response to this has been a general recognition of a need for more detailed monitoring and more detailed analytical research. In Switzerland, this took the form of the Sanasilva project (Schlaepfer, 1992). The results from this (Figure 18.1) illustrate that annual changes in forest condition may occur and that there has been an upward trend in the amount of unexplained defoliation observed in Switzerland between 1985 and 1992.

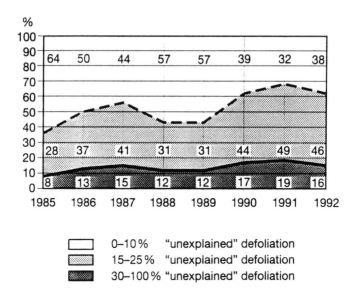

Figure 18.1 Development of 'unexplained' defoliation in Switzerland between 1985 and 1992. Values represent the proportion of trees in a given defoliation category, weighted to take into account the basal area of the trees. The sample size was approximately 7,000 trees in each year.

This trend has not occurred consistently across Switzerland, with a decrease in the proportion of defoliated (> 25 per cent) trees occurring on the south side of the Alps and no long-term changes being evident in the forests of the Swiss Plateau. In contrast, the proportion of defoliated trees in the Jura Mountains has more than trebled over the survey period.

Currently, most programmes are concerned only with crown condition and, within this context, only assess the transparency and overall discoloration of individual tree crowns (Anon, 1992). Assessments are normally made from the ground, but in some cases are supplemented with colour infrared air photographs. The inventories of 'forest health' have provided a very limited data base from which to work. Sites are poorly described and there is insufficient information on the factors that could be affecting defoliation and discoloration, particularly in relation to the history of the plots. There are historical reasons for this. When the programmes were set up, the overriding concern was that acid deposition was the cause of forest mortality.

The situation is now recognized as being much more complex. Not only have other forms of pollution, particularly ozone, been recognized as being of potential importance, but the role of factors such as drought stress and extreme frosts has become increasingly apparent (Innes, 1993a). In addition, it is now widely recognized that different types of decline are characterized by different sets of symptoms and that these can be triggered and accelerated by many different factors. However, the causes of the declines and subsequent recoveries remain difficult to determine because of the lack of reliable field data and the basic lack of understanding of 'normal' tree growth. At the same time, there has been increasing pressure to undertake long-term ecological monitoring, particularly in remote areas (e.g. Fox and Bernabo, 1987; Bruns et al., 1991; Silsbee and Peterson, 1991).

One possibility is to undertake experimental research to verify the results of the field inventories. However, there are major problems with the extrapolation of results from experimental situations to the field. These mainly revolve around the differences between the responses of mature trees and seedlings to stress (e.g. Borchert, 1976; Edwards et al., 1993). In addition, it is rarely possible to undertake experimental work with mature trees. These problems reflect the different aims that often exist between laboratory and field studies. In laboratory studies, work is usually directed towards establishing mechanisms, whereas the field requirements are for the explanation of existing symptoms. The two aims require different approaches and problems have arisen when attempts have been made to apply the results of laboratory studies to the field.

These two difficulties have resulted in differences in the ways that laboratory-based and field-based scientists approach the problems of forest declines. Experimental work strongly suggests that under certain conditions, several different pollutants can influence the vigour of young trees at pollutant concentrations close to those seen in forest areas (Ashmore et al., 1990). However, this is not supported by field observations, which have yet to demonstrate a correlation in space or in time between the occurrence of poor crown condition and specific pollutants.

Experimental work has also indicated that many factors, including pollution, can influence the processes operating in forest ecosystems. In many cases, effects on the other ecosystem components are easier to demonstrate than effects on mature trees. For instance, soil columns can be collected reasonably easily in the field and subjected in the laboratory to experimental manipulations under highly realistic conditions. The results indicate that several processes occurring today may have been induced by man's activities. There is a clear case for the careful integration of observational and experimental studies of forest ecosystems.

The monitoring programme will provide an opportunity for integrating various approaches to environmental problems (Figure 18.2). Four main approaches can be identified: experimental research, modelling, special surveys, and monitoring, with monitoring forming a central theme and an essential part of hypothesis construction and testing. Monitoring provides the baseline data which are essential for the calibration of models. Similarly,

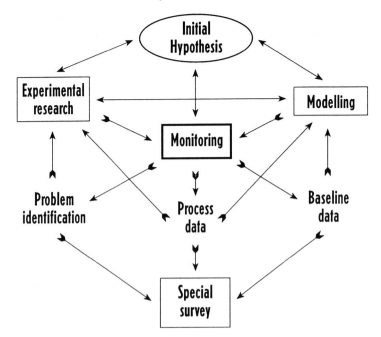

Figure 18.2 Hypothetical relationships between monitoring and other scientific research methods. Monitoring should be clearly distinguished from special surveys aimed at resolving particular problems.

it enables potential or actual problems to be identified that are then the subject of experimental research. In addition, process data are supplied (for example on movement of water through the ecosystem) which can be used in both modelling and experimental studies. The information collected from modelling may also result in the identification of a need for more specialized surveys which, because of their cost, can only be undertaken on a limited number of sites or over a restricted time period. Experimental research, monitoring and modelling can all result in the modification of the original hypothesis.

CHANGES IN FOREST ECOSYSTEMS

There is a large literature available on the subject of forest condition (Innes, 1993a). However, much less is known about the long-term changes that may be occurring in forest ecosystems. For example, studies in many countries have indicated that the chemical composition of forest soils is changing (Billett *et al.*, 1988; Hallbäcken 1992). The proportion of this change in soil chemistry that is attributable to pollution and the proportion that is due to the growth of the trees remains speculative. Generally, forest soils have been acidifying, but this may be due to a combination of past mismanagement (particularly litter raking), acidification caused by base cation uptake by biomass (and subsequent losses through harvesting) and acidic deposition.

External factors are also affecting the forest ecosystems. Three examples will suffice. There is widespread evidence of increasing growth rates of trees in arctic and alpine areas (Innes, 1991). The cause is unknown, but appears to be related to increased atmospheric concentrations of carbon dioxide, increased nitrogen deposition, warmer temperatures, better management, or a combination of these factors. This clearly has direct relevance to those concerned with the maintenance of alpine forest ecosystems.

The second example concerns forest bird communities. Major changes in forest bird communities were seen in Europe in the 1960s following the introduction of certain insecticides. Raptors in particular suffered major population declines and similar problems occurred elsewhere (Newton, 1988; Newton and Galbraith, 1991). This would not have been discovered without population monitoring. More recently, there has been speculation that changes in the wintering habitats of many European and North American summer visitors have caused major population changes in some species (Berthold et al., 1986; Terborgh, 1989; Baillie and Peach, 1992). The lack of good monitoring data from forests means that this will remain speculation until studies are initiated.

The ground flora of many forest ecosystems is known to be changing (e.g. Ellenberg, 1979; Falkengren-Grerup and Tyler, 1991). Some of these changes may be due to the structural evolution of the forest canopy. Other changes may be due to external forcing factors, such as acidic deposition. In Switzerland, Kuhn (1990) has documented changes in the ground flora of a number of forests over the last 40 years. While management practices and natural regeneration clearly played a part in these changes, other factors, particularly the deposition of nitrogen, have influenced the vegetation composition. However, to help the determination of the precise causes of the changes, Kuhn (1990) argued that much more intensive monitoring would be required.

MONITORING SYSTEMS IN OTHER COUNTRIES

Many countries have already installed some form of environmental monitoring. For example, 20 intensive forest monitoring stands have been established in Bavaria (Nüsslein, 1992; Preuhsler et al., 1992), 20 in Norway (Jensen, 1992; Nelleman, 1992) and about 100 in France (Barthot, personal communication). These represent the exception rather than the rule. Mostly, sample plots or programmes are related to specific phenomena (such as the water quality of lakes) and an integrated, ecosystem approach has rarely been adopted (Thomsen, 1991). Sometimes, inadequate attention has been given to the setting up of long-term monitoring projects, leading a National Academy of Sciences of the United States of America committee to state in 1989 that 'The failure to commit adequate resources of time, funding and expertise to up-front program design and to the ultimate interpretation and reporting of information will result in the failure of the entire program'. These criticisms do not necessarily apply to any of the forest monitoring programmes mentioned above.

In terms of its scope, the level of intensive monitoring that will be undertaken in Swiss forests is similar to the level of monitoring undertaken by the International Co-operative Programme on Integrated Monitoring of the United Nations Economic Commission for Europe. The UNECE programme has recently completed its pilot phase (Nihlgård and Pylvänäi-nen 1992), with the result that many of the costings, instruments, techniques, etc. have been evaluated. Full advantage will be taken of this pilot study, together with the information gathered during the pilot phases of many of the other monitoring programmes currently being established.

A NEW MONITORING PROJECT

The project described in this paper is based on the premise that to understand the functioning of forests, it is necessary to examine the whole forest ecosystem. Not all aspects of forest ecosystems are directly related to the health of the trees. However, it is now increasingly recognized that forest scientists and others should be examining the *whole* ecosystem, and not just part of it.

In Switzerland, a new forest ecosystem monitoring programme is being installed over the period 1993–1995. It seeks to elaborate some of the possible cause–effect relationships between the condition of forest ecosystems and the environment. In addition, it seeks to determine the nature and extent of any changes in forest ecosystem processes over time and the implications of these for the condition of the trees. Conversely, effects of the trees on the ecosystem will also be evaluated. It is particularly concerned with those processes affected directly or indirectly by man, although other processes must be regarded as being equally or even more important. The main approach will be through the long-term evaluation of ecosystem processes as they relate to trees. Although other aspects of the forest ecosystem are of importance, the programme will concentrate (initially at least) on those processes known or believed to be affecting the condition of the trees.

AIMS OF THE PROGRAMME

The two most important objectives of the intensive monitoring programme are:

1   The integral and comprehensive assessment of the current status and future changes in forest ecosystems.
2   Recording the development, growth, pressures, and health status of the most important forest types in Switzerland.

During the initiation phase (1993–1995), the project is seeking answers to the following questions:

- What are the most important characteristics of forest ecosystems?
- How can these characteristics be recognized and how do they inter-connect?
- How do factors such as soil, ground flora, fauna, air pollution, climate, and stand characteristics interact in forest ecosystems?
- In which forest types should the plots be installed?
- What other factors might influence the choice of monitoring plots?
- With which characteristics can the pressures and status of forests and forest ecosystems best be evaluated and understood?

It is not possible to develop an ecosystem monitoring programme without some idea of the longer-term aims of the project. These can be divided into broad groups; aims that are likely to be satisfied in the medium-term (1–10 years) and aims that will take longer.

Medium-term aims

The medium-term aims of the programme can be directly related to perceived needs at the present time. In broad terms, the aims of the programme can be summarized as:

- to enhance, maintain, and monitor forest status through an under-standing of the processes operating in forest ecosystems;
- to monitor the status and trends of ecological processes in forest ecosystems;
- to develop a better understanding of the precise nature of 'forest health' and to establish how it can best be characterized;
- to develop new methods for anticipating emerging ecological problems; and
- to provide basic information for the calibration and verification of models related to practical forest ecosystem problems, including growth models, critical loads models, and models predicting the impact of global change.

Long-term aims

No satisfactory basis exists for developing long-term aims for a monitoring project. However, the following general aims can be proposed:

- to identify baseline conditions in forest ecosystems and to quantify long-term changes in tree and forest status;
- to increase the understanding of the mechanisms leading to change within forest ecosystems;
- to understand the processes by which any changes are achieved; and
- to provide an opportunity for the combination of experimental-analytical science and field observations in known field environments.

DIFFERENT LEVELS OF MONITORING

It is possible to undertake forest monitoring at a number of different scales. The simplest is the international survey of crown condition organized by the United Nations Economic Commission for Europe. This involves assessments of crown transparency and discoloration of a limited number of trees per plot. In the annual survey of forest condition in Switzerland, more detailed monitoring is undertaken, with several indices of tree condition (crown transparency, crown discoloration, causes of damage, social status, etc.) being conducted. In addition, it is proposed that the observations will be complemented by site descriptions (soils, ground vegetation, climate, and air pollution). This is probably the minimum scale that can be used to *infer* cause–effect relationships which must then be tested in experimental research. A more intensive form of monitoring involves assessing changes in components of the forest ecosystem (such as soils and flora) in addition to changes in crown condition. Obviously, the number of potential indices that could be monitored is almost infinite, and a selection process must be adopted to determine which indices would be the most useful.

Three levels of monitoring were considered during the formulation of the programme, each more detailed than the last and all being more detailed than the assessments made in the annual forest health inventories. The intensive monitoring plots should be seen as medium- to high-intensity monitoring sites. The plots used in the annual health inventories should then be seen as low-intensity plots. Some intensive monitoring plots already exist in Switzerland, sponsored by the cantonal authorities. It will be important to ensure that there is no duplication of effort over these plots.

The most intensive monitoring is likely to involve integrated ecosystem monitoring, with a very small number of relatively large (> 1,000 ha) observational units being assessed. These larger sites will probably not be restricted to forests, but would include a variety of other ecosystem types. The aim is to have a tiered monitoring network that will enable a variety of different programmes to be included. These include the UNECE integrated catchment monitoring sites, the UNECE forest programme Level II and III sites, the proposed Global Terrestrial Observation System sites and the proposed monitoring sites of the International Union of Forest Research Organisations. These programmes all collect data at a variety of scales and,

Table 18.1 Different levels of monitoring

Level 1	Level 2	Level 3
Disturbance	Air quality	Aquatic biology
Forests	Bulk precipitation	Bats
crown class	Epiphytes	Breeding birds
foliar composition	Heavy metal deposition	Butterflies
fruiting	Increment analysis	Contamination
regeneration	Litterfall	Gauging station
structure	Meteorology 2	Meteorology 3
tree damage	Root disease evaluation	Moths
tree growth	Soils: biological activity	Other insects
tree health	Throughfall	Stream chemistry
tree phenology	Tree foliage (hands-on)	Water quality
Ground flora	Tree genetic structure	
Meteorology 1		
Site description	**+ Level 1**	**+ Levels 1 and 2**
Soils analyses		

Note: All the items listed under Level 1 would be monitored at the most basic level. Level 2 would include all the items listed under Level 2 together with those listed under Level 1. Three different types of meteorology have been included, representing data taken from a nearby station at Level 1 (meteorology 1), use of an on-site Stephenson screen or similar device at Level 2 (meteorology 2) and use of an on-site fully automatic weather station at Level 3 (meteorology 3).

as with the established monitoring plots in Switzerland, it is important to ensure that as little duplication as possible occurs.

An immediate reaction to the stated objective of the monitoring plan might be that every aspect of the forest ecosystem should be monitored. This would clearly be impossible. However, the next approach might be to measure everything that can be practically measured. Such an approach would have severe limitations as much time could be spent collecting irrelevant information. Instead, an overall framework based on modelling has been developed. The basic requirements for the modelling at each level are given in Table 18.1. Level 1 essentially involves the monitoring of the basic silvicultural system within the forest. Emphasis is on the condition of the trees and little attention is given to the rest of the forest ecosystem. At Level 2, emphasis is still on the trees, but more attention is given to the processes that might affect the condition of the trees. In addition, pollution inputs are assessed. At Level 3, a number of components of the forest ecosystem unrelated to the trees are included in the assessments.

Following discussions between the Swiss Federal Institute for Forest, Snow and Landscape Research and the Eidg. Forstdirektion (the Federal Forest Administration of the Bundesamt für Umwelt, Wald und Landschaft – Federal Office for Environment, Forestry and Landscape), a decision was reached to limit the monitoring to Level 1 + 2. This means that some aspects of the forest ecosystem will be excluded from the programme. However, the plots are being established in such a way as to permit the inclusion of parameters from Level 3 (particularly vertebrates and invertebrates) in the future and every encouragement will be given to independent assessments on the monitoring plots.

PROGRAMME BASIS

To begin to understand the complex relationships that are present in forest ecosystems and the changes that these are subject to, a model-based approach is essential. Modelling forms an important part of both the design and running of a monitoring programme (Wiersma, 1990; Wiersma *et al.*, 1991) and it provides the following advantages:

- the different links within the forest ecosystem would be gradually pieced together;
- a system of black (no information), grey (little information) and white (process fully understood) boxes would be possible, allowing for differential data quality/availability;
- measurements of fluxes between boxes would enable critical areas to be identified and quantified, without necessitating a detailed knowledge of the contents of the boxes. However, knowledge of the processes operating within the boxes is essential if a true understanding of the ecosystems is to be gained; and
- the model could be used in the monitoring programme initialization phase to assess current knowledge, the gaps that exist and the areas where measurements were a priority.

One possible way to develop this is through a forest ecosystem model. This approach has already been proposed (e.g. the SILVISTAR model of Koop (1989)). An alternative approach is to take an international ecosystem model such as FORCYTE (Kimmins *et al.*, 1990) as a basis. However, as this model is aimed at predicting forest growth, some modifications to it are necessary. The modelling has several aims, including the prediction of the ecosystem responses under specific conditions and the testing of our understanding of the basic processes operating within the ecosystem. In the latter sense, the modelling will also help to define further research needs.

QUALITY CONTROL CRITERIA

A particularly important aspect of any monitoring is the quality control procedures (Cline and Burkman, 1989). It is inadequate attention to this problem that led to major difficulties with the interpretation of international statistics of forest health (Innes *et al.*, 1993). A standard operating procedure, embracing measurement quality objectives, will be developed for each aspect of the ecosystem being monitored. Analytical procedures will be backed up by standard laboratory quality assurance techniques. Where practical, a portion of every physical sample collected in the field will be archived for future reference. Full details of the quality assurance programme are still being determined and it is likely that the programme will evolve along with the main monitoring programme.

NUMBER AND NATURE OF MONITORING PLOTS

Given the nature of this programme, the primary limitation on the number of plots is financial. From a forestry perspective, it is important to ensure that all the main forest types are represented in the sampling design. From an ecological perspective, all the most important forest community types should be included. Kuhn (1992) identified the following communities as being of importance, and plots will be located in each:

> *Galio silv. – Carpinetum*
> *Luzulo silvaticae – Fagetum*
> *Galio odorati – Fagetum* or *Milio – Fagetum*
> *Cardamino – Fagetum*
> *Aceri – Fraxinetum* or *Ulmo – Fraxinetum*
> *Bazzanio – Abietetum*
> *Dryopterido – Abietetum* or *Calamgrostio vill. – Abietetum*
> *Sphagno – Piceetum* or *Veronico latifoliae – Piceetum*
> *Erico – Pinetum*
> *Larici – Pinetum cembrae*
> *Insubr. Eichen – Birkenwald (Kastanie)*

The primary aim of the design is to ensure that these forest types are adequately represented. The aim of the plot location is to achieve a replicated factorial design, with forest community type and stand management being the principal variables and with the total number of plots being approximately 80. As all management types do not occur in all forest types, extra plots will be located in newly established forest in order to monitor changes associated with forest succession. These plots will include managed and unmanaged, protected, sites. In locating the monitoring plots, it is also important to remember that the programme is a long-term one and has been set up in response to a widespread feeling of the need for such a programme. Consequently, regional considerations will be a criterion for plot location.

The most important weighting factor in the regional selection of forests is the presence of mountain forests. There are two reasons for this. Firstly, the level of concern for mountain forests is generally higher than for low-lying forests. Secondly, mountain forests are particularly important in Switzerland, and the programme could therefore make a unique contribution to any pan-European programme of intensive monitoring plots. Any plot design will have to be clearly justifiable to all groups involved in the running of the monitoring programme (particularly the cantonal forest authorities).

An additional criterion will be the presence of existing monitoring plots. A number of different monitoring schemes already exist (see above) and it is important that the information from these be incorporated into the new network, where this can be done without compromising the overall site requirements. This is a problem facing the developers of many monitoring projects and should not be underestimated in importance. It is vital that as full use as possible be made of existing sites for which there is information. Although the monitoring is seen as long-term, there will be increasing pressure to produce answers in the short term, and any prior information that can be incorporated into the programme will be of considerable value. As a first step, the three intensive monitoring sites from the National Research Programme 14+ 'Forest damage and air pollution in Switzerland', Alpthal, Davos and Lägeren, are being incorporated into the programme as pilot monitoring plots.

Plot design

A plot area of 30 ha will be used to monitor animal and bird populations. Each plot will consist of over 80 per cent forest, although clearings are required for, for example, meteorological measurements. Within the 30 ha, there will be a core of area of 1 ha, in which intensive observations of phenomena such as litterfall, tree health and soil processes will be undertaken. Where possible, the plots will consist of hydrological catchments although it is likely that this criterion will only be met a limited number of cases.

Each core area will consist of a series of one or more protected subplots which will be used for non-destructive sampling. There are a number of different possible sampling designs, ranging from a single central core area surrounded by a destructive sampling zone to a number of small cores areas spread through a destructive sampling zone. The relative merits of these are being assessed as part of the pilot programme and the possibility of a mixed sampling design (depending on the nature of the forest ecosystem) is being investigated. In addition, a buffer zone may be required between destructive and non-destructive sampling zones.

In mountain areas, the possibility of looking at altitudinal transects is being considered. These will improve cost-effectiveness and will also provide additional valuable information related to global warming. There are a number of problems with the setting up of monitoring plots in mountain areas, and these are currently being addressed. They include difficulties of winter access, the need for the ruggedness of any instrumentation and the lack of forest homogeneity over relatively short distances.

VALUE OF THE PROGRAMME TO FOREST PRACTICE, RESEARCH AND POLICY MAKING

Forestry practice

Within Switzerland, two broad forest types can be recognized. In low-lying areas, forests primarily exist for wood production. Research here has therefore focused on various means of increasing the quality and volume of timber, often to the detriment of other parts of the forest ecosystem. Insufficient attention has been given to the forest as an ecosystem, with the result that problems have arisen. For example, poor matching of site to species or provenance has created problems for the cycling of nutrients. This is particularly evident in some sites where Norway spruce or another species of conifer has replaced beech.

In mountain areas, the primary function of forests today is one of protection. The presence of forest cover on steep slopes reduces potential problems of soil erosion, reduces the likelihood of severe flooding (see, for example, Leuppi and Forster, 1990) and provides protection against avalanches (e.g. Marx, 1986). The forests are consequently very important but recent trends reported in the Swiss forest health monitoring programme suggest that they may be unstable (*Sanasilva Waldschadenbericht*, 1991). Consequently, any information on the factors affecting their status and dynamics will be of importance to their silviculture.

The intensive monitoring project will help to reach the multipurpose objectives of forest management by providing baseline information on the day-to-day functioning of the main forest ecosystem types in Switzerland. Emphasis is being placed on protection forests in mountain areas as these are both critical and are better developed in Switzerland than anywhere else (Küttel, 1990). The information will enable breaks in the forest ecosystem connections to be identified and, in the longer term, will suggest possible remedial actions.

Forestry research

One of the greatest problems facing forestry research has been the lack of good basic data on forest ecosystems and tree ecophysiology in given field situations. Occasionally, surveys have been undertaken, but these do not provide any information on changes over time. The lack of basic information represents a potential shortcoming of the many models of forests and forest ecosystems. This is important, as such models are being used increasingly to predict future forest dynamics, both in relation to environmental problems such as climatic change and in relation to forest management issues, despite the many scientific uncertainties associated with them. An example of the latter is when the rotation age of a species is increased to benefit wildlife

(Petty and Avery, 1990). Forest models could, potentially, predict the implications of such a change on the entire forest ecosystem.

The programme goes much further than the collection of baseline data. By monitoring both the state of the forest ecosystem and some of the more important processes within it, information will be gained on the impact of external influences on the forest ecosystem, such as climate and air pollution. These data will be incorporated into the models of future forest species composition and production, enabling a better assessment of the likely impact of individual external influences on forest ecosystems.

The current network of pollution monitoring sites in Switzerland is limited in its extent and geographical coverage. This means that projects such as the annual inventory of forest health has had difficulties in estimating site-specific pollution concentrations (this is also true of climatic modelling). The establishment of active (direct measurement) and passive (indirect measurement) pollution and climatic monitoring stations in forest environments will add greatly to the coverage of these measurements in Switzerland, providing invaluable data to a number of groups involved with the more detailed assessment of the environment.

This project will also contribute to the assessment of carbon dynamics in forest ecosystems. This is a critical area in current forest research in view of the potential of forests as short-term carbon sinks (e.g. Post *et al.*, 1982; Birdsey, 1990; Grigal and Ohmann, 1992). The project will provide the details of soil carbon dynamics and biomass turnover that are missing from many current models.

Public interest

There is a growing interest in determining the status of forest ecosystems. In the past, foresters have been viewed as protecting the environment. However, with increasing public concern over issues such as biodiversity, the importance of forestry in environmental impoverishment has been stressed. This has become particularly apparent in countries such as the USA, Canada, and Great Britain, where foresters and conservationists are now opposed to each other (Innes, 1993b). The intensive monitoring project provides an opportunity to redress this dichotomy, since its emphasis will be on forest ecosystems, rather than on timber production. However, it will also provide important information on the factors affecting forest development, another important area of the public interest.

National and international policy

Forests are frequently viewed as ecosystems that are extremely sensitive to external forcing factors. Consequently, they are of considerable interest for impacts modelling, particularly in relation to pollution. Critical loads and levels have been developed for a variety of forests, but this information is unreliable and in need of verification. The intensive monitoring project will provide the baseline information essential for such studies.

The exposure of forests to various different types of pollution is currently extremely complex, with a variety of different trends in operation (Hochstein and Hildebrand, 1992). Consequently, reliable monitoring and understanding of processes is essential if sensible policy decisions are to be reached in relation to forests.

There has been a recent upsurge in interest in the possibility that forests represent a potential sink for atmospheric carbon dioxide. It has been proposed that a vigorous afforestation programme will help to offset the increases in atmospheric CO_2 concentrations that are currently occurring. While the figures that these claims are based on are open to question, it is clear that the existing forests of the world are a very important CO_2 reservoir. However, the interactions between the forest and soil carbon pools are poorly understood and there are a number of gaps in our knowledge of how trees and forests will respond to increasing CO_2 concentrations. Monitoring of forest ecosystem processes will provide important information will enable policy makers to place correctly the importance of forests in the global carbon cycle.

CONCLUSIONS

A number of countries have begun to implement programmes of forest ecosystem monitoring. This follows the realization that the annual inventories of forest health have provided little information on the causes of phenomena such as 'forest decline'. Given the predictions over future climate change, it is important that an adequate monitoring system for forests is installed at the earliest possible opportunity – as is currently being done in Switzerland.

The monitoring system that has been designed will cover the whole of Switzerland, but will have emphasis placed on mountain forests. These are particularly important in view of their protective function. If the forests were to be damaged, the risk of processes such as avalanches and soil erosion would increase, creating additional problems. In order to understand the factors affecting changes in forest ecosystem processes, detailed monitoring of the processes is required, together with experimental and modelling work. The monitoring system will provide essential baseline data and will provide the data to generate plausible cause–effect hypotheses that can then be tested by experimental techniques.

A number of problems related to environmental monitoring can be identified that are specific to or particularly apparent in mountain forests. These include the shape and size of the sample plot, the adverse weather conditions (necessitating specialized monitoring equipment) and the difficulties of access. All these need to be resolved before the monitoring can be fully established. In Switzerland, a three-year pilot phase is underway, during which the techniques and equipment required for the project will be evaluated. This development work is being undertaken in close collaboration with others interested in environmental monitoring in mountain areas so that maximum standardization can be achieved.

The project will produce large amounts of data. Steps are being taken to ensure that the data base will be widely available, in line with other similar projects related to environmental change. Further details of the sites, procedures, and data base are available from the author.

REFERENCES

Anon (1992) *Forest Condition in Europe*, United Nations Economic Commission for Europe, Geneva; and Commission of the European Communities, Brussels.

Ashmore, M.R., Bell, J.N.B. and Brown, I.J. (1990) *Air Pollution and Forest Ecosystems in the European Community*, Air Pollution Research Report 29, Commission of the European Communities, Brussels.

Baillie, S.R. and Peach, W.J. (1992) 'Population limitation in Palaearctic-African migrant passerines', in H.Q.P. Crick and P.J. Jones (eds), *The Ecology and Conservation of Palaearctic-African Migrants. Ibis* 134, Supplement 1: 120–32.

Berthold, P., Fliege, G., Querner, U. and Winkler, H. (1986) 'Die Bestandsent-wicklung von Kleinvögeln in Mitteleuropa: Analyse von Fangzahlen', *Journal of Ornithology*, 127: 397–437.

Billett, M.F., Fitzpatrick, E.A. and Cresser, M.S. (1988) 'Long-term changes in the acidity of forest soils in north-east Scotland', *Soil Use and Management*, 4: 102–7.

Birdsey, R.A. (1990) 'Inventory of carbon storage and accumulation in U.S. forest ecosystems', in *Proceedings IUFRO World Congress, Montreal, 5–11 August 1990*.

Borchert, R. (1976) 'Differences in shoot growth patterns between juvenile and adult trees and their interpretation based on systems analysis of trees', *Acta Horticultura* 56: 123–30.

Bruns, D.A., Wiersma, G.B. and Rykiel, E.J. (1991) 'Ecosystem monitoring at global baseline sites', *Environmental Monitoring and Assessment*, 17: 3–31.

Brzeziecki, B., Kienast, F. and Wildi, O. (1993) 'Simulated map of the potential natural forest vegetation of Switzerland', *Journal of Vegetation Science*, 4: 499–508.

Bugmann, H. and Fischlin, A. (1994) 'Comparing the behaviour of mountainous forest succession models in a changing climate', this volume.

Cline, S.P. and Burkman, W.G. (1989) 'The role of quality assurance in ecological research programs', in J.B. Bucher and I. Bucher-Wallin (eds), *Air Pollution and Forest Decline*, Eidgenössische Anstalt für das forstliche Versuchswesen, Birmens-dorf, pp. 361–5.

Edwards, G.S., Wullschleger, S.D. and Kelly, J.M. (1993) 'Growth and physiology of northern red oak: preliminary comparisons of mature tree and seedling responses to ozone', *Environmental Pollution*, 83: 215–22.

Ellenberg, H. (1979) 'Veränderungen der Flora Mitteleuropas unter dem Einfluss von Düngung und Immissionen', *Schweizerische Zeitschrift für Forstwesen*, 136: 19–39.

Falkengren-Grerup, U. and Tyler, G. (1991) 'Dynamic floristic changes of Swedish beech forest in relation to soil acidity and stand management', *Vegetatio*, 95: 149–58.

Fox, D.G. and Bernabo, J.C. (1987) *Guidelines for Measuring the Physical, Chemical, and Biological Condition of Wilderness Ecosystems*, General Technical Report RM–146, United States Department of Agriculture Forest Service, Fort Collins.

Grigal, D.F. and Ohmann, L.F. (1992) 'Carbon storage in upland forests of the Lake States', *Soil Science Society of America Journal*, 56: 935–43.

Hallbäcken, L. (1992) 'Long term changes of base cation pools in soil and biomass in a beech and a spruce forest of southern Sweden', *Zeitschrift für Pflanzenernährung und Bodenkunde*, 155: 51–60.

Hochstein, E. and Hildebrand, E.E. (1992) 'Stand und Entwicklung der Stoffeinträge in Waldbestände von Baden-Württemberg', *Allgemeine Forst- und Jagdzeitung*, 163: 21–6.

Innes, J.L. (1991) 'High-altitude and high-latitude tree growth in relation to past, present and future global climate change', *The Holocene*, 1: 168–73.

—— (1993a) *Forest Health. Its Assessment and Status*, Commonwealth Agricultural Bureau, Oxford.

—— (1993b) '"New perspectives in forestry": a basis for a future forest management policy in Great Britain?' *Forestry*, 66: 395–421.

Innes, J.L., Landmann, G. and Mettendorf, B. (1993) 'Consistency of observations of forest tree defoliation in three European countries', *Environmental Monitoring and Assessment* 25: 29–40.

Jensen, A. (1992) 'Terrestrisk naturovervåking. Overvåking av jord og jordvann 1991', *Rapport fra Skogforsk*, 9/92.

Kimmins, J.P., Scoullar, K.A. and Apps, M.J. (1990) *FORCYTE-11 User's Manual for the Benchmark Version*, Forestry Canada, Ottawa.

Koop, A. (1989) *Forest Dynamics. SILVI-STAR: A Comprehensive Monitoring System*, Springer Verlag, Berlin.

Kuhn, N. (1990) *Veränderung von Waldstandorten. Ergebnisse, Erfahrungen und Konsequenzen, mit einem Konzept für die Dauerbeobachtung von Waldbeständen*, Berichte 319, Eidgenössische Anstalt für das forstliche Versuchswesen, Birmensdorf.

—— (1992) 'Dauerbeobachtung in Schweizer Waldbeständen. Betreung von Waldreservaten', unpublished MSS, Swiss Federal Research Institute for Forest, Snow and Landscape Research, Birmensdorf.

Küttel, M. (1990) 'Der subalpine Schutzwald im Urserental – ein inelastisches Ökosystem', *Botanica Helvetica*, 100: 183–97.

Leuppi, E. and Forster, F. (1990) 'Zur Frage der Wirksamkeit des Waldes für den Hochwasserschutz – ein Beispiel aus dem oberen Reusstal', *Schweizerische Zeitschrift für Forstwesen*, 141: 943–54.

Marx, J. (1986) 'Randbedingungen für die Walderschliessung längs der Gotthard-Nordrampe', *Schweizer Zeitschrift für Forstwesen*, 137: 545–55.

Nelleman, C. (1992) 'Vitalitetsregistreringer på faste intensive overvåkingsflater 1986–91', *Rapport fra Skogforsk*, 20/92.

Newton, I. (1988) 'Determination of critical pollutant levels in wild populations, with examples from organochlorines in birds of prey', *Environmental Pollution*, 55: 29–40.

Newton, I. and Galbraith, E.A. (1991) 'Organochlorines and mercury in the eggs of Golden Eagles *Aquila chrysaetos* from Scotland', *Ibis* 133: 115–20.

Nihlgård, B. and Pylvänäinen, M. (1992) *Evaluation of Integrated Monitoring in Terrestrial Reference Areas of Europe and North America. The Pilot Programme 1989–1991*, Environment Data Centre, National Board of Waters and the Environment, Helsinki.

Nüsslein, S. (1992) 'Untersuchungen bei den Waldklimastationen in Bayern', *Allgemeine Forst Zeitschrift*, 10/1992: 534–7.

Petty, S.J. and Avery, M.I. (1990) *Forest Bird Communities*, Forestry Commission Occasional Paper 26, HMSO, London.

Post, W.M., Emanuel, W.R., Zinke, P.J. and Stangenberger, A.G. (1982) 'Soil carbon pools and world life zones', *Nature*, 298: 156–9.

Preuhsler, T., Gietl, G., Grimmeisen, W., Kennel, M. and Lechler, H.H. (1992) 'Forschungsprojekt Waldklimastationen in Bayern', *Allgemeine Forst Zeitschrift*, 10/1992: 529–33.

Sanasilva Waldschadenbericht (1991) Swiss Federal Institute for Forest, Snow and Landscape Research, Birmensdorf, and BUWAL, Berne.

Schlaepfer, R. (1992) 'Waldschadenforschung in der Schweiz: Eine Synthese', in *Waldschadenforschung in der Schweiz: Stand der Kenntnisse*, Forum für Wissen 1992, Eidgenössische Forschungsanstalt für Wald, Schnee und Landschaft, Birmensdorf, pp. III–VIII.

Silsbee, D.G. and Peterson, D.L. (1991) *Designing and Implementing Comprehensive Long-term Inventory and Monitoring Programs for National Park System Lands*, Natural Resources Report NPS/NRUW/NRR–91/04, United States Department of the Interior, Denver.

Terborgh, J. (1989) *Where Have All the Birds Gone?*, Princeton University Press, Princeton.

Thomsen, M. (1991) 'Long-term monitoring sites', in J.L. Innes (ed.), *Interim Report on Cause–Effect Relationships in Forest Decline*, United Nations Economic Commission for Europe, Geneva.

Wiersma, G.B. (1990) 'Conceptual basis for environmental monitoring programs', *Toxicological and Environmental Chemistry*, 27: 241–9.

Wiersma, G.B., Otis, M.D. and White, G.J. (1991) 'Application of simple models to the design of environmental monitoring systems: a remote site test case', *Journal of Environmental Management*, 32: 81–92.

Part III

SOCIO-ECONOMIC ASPECTS OF CLIMATE CHANGE IN MOUNTAIN REGIONS

19

CLIMATE RISK CONCERN IN AN ALPINE COMMUNITY

On the social embeddedness of risk-perception

G. Dürrenberger, H. Kastenholz, and R. Rudel

INTRODUCTION

Global warming caused by human activities has become a main topic in scientific as well as in policy debate (Schneider, 1989; Cline, 1991; IPCC, 1991; Stern *et al.*, 1992). The facts that concentration of CO_2 increased by about 25 per cent and of CH_4 by 100 per cent since the begining of the industrial era, and that these increases have been caused by human activities are not controversial. The same is true for the assessment of the effects of CFCs. There is general agreement that CFCs have contributed about 25 per cent to the total radiative effect in the last ten years. It is also a recognized fact that over 50 per cent of greenhouse gas warming stems from CO_2, that is, from the burning of fossil fuels and deforestation. What is controversial, however, is how the enhanced greenhouse effect will influence temperature, rain-patterns, storm activities, sea level, permafrost, soil moisture, and other parameters (for a review see Dickinson, 1989).

Forecasts of potential climatic impacts of, for instance, a doubled CO_2 concentration in the atmosphere, are highly uncertain (Kates *et al.*, 1985; EPA, 1990; IPCC, 1991). Important processes such as changes in ocean-circulation are not yet fully understood and regional climate modelling is still in its infancy (Giorgi and Mearns, 1991; Malone and Yohe, 1992). Against such a background, the assessment of societal impacts of global warming is even more problematic. Current judgements range from beneficial to disastrous (Schneider, 1991; Ausubel, 1991; Fajer and Bazzaz, 1992). This uncertainty is unlikely to be reduced in the near future, which has led to a 'wait-and-see attitude' in policy debate: efficient action should not be taken before reliable scientific predictions about the dynamics of climate change are available. On the other hand waiting could substantially increase future costs of protection measures against the effects of greenhouse warming. Hence, it is sensible to hedge against those risks.

A major problem society faces is the fact that no reliable assessment of potential monetary losses (or profits) attributable to global warming can be made (Tobey, 1992). This hampers the process of determining the optimal levels of environmental taxes aimed at an internalization of external costs of, for instance, CO_2 emissions (see Green, 1993). The implementation of measures, therefore, depends to a considerable extent on people's perception of the risks of the greenhouse effect, and on the interests associated with potential measures. The latter are likely to nourish opposing risk-preferences, for instance between developed and developing countries or between energy-intensive and energy-extensive industries. Negotiating targets of emission limits becomes a very difficult task, on the international as well as on the national level (Haas, 1990; Morrisette *et al.*, 1991).

Because in democratic societies majorities are needed to implement political measures, the formation and mobilization of public support for a specific measure is cardinal. Hence, it will be an important scientific objective for research in the field of human dimensions of climate change to understand how climate risk awareness develops into climate risk concern and into climate-relevant political will (Löfstedt, 1991; Kempton, 1991). In our paper, we will focus on the first two aspects, that is, on determinants of climate risk concern. A small step towards the third aspect is made by relating concern to actions.

In the first part of the paper, we will very briefly address work in the field of human dimensions of climate change. In doing so, we will focus on a theoretical deficiency of the risk-perception approach: its atomistic view of human agency. We will argue that risk perception is strongly shaped by the social milieux in which people are embedded. This is especially true for alpine regions, which are less strongly influenced by modern individualistic lifestyles. Against this background, in the third section we will present our survey on climate risk awareness conducted in a Swiss mountain region, the Surselva, and discuss some methodological issues. In the fourth section we will discuss main findings. We will end the paper by drawing some conclusions with regard to further research needs in the field of human dimensions of climate change.

TOWARDS A NON-ATOMISTIC VIEW OF RISK PERCEPTION

We will broadly distinguish between three approaches that deal with human dimensions of climate change. First, impact studies (undertaken mainly by economists), second, perception studies (mostly advanced by psychologists and geographers), and third, sociological accounts (carried out by anthropologists, sociologists and political scientists).

1. During the last decade or so, environmental impact-assessment has developed into an influential business (Dietz and Rycroft, 1987). Two

different types of impact studies can be distinguished:

First, studies that try to assess the physical impacts of environmental risks, that is, the magnitude of damage caused by possible future events associated with a warmer atmosphere. One kind of study relates to health, and is mainly oriented at assessing possible health-risks associated not only with higher levels of tropospheric ozone and radiation, but also with changed spread-patterns of diseases (see, e.g., Gellert, Neumann and Gordon, 1989; Calabrese, Gilbert and Beck, 1990; WHO, 1990). A large body of impact knowledge has also been gathered with respect to natural hazards that relate to climate change, on the global (see, e.g., Ausubel, 1991; IPCC, 1991) as well as on the regional level (Glantz, 1988; Schmandt, 1991; Riebsame and Magalhaes, 1991). Most of this work has been undertaken by modellers and natural scientists. In contrast to technical risk assessment, the analysis of climatic risks is much more complex. Uncertainty is considerable and the general opinion is that it cannot be significantly reduced in due course. This fact is of the utmost importance for the second type of impact-study.

There is an increasing demand for studies related to the costs and benefits of measures that protect us against negative environmental impacts, for instance in order to be able to insure against potential financial losses resulting from future climate disasters such as, for example, a decrease in crop-production due to reduced precipitation (see Peele, 1987; Parry et al., 1988). On the other hand, a growing number of inquiries deal with the assessment of costs arising from measures to prevent such losses, for example costs linked to the introduction of a CO_2 tax that reduces the emission of this greenhouse gas (Manne and Richels, 1991; Nordhaus, 1991; Rothen, 1993) and, hence, the probability that critical shifts in precipitation patterns will take shape.

2. In those economic models a human actor is conceived as an atomistic being with a given preference structure and limited (monetary) resources, and society is treated as the aggregate outcome of myriads of individual actions performed by thousands of isolated rational actors (Granovetter, 1985). According to this atomistic conception of human actors, in a world of given outcomes an individual can choose between those actions that best satisfy his or her preferences, that is, that maximize self-interest. In the case of uncertain outcomes, risk-preferences can be introduced into the model. An impressive stock of knowledge that deals with such preferences has been accumulated by psychological studies (Kahnemann and Tversky, 1979; Fischhoff et al., 1981; Covello et al., 1983). In a nutshell, it has been shown that in analogue situations people often behave differently: some are willing to take a risk, others behave risk-aversely. This can be illustrated by the following two game-situations.

A person can choose between two alternatives: (a) to get a certain amount of money for free and (b) to take part in a lottery with the same expectation value but the chance to win an even greater amount of money, for instance, to win four times as much money with a probability of 0.25 and with the associated risk of winning nothing at all with a probability of 0.75. It has been empirically shown, that if the money at stake is high, most people will restrain from gambling and take the lump sum. On the other hand, if the situation is framed in terms of losses, persons seem to be more willing to engage in the lottery, that is, to take the risk.

If a person is faced with a range of possible outcomes, each of which promises the same utility, the criterion of utility-maximization is not relevant for selection. An additional criterion is needed. However, this implies that preferences can no longer be deduced from actual behaviour, which jeopardizes fundamental assumptions of the atomistic rational-choice approach (Sen, 1982; Kirman, 1989; Granovetter, 1992). If preferences and behaviour diverge, the study of the inter-relationship between (individual) dispositions and (social) forces becomes important.

3. Actions may not only be shaped by preferences, but also by social factors such as rules or values. Anthropologists and sociologists have tried to identify the social factors that influence risk-behaviour (Douglas and Wildavsky, 1982; Heimer, 1988; *Journal of Cross-Cultural Psychology*, 1991; Short and Clarke, 1992). In this view, a person is embedded in a social environment that shapes individual preferences and behaviour.

By introducing the social realm, a fundamental problem of the atomistic model can be resolved – the problem of the formation of altruistic actions. Analytically, this problem can be described in terms of the prisoner's dilemma known from the theory of games: in the absence of shared social rules that settle conflicts and guide action to the advantage of the community as a whole, social actors behave egoistically in ways which lead to sub-optimal results, at least in the short run (Axelrod, 1984).

The fact that humankind is actually going to destroy the planet's climate is an alarming, yet clear indication that society is trapped in a prisoner's dilemma of global magnitude. Because the environmental costs associated with the emission of greenhouse gases are highly uncertain and are not only carried by the polluter but by all humankind, a nation has no strong economic incentives to reduce emissions. If every single nation is following this rationale, the international community is trapped in a global prisoner's dilemma that leads to what has become known as a 'tragedy of the commons' (Hardin, 1968; WCED, 1987).

From a theoretical point of view, the resolution of this dilemma entails the replacement of the atomistic conception of rational individuals who act solely in pursuit of self-interest, by an account which embeds such behaviour in social structures, that is, in rules and trust-networks that allow co-operative social behaviour (with respect to risks, see Renn and Levine, 1988; Argyle, 1991; Short, 1991). In such a view, research on human dimensions of climate change should not only concentrate on impact studies but also on investigations into the social construction (and reconstruction) of climate risk perceptions and, more generally, of environmental risk awareness. We should take seriously the fact that people do not simply behave according to a well-defined hierarchy of personal preferences, but also follow social expectations, roles, and rules.

Against this background, the probability of a person being concerned by climate risks should considerably depend on social factors that have the power to shape his or her perception. A person embedded in an environmentally conscious milieu, for instance, may be significantly more risk-averse *vis-à-vis* environmental risks than a person who is primarily shaped by an entrepreneurial culture oriented toward risky actions (see, for example, Krimsky and Golding, 1992). In this paper, we will not focus on such subtle differences but on the general hypothesis that risk perception is strongly shaped by social and cultural factors, and not only by personal factors in a narrow sense. By stating this, we are well aware of the fact that the latter expose social characteristics, too. However, we discriminate between first-order and second-order indicators. Gender, age, status, and knowledge are treated as second-order indicators: networks, rules and values as factors of the first order.

Our first hypothesis is that 'inter-personal factors' such as social networks (Wellman and Berkowitz, 1988), rules, and values (Burns and Dietz, 1992) to which a person is exposed are better determinants of climate risk concern than 'personal factors' such as demographic attributes or the state of knowledge (see also Samdahl and Robertson, 1989). Our second hypothesis is that the influence of social networks, rules, and values even increases when we shift from concern to actions, especially when we focus on political actions. This hypothesis relates to the very core of social theory: actions cannot be meaningfully conceptualized as isolated behaviour of an atomistic actor but only as inter-personal (or inter-subjective) behaviour of a social being (Harré, 1979).

In the following section we will describe our sample, the variables, and the methodology used to test our hypotheses of the social embeddedness of risk concern and environmentally relevant actions. In the fourth section, we will present and discuss the findings.

DESCRIPTION OF SURVEY AND METHODOLOGY

To study climate risk awareness, we have distributed questionnaires in an alpine region in Switzerland, the Surselva, a mountain area in the western part of Canton Grison (see Kastenholz, 1992; Jaeger *et al.*, 1993). Climate risks have become relevant for this region for two major reasons. First, heavy storms have repeatedly hit the mountain forest. This has stimulated regional discussions about climate change and storm activity. Second, winter tourism, a main economic asset of the Surselva, has been eroded by declining skiing conditions due to a lack of snow in recent winters. This fact, too, was discussed in the context of climate change: it was argued that sluggish business cannot be regarded as an instance of ordinary entrepreneurial risk and, hence, decreased income should be compensated by dole. Indeed, the Swiss Federal Government accepted this argument and granted unemployment pay in 1991. The Surselva has about 23,000 inhabitants, 50 per cent working in the service sector, 30 per cent in industry, and 20 per cent in agriculture and forestry. We have distributed a standardized questionnaire to a random sample of people over 16 years old (n = 230). The sample is representative for the region.

In order to test our first hypothesis, we built two models with risk concern as dependent variable. The first model, known as the 'atomistic model', is focused on personal attributes, that is, demographic characteristics (gender, age, status) and knowledge. The second model, named the 'social model', is focused on inter-personal variables such as social networks, family structure, rules, and cultural values. As we discriminated between climate risk concern and environmental risk concern we could calculate two versions of each model, one climate-oriented and one environment-oriented.

The second hypothesis was tested by the same method. As dependent variable, we replaced concern by action. Again, we discriminated between a climate-oriented version and an environment-oriented version. In addition, in the case of climate-relevant actions we also differentiated between, on the one hand, political actions and, on the other hand, actions performed in the private and vocational spheres.

All variables were treated as logical predicates. A logical predicate is a formalized statement like 'She is sick' or 'I'm concerned by climate risks.' Whether a specific statement is true or not strongly depends on the context. Sometimes, one cannot clearly judge and the statement remains undetermined. In this paper, however, we have defined our predicates in a boolean sense, that is, a formalized statement can be 'true' or 'false' (for a three-valued logic of formalized statements see Blau, 1977; for a detailed discussion of the difference between formalized statements and quantitative statements see Jaeger *et al.*, 1993).

We now turn to a short description of the variables. First, the dependent variables. We defined the predicate 'climate risk concern', that is, the fact that somebody is distressed by the possibility that climate change may lead to severe environmental disruptions and social problems, on the basis of four items (questions emphasizing this personal distress). If a person expressed concern in at least three of the four questions (on a five-point Likert-type answer format ranging from 'strongly agree' to 'strongly disagree') the predicate value was set 'true', that is, the person was regarded as concerned about climate risks. Generally, the same methodology was applied to the other dependent variables. The only thing that changed was the number of items. In the following discussion of the predicates, after indicating the number of items used to define a predicate, we put in brackets the minimum number of questions that had to be answered positively for the predicate in question to be judged 'true'.

In the case of the predicate 'environmental risk concern', three items (two) were used. For the predicate 'climate-relevant actions' we used nine (six) items encompassing actions towards energy saving, steps to become informed about climate risks, or commitments to political activities. The predicate 'environment-relevant actions' was handled in exactly the same way. The nine questions concerned steps towards the recycling of household garbage, actions aimed at reducing car-use, or active support of political activities in favour of the environment. In the case of climate-relevant actions, we also diffentiated into two subgroups: 'political actions' and 'personal actions'. The former consisted of three items (two), the latter of the remaining six (four).

Some of the independent variables were based on a single item, namely gender, age, family, or social networks. Gender is self-explanatory, age was categorized into two groups (younger than 40 years, 40 years and older). Family status was treated as a predicate that discriminates between people who live with and educate children and people who do not. The predicate 'climate-relevant networks' ('environment-relevant networks'), finally, tried to assess the involvement of persons in social milieux in which climate problems (environmental problems) are issues that matter. We asked how often a person was engaged in conversations about climate problems (environmental problems). If a person was engaged in such conversations at least once a week, we assigned the value 'true' to the predicate.

Four variables were defined on the basis of several items, namely status, knowledge, rules, and cultural values. The predicate 'status' was categorized on the basis of two items into the four groups 'high', 'middle', 'low', and 'undetermined'. The two items referred to formal education and current business position. The variable 'knowledge' was built up of three items (two) that related to the mechanism and nature of the greenhouse effect. The predicates 'climate-relevant rules' and 'environment-relevant rules' were composed of two (two) and four (three) items that registered a person's exposure to social and moral rules oriented at climate and environmentally relevant actions. Finally, we looked at cultural values. We conceived them as more or less coherent sets of rules shared by a large proportion of a population. For our survey, one such set considered to exert strong influence on people's perceptions and actions was of special interest: the so-called 'New Environmental Paradigm' or NEP (Dunlap and Van Liere, 1978). We used five items (four) which are very close (albeit not identical) to some of the twelve questions originally used by Dunlap and Van Liere.

To test our hypotheses, we used logit-models which are most appropriate to the analysis of categorical data. In a logit-model, a predicate's probability of occurrence is modelled by a logit function in dependence on the probability of occurrence of other predicates (for the mathematics of logit

G. DÜRRENBERGER, H. KASTENHOLZ AND R. RUDEL

models see Cramer, 1991). To test the first hypothesis, we built two models. In the 'atomistic model' the predicates 'gender', 'age', 'status', and 'knowledge' were used. All of those predicates primarily characterize persons and do not directly stand for the social and cultural environment in which a person is embedded. This is the case when we refer to the predicates 'family', 'climate-relevant networks' ('environment-relevant networks'), 'climate-relevant rules' ('environment-relevant rules') and NEP. Accordingly, those variables were entered into the 'social model'. As dependent variables we used the predicates 'climate risk concern', 'environmental risk concern', 'environment-relevant actions', 'climate-relevant actions', climate-relevant 'political actions', and climate-relevant 'personal actions'.

The models were calculated with the standard software 'JMP' on an Apple Macintosh computer. For reasons of clarity, we restrict the presentation of the findings to a minimum of statistical details, that is, to the model fit, the probability of the ChiSquare of the model, the parameter estimates and the probabilities of the ChiSquare of the variables.

RESULTS AND DISCUSSION

Before presenting the results of the models, we look at the frequency distributions of risk concern. People are strongly concerned by environmental and climate risks. About half of the population is dismayed by those risks. In a similar, albeit not identical, survey conducted in an urban region – the region of southern Ticino – we have found comparable levels of environmental concern. Although this region is not as directly affected by the climate change issue as is the Surselva (the Ticino is heavily burdened by air pollution, NO_x and tropospheric ozone, that stems to a very large part from transit and regional cross-border traffic), concern with regard to climate risks has been somewhat greater (up to 60 per cent) in the Ticino than in the Surselva. A reason for that might be that air pollution and climate risks are sometimes perceived as similar phenomena (see Löfstedt, 1991).

An indication of climate problems not being separated from other ecological problems is given by the fact that in a logit-model with climate and environmental concern as the only variables, a highly significant impact (significance level below 0.0001) of environmental risk concern on climate risk concern (model fit was 0.12 in both samples) can be found. Those concerned by climate risks are concerned by environmental problems, too – and vice versa. Indeed, it seems that climate risk awareness is a special aspect of a much broader awareness of environmental problems (Henderson-Sellers, 1990; Kempton, 1991).

Let us now turn to our two models. We will first present the results of the models that refer to risk concern. Then we will look at the corresponding models with climate- and environment-relevant actions as dependent variables.

308

Table 19.1 Atomistic models of risk concern

	Climatic risk concern	Environmental risk concern
Gender (men)	–0.12 (0.44)	0.10 (0.53)
Age (elderly)	0.12 (0.39)	0.00 (0.95)
Status (high)	–0.44 (0.24)	–0.01 (0.99)
Status (low)	–0.48 (0.23)	–0.53 (0.18)
Status (middle)	0.41 (0.10)	0.36 (0.14)
Knowledge (high)	0.39 (0.01)	0.17 (0.24)
Model Fit	0.04	0.02
ProbChiSquare	0.04	0.51

Table 19.2 Social models of risk concern

	Climatic risk concern	Environmental risk concern
Family (false)	0.17 (0.25)	0.31 (0.04)
Networks (false)	–0.19 (0.23)	–0.40 (0.01)
Rules (false)	–0.34 (0.03)	–0.47 (0.00)
NEP (false)	–0.47 (0.00)	–0.10 (0.50)
Model Fit	0.06	0.08
ProbChiSquare	0.00	0.00

As shown in Table 19.1, the atomistic models of risk concern display very poor results (the figures in the table represent the parameter estimates and, in brackets, the probabilities of the ChiSquare of the variables). In fact, the probabilities of the ChiSquares of the models show such high figures that the variables cannot be regarded as sensible predictors of climate and environmental risk concern. The same holds for 'gender', 'age', and 'status'. A partial exception is the predicate 'knowledge' that is significant at a level of 1 per cent in the climate-focused model. The frequency distribution of the predicate shows that about 30 per cent of the people have an adequate knowledge of the nature of the greenhouse effect. This indicates that knowledge can be important for the formation of risk concern, though it is, however, not essential. Hence, the proliferation of scientific information and expert knowledge via the media to the general public is just one factor that increases risk awareness among laypersons.

Table 19.2 shows the two versions of the social model of risk concern. In contrast to the atomistic model, the social variables give a much better model fit and can be regarded as significant predictors of risk concern. Thus, the data strongly confirm our hypothesis that concern about climatic and environmental risks is not primarily shaped by personal attributes but by the social milieu of a person.

Some differences between the version focused on climate risks and the model focused on environmental risks deserve additional comments. First, environmental risk concern can be better predicted by the variables used than climate risk concern. This implies that the social significance of environmental degradation is higher than the social relevance of climate problems. A plausible explanation for this phenomenon is the fact that environmental degradation can, in most cases, be experienced sensorily whereas climate risks are perceivable by means of scientific evidence only. Another reason might be the fact that, compared with environmental problems, climate problems are relatively new issues.

Second, Table 19.2 shows that the two versions are complementary in a way that suits our first hypothesis. Those variables that directly refer to the social milieu of a person (family, networks, rules) are better predictors of environmental concern than of climate concern. This is largely due to the fact that environmentally conscious social groups are more widespread than climate conscious groups. In fact, only 28 per cent (29 per cent) of the people in our sample are exposed to climate-relevant rules (networks) whereas in the case of environment-relevant rules (networks) the corresponding figures amount to 43 per cent (41 per cent). In the case of the predicate 'family', significance in predicting risk concern is only 75 per cent in the climate-focused model but over 95 per cent in the environment-focused model. What is especially striking, however, is the fact that people with families are less concerned about environmental problems than people who live without children. A possible explanation of this finding could relate to repression. Yet this conclusion is premature because, as we will see, people with families score higher in the models oriented at action. This indicates that concern also depends on activities: People who engage in ecological actions are, so to speak, prejudiced in favour of the environment, whereas people who refrain from such activities are more fatalistic about the future state of nature.

Third, this assessment is partly confirmed by the variable NEP. Van Liere and Dunlap (1983) have shown that environmental concern is integrated into a rather broad belief-system which they have called the 'New Environmental Paradigm'. This paradigm includes basic beliefs about, for instance, limits to growth or human responsibility for nature. It generally conflicts with established cultural orientations towards technical progress and material wealth. Catton and Dunlap (1978) have called the latter 'Human Exceptionalist Paradigm', or HEP, because humankind is not considered to be an intrinsic part of nature. Concerning our models, the significance of the NEP substantially differs between the environmental and the climate version. People who have internalized the NEP are clearly more concerned about climate risks than people who do not stick to the new environmental paradigm. This influence is highly significant.

With respect to environmental risks, however, the influence of the NEP vanishes (though this is not the case in the urban sample). We have no

Table 19.3 Social models of ecologically relevant actions

	Climate actions	*Environ. actions*
Family (false)	−0.34 (0.04)	−0.29 (0.08)
Networks (false)	−0.41 (0.01)	−0.79 (0.00)
Rules (false)	−0.32 (0.04)	−0.55 (0.00)
NEP (false)	−0.22 (0.15)	−0.23 (0.14)
Model Fit	0.07	0.16
ProbChiSquare	0.00	0.00

straightforward explanation of this fact but we suspect, first, that the strong influence of the other variables led to a decrease in the relative importance of the NEP in the model. Second, in the absence of robust-climate relevant social networks and rules, more general rule-systems such as the NEP are likely to exert a strong influence on people's climate risk perceptions. In the case of environmental risks, consolidated rules exist that point at environmentally conscious activities and significantly shape concern.

We now turn to our second hypothesis (see also Jaeger *et al.*, 1993), the hypothesis that the predictive quality of the social model increases when we replace concern by action. This is indeed the case (Table 19.3). Model fit has increased in both versions, and the version with 'environmentally relevant actions' as a dependent variable has by far the best fit of all models discussed. In both models the NEP is no longer significant, either in the case of the environment focused model or in the case of the climate oriented version. However, the predicates 'family', 'networks', and 'rules' that represent a somewhat more direct influence of the social environment on people's activities show strong impacts.

To confirm our assessment we built an atomistic model with 'climate-relevant actions' and 'environment-relevant actions' as dependent variables. The independent variables already used in our first model were supplemented by the variables 'climate risk concern' and 'evironmental risk concern' respectively (Table 19.4). Model fit of the environment focused model is clearly weaker compared with the social model. In the case of the climate focused model, fit has somewhat increased. However, this is largely due to the fact that an additional variable (which moreover strongly affects the dependent variable) has entered into the model.

In the atomistic model, 'age' and 'concern' are the only two variables which significantly influence ecologically conscious actions. In the case of 'concern', this should not cause much surprise because, as discussed above, risk concern is, so to speak, pre-shaped by the social environment. As to 'age', however, there is no ready explanation at hand. Why should older people act more in favour of the environment than younger persons? We have seen that concern does not differ very much between the older and the

Table 19.4 Atomistic models of ecologically relevant actions

	Climate actions	Environ. actions
Gender (men)	0.27 (0.12)	0.01 (0.95)
Age (elderly)	0.53 (0.00)	0.43 (0.00)
Status (high)	0.42 (0.29)	0.50 (0.18)
Status (low)	−0.99 (0.05)	−0.61 (0.16)
Status (middle)	0.49 (0.08)	−0.04 (0.88)
Knowledge (high)	0.17 (0.28)	0.21 (0.17)
Concern (high)	0.29 (0.05)	0.50 (0.00)
Model Fit	0.08	0.08
ProbChiSquare	0.002	0.005

Table 19.5 Social models of climate-relevant political and personal actions

	Political actions	Personal actions
Family (false)	−0.27 (0.13)	−0.40 (0.02)
Networks (false)	−0.32 (0.06)	−0.37 (0.02)
Rules (false)	−0.46 (0.01)	−0.28 (0.08)
NEP (false)	−0.30 (0.07)	−0.18 (0.23)
Model Fit	0.08	0.07
ProbChiSquare	0.00	0.00

younger generation. A convincing reason might relate to lifestyles. The older generation has experienced nature as an integral part of rural community life. For instance, managing commons (see Netting, 1981) has strengthened ties not only to the community but also to the environment. The younger generation, on the other hand, has been attracted by a modern lifestyle that stresses individualistic and urban values. This lifestyle has become influential in the region with the recent growth of the tourist industry.

This assessment is partly confirmed by the fact that in the urban sample environmental concern is primarily shaped by gender and the NEP whereas the role of social networks is less important. Women are more aware of environmental risks than men and people having internalized the NEP are more likely to be concerned about environmental problems than people sticking to modernist values. This seems to indicate that the urbanite's lifestyle is more individualistic and can, hence, be more effectively formed by general values, or, to put it the other way round, it is less strongly shaped by close social relations. However, as has been mentioned at the beginning of the third section we have to be very cautious with this assertion because of differences between the two samples and in the variables that entered into the models.

In order to assess the significance of the social variables for the formation of political will to take actions with regard to climate risks, we have built a model that differentiated the predicate 'climate-relevant actions' into climate-relevant 'political actions' and climate-relevant 'personal actions', which encompass private activities as well as actions undertaken in business life (Table 19.5). This more detailed analysis gives an interesting insight for our argument about the relevance of the social environment for climate-conscious actions.

Table 19.5 shows that the two versions with political actions and personal actions do not differ substantially. Model fit is similar and the 'networks' and 'rules' variables are significant for political as well as for personal actions. 'Family' is more important in the version with 'personal actions' as dependent variable whereas the influence of the NEP is more accentuated in the model with 'political actions', that is, the NEP seems to influence the willingness to take political action. In the private and business sphere, however, this cultural orientation is less important with regard to action. This supports the assessment that the NEP is an important factor with regard to the build-up of public concern and action. Yet what is most striking in our analysis is the fact that political and personal actions cannot be neatly separated from each other. Hence, the fostering of climate-relevant actions in the personal sphere is likely to contribute to the formation of political actions, for example the willingness actively to support measures aimed at the management of climate risks.

CONCLUSION

Our data strongly support a non-atomistic view of risk perception. Predicates such as age, gender, and status, which refer to personal attributes, are weak predictors of risk concern. On the other hand, the fact that a person is embedded in social networks and exposed to moral rules and values that favour ecological actions is a clear and significant indication that he or she is concerned about ecological problems as well as that she or he will take ecologically relevant action. In the case of climate risks, rules and values exert a prime influence on people's perceptions whereas in the case of environmental problems the influence of general environmental values fades away and the importance of social networks and rules strongly increases.

A plausible explanation of this fact might be that the greenhouse effect is an abstract, only scientifically conceivable phenomenon which has not yet become integrated into the social networks. The formation of climate risk concern is limited to 'pioneers'. Gradually, pioneers may become a social reality, which not only reinforces the pioneers' assessment of climate risks but also becomes a relevant environment for common actors. Climate risk concern then has the status of a social affair (Roqueplo, 1992).

Thus, we can postulate the following process of building up social awareness of climate risks. (i) Information capable of mobilizing environmental values embodied in social milieux builds up concern in individuals who are embedded in those milieux. (ii) Concerned individuals establish new social rules of climate-conscious behaviour. (iii) Awareness of people embedded in such rule-systems is likely to increase. Value-biases favouring environmental actions, as in the case of the NEP, will clearly support this process.

If we shift from concern to action, the influence of action-oriented rules and networks is even more important. Such rules and networks seem to be the crucial forces that turn awareness into action. Our models support the view that actions are much more strongly shaped by social factors than by personal factors such as demographic variables and knowledge.

This indicates that social milieux strongly influence perceptions of climate risks and the willingness of people to take climate-relevant environmental action. However, further empirical work has to be done, for instance with respect to different social groups such as entrepreneurs, politicians, grassroot-movements, the sciences, etc. Are there major differences in the formation of risk awareness between those groups? Do possible differences hinder or facilitate processes of social learning aimed at a reduction of climate risks?

An important arena to study such processes is the region. People are much more willing to take action when such action is perceived as beneficial for their own lives. Therefore, future research on the human dimensions of climate change should increasingly focus on how global climate risks translate into regional phenomena of social and individual relevance. Innovative alpine regions with their historical tradition of resolving problems associated with the management of commons may become important arenas of such regional risk-research.

Indeed, the study of the human dimensions of global environmental change must beware of the pitfall of looking for global strategies only. Global strategies focus mainly on international diplomacy that confines itself to negotiating regulatory targets. Clearly, the topic of regulation is an important issue in regional risk-debates. However, we consider the regional arena as a promising platform for the development of social innovations ultimately needed to reduce climate risks. On the national and international level, social milieux are more rigidly separated from each other. Hence, fruitful risk-communication that leads to innovative solutions is more difficult to establish. Within regional milieux, however, environmentally relevant development options may be more easy to identify and to seize by the forelock. Hence, innovation-oriented regional research may sensibly complement the study of regulation-oriented global and national strategies.

In theoretical respects, this implies a certain shift from atomistic reasoning and modelling towards the study of social networks and rule-systems in order to develop a more comprehensive understanding of processes of, for instance, social learning and social change. In tackling global environmental problems like climate change, such an understanding is badly needed to foster social processes like the development of new technologies that are environmentally less harmful, shifts in consumer preferences towards less energy-consuming products, or the formation of political will to take regulatory action against further greenhouse warming.

ACKNOWLEDGEMENTS

We wish to thank C. Jaeger, O. Renn, S. Rothen and B. Truffer for helpful discussions on topics raised in this paper.

The paper is part of a research project on regional strategies against global climate risks, funded by the Swiss National Foundation for Scientific Research under grant #4031–033525.

REFERENCES

Argyle, M. (1991) *Cooperation. The Basis of Sociability*, Routledge, London.
Ausubel, J.H. (1991) 'A second look at the impacts of climate change', *American Scientist*, 79: 210–21.
Axelrod, R.M. (1984) *The Evolution of Cooperation*, Basic Books, New York.
Blau, U. (1977) *Die dreiwertige Logik der Sprache*, de Gruyter, Berlin.
Burns, T. R. and Dietz, T. (1992) 'Cultural evolution: social rule systems, selection and human agency', *International Sociology*, 7: 259–83.
Calabrese, E.J., Gilbert, C.E. and Beck, B.D. (eds) (1990) *Ozone Risk Communication and Management*, Chelsea, Michigan.
Catton, W.R. and Dunlap, R.E.(1978) 'Environmental sociology: a new paradigm', *The American Sociologist*, 13: 41–9.
Cline, W.R. (1991) 'Scientific basis for the greenhouse effect', *The Economic Journal*, 101: 904–19.
Covello, V.T., Flamm, G., Fodericks, J. and Tardiff, R. (eds) (1983) *Analysis of Actual vs. Perceived Risks*, Plenum, New York.
Cramer, J.S. (1991) *The Logit-Model: An Introduction for Economists*, Routledge, London.
Dickinson, R.E. (1989) 'Uncertainties of estimates of climatic change: a review', *Climatic Change*, 15: 5–13.
Dietz, T.M. and Rycroft, R.W. (1987) *The Risk Professionals*, Russell Sage Foundation, New York.
Douglas, M. and Wildavsky, A. (1982) *Risk and Culture*, University of California Press, Berkeley.
Dunlap, R.E. and Van Liere, K.D. (1978) 'The "New Environmental Paradigm"', *Journal of Environmental Education*, 9, Summer: 10–19.
EPA (1990) *The Potential Effects of Global Climate Change on the United States*, Hemisphere Publishing, New York.
Fajer, E.D. and Bazzaz, F.A. (1992) 'Is carbon dioxide a "good" greenhouse gas? Effects of increasing carbon dioxide on ecological systems', *Global Environmental Change*, 4: 301–10.
Fischhoff, B., Lichtenstein, S., Slovic, P., Derby, S.L. and Keeny, R.L. (1981) *Acceptable Risk*, Cambridge University Press, New York.
Gellert, G.A., Neumann, A.K. and Gordon, R.S. (1989) 'The obsolence of distinct domestic and international health sectors', *Journal of Public Helath Policy*, 10: 421–4.
Giorgi, F. and Mearns, L.O. (1991) 'Approaches to the simulation of regional climate

change: a review', *Rev. Geophys.*, 29: 191–216.

Glantz, M.H. (ed.) (1988) *Societal Responses to Regional Climatic Change*, Westview Press, Boulder.

Granovetter, M. (1985) 'Economic action and social structure: the problem of embeddedness', *American Journal of Sociology*, 91: 481–510.

—— (1992) 'Economic institutions as social constructions: a framework for analysis', *Acta Sociologica*, 35: 3–11.

Green, Ch. (1993) 'Economics and the "greenhouse effect"', *Climatic Change*, 22: 265–91.

Haas, P.M. (1990) 'Obtaining international environmental protection through epistemic consensus', *Millenium*, 19: 347–63.

Hardin, G. (1968) 'The tragedy of the commons', *Science*, 162: 1243–8.

Harré, R. (1979) *Social Being*, Basil Blackwell, Oxford.

Heimer, C.A. (1988) 'Social structure, psychology, and the estimation of risk', *Annual Review of Sociology*, 14: 491–519.

Henderson-Sellers, A. (1990) 'Australian public perception of the greenhouse issue', *Climatic Change*, 19: 69–96.

IPCC (1991) *Climate Change. Scientific Assessment*, Cambridge University Press, Cambridge.

Jaeger, C., Dürrenberger, G., Kastenholz, H. and Truffer, B. (1993) 'Determinants of environmental action with regard to climatic change', *Climatic Change*, 23: 193–211.

Journal of Cross-Cultural Psychology (1991) *Risk and Culture*, Special Issue, 22.

Kahnemann, D. and Tversky, A. (1979) 'Prospect theory: an analysis of decision under risk', *Econometrica*, 47: 263–91.

Kastenholz, H. (1992) 'Bedingungen umweltverantwortlichen Handelns in einer Schweizer Bergregion. Eine empirische Studie unter der besonderen Berücksichtigung anthropogen verursachter Klimaveränderungen', Ph.D. Dissertation, ETH Zürich.

Kates, R.W., Ausubel, J.H. and Berberian, M. (1985) *Climate Impact Assessment. Studies of the Interaction of Climate and Society*, John Wiley & Sons, New York.

Kempton, W. (1991) 'Lay perspectives on global climate change', *Global Environmental Change*, 3: 183–208.

Kirman, A. (1989) 'The intrinsic limits of modern economic theory: the Emperor has no clothes', *The Economic Journal*, 99: 126–39.

Krimsky, S. and Golding, D. (1992) *Social Theories of Risks*, Praeger, New York.

Löfstedt, R.E. (1991) 'Climate change perceptions and energy-use decisions in northern Sweden', *Global Environmental Change*, 4: 321–4.

Malone, T. and Yohe, G. (1992) 'Towards a general method for analysing regional impacts of global change', *Global Environmental Change*, 2: 101–10.

Manne, A.S. and Richels, R. (1991) 'Global CO_2 emission reductions: the impacts of rising energy costs', *The Energy Journal*, 12: 87–107.

Morrisette, P.M., Darmstadter, J., Plantinga, A.J. and Toman, M.A. (1991) 'Prospects for a global greenhouse gas accord', *Global Environmental Change*, 3: 209–23.

Netting, R. McC. (1981) *Balancing on an Alp: Ecological Change and Continuity in a Swiss Mountain Community*, Cambridge University Press, Cambridge.

Nordhaus, W.D. (1991) 'To slow or not to slow: the economics of the greenhouse effects', *The Economic Journal*, 101: 920–37.

Parry, M.L., Carter, T.R. and Konijn, N.T. (eds) (1988) *The Impacts of Climate Variations on Agriculture*, 2 vols, Kluwer, Dodrecht.

Peele, B.D. (1987) 'Insurance and the greenhouse effect', in G.I. Pearman (ed.), *Greenhouse: Planning for Climate Change*, Brill, Leiden.

Renn, O. and Levine, D. (1988) 'Trust and credibility in risk communication', in H. Jungermann, R.E. Kasperson and P.E. Wiedemann (eds), *Risk Communication*, KFA Jülich, Jülich.

Riebsame, W.E. and Magalhaes, A.R. (1991) 'Assessing the regional implications of climate variability and change', in J. Jaeger and H.L. Ferguson (eds), *Climate Change: Science, Impacts and Policy*, Cambridge University Press, Cambridge.

Roqueplo, Ph. (1992) 'Le statut social des phénomènes d'environnement. Exemples des pluies acides et de l'effet de serre', in Commission Nationale Suisse pour l'UNESCO (ed.), *Séminaire 'Environnement et Société: la Contribution des Sciences Sociales'*, Rapport Finale, UNESCO, Berne.

Rothen, S. (1993) *Kohlendioxid und Energie. Eine Untersuchung für die Schweiz*, Rüegger, Chur.

Samdahl, D. and Robertson, R. (1989) 'Social determinants of environmental concern: specification and test of the model', *Environment and Behavior*, 21: 57–81.

Schmandt, J.A. (1991) 'The regions and global warming: impacts and response strategies', *Zeitschrift für Umweltpolitik und Umweltrecht*, 14: 133–57.

Schneider, S.H. (1989) 'The greenhouse effect: science and policy', *Science*, 243: 771–81.

—— (1991) 'Why global warming schould concern us', *Global Environmental Change*, 4: 268–71.

Sen, A. (1982) 'Rational fools: a critique of the behavioural foundations of economic theory', in A. Sen (ed.), *Choice, Welfare and Measurement*, Blackwell, Oxford.

Short, J.F. (1991) *Trust, Acceptable Risk, and the Law*, Washington State University, Pullman.

Short, J.F. and Clarke, L. (eds) (1992) *Organizations, Uncertainties, and Risk*, Westview Press, Boulder.

Stern, P.C., Young, O.R. and Druckman, D. (1992) *Global Environmental Change: Understanding the Human Dimensions*, National Academy Press, Washington DC.

Tobey, J.A. (1992) 'Economic issues in global climate change', *Global Environmental Change*, 3: 215–28.

Van Liere, K.D. and Dunlap, R.E. (1983) 'Cognitive integration of social and environmental beliefs', *Sociological Inquiry*, 2/3: 333–41.

WCED (1987) *Our Common Future*, Oxford University Press, Oxford.

WHO (1990) *Potential Health Effects of Climatic Change*, WHO, Geneva.

Wellman, B. and Berkowitz, S.D. (1988) *Social Structures: A Network Approach*, Cambridge University Press, Cambridge.

20

ENVIRONMENTAL PERCEPTION, CLIMATE CHANGE, AND TOURISM

A.S. Bailly

In 1990, a book written by L.K. Caldwell *Between Two Worlds: Science, the Environmental Movement and Policy Choice* was published by Cambridge University Press to deal with the relationships between science and the world of people. Caldwell shows how science can help shape new relationships between people and the environment when the information is clearly transmitted – which is not always the case. He reminds us that when the popular environmental movement was looking to science for guidance, most natural sciences were unprepared to explain the interactions between population expansion, technological advance, and environmental changes in such a way that people could understand, so as to bear effectively on public behaviour and decisions. The fundamental question of environmental policies today is whether the realities of this earth, as the sciences reveal them, can be translated into representation and behaviour appropriate to the continuation of life on earth. As Schneider (1990) writes: 'My principal purpose ... is to make basic issues and policy implications of global atmospheric change more accessible to the public', without loss of scientific validity and soundness so that the public can be concerned with the future of this planet.

This paper is concerned with environmental perception, an aspect that many climate researchers would regard as non-scientific, since it deals more with the concept of 'climate sensitivity' than with those of climatology and meteorology. The idea of 'climate sensitivity' is to consider the behavioural approach as the base for the analysis of the decision-making process: we deal with the diffusion of information about climate change and its impact on human behaviour. One of the main questions for economic practices – tourism in mountain environments being one – is not climatic change as a meteorological process, but the way people perceive it. The cognition process is too often ignored, even if people need to be more informed on the implications of past, present, and future climate change for their socio-economic decisions.

One of the questions to be raised is how, in a given climate, the individual constructs his own reality in linking together the functional and the symbolic; then how our perception of climate is related to our experience and how the imaginary and the real are connected for each decision.

Maunder (1986) identifies three levels of sensitivities: 'how much variation is needed before it becomes significant ... in affecting decisions; are such significant ... climate variations the same ... from one season to the next; and what specific effect does a significant ... climate variation have on particular activities' such as tourism?

ENVIRONMENTAL PERCEPTION AND GEOGRAPHY OF REPRESENTATIONS

The classical conception of geography – based on Cartesian precepts of evidence (for example, the observer's independent certainty), reductionism (a breaking down into sets of simple elements), causality (the presupposition of an order connecting linear causalities) and exhaustiveness (the certainty that nothing essential has been omitted) – has been thrown into question by the holistic approach of the geography of representations;[1] for how can our scientific practices be separated from our interior existence with its affective and emotional aspects? Is it not just as important that the researcher understand the subtle and complex links which unite the real and the imaginary (Bailly and Debarbieux, 1991) above and beyond the observation of facts, even for climate change?

To the precept of evidence, one must oppose that of subjectivism, allowing us, as it does, to account for human irrationality; the precept of reductionism must be replaced by that of complexity, for the more a phenomenon under scrutiny is reduced, the more complex it becomes and the greater the number of dimensions one finds; probabilism is to be opposed to causalism, as it is impossible to foresee everything (for nature is random); exhaustiveness must be replaced by the ideological, that is, the partial representation of phenomena based on our explicit and implicit choices.

The geographer, like all researchers in the human sciences, finds himself faced with a fabulous and complex world, a chaotic swirl of existential experiences; should he or she want to understand its creation and evolution, and find order in chaos, it will be necessary to accept the random character of change and the apparent irrationality of human behaviour. He or she will also have to take an interest in the symbolism of phenomena, in their subjective connotations. In this sense the study of climate is not only a collection of empirically observable facts, but also the study of their meanings.

The geography of representations is based on the theories of cognitive psychologists (Piaget and Inhelder, 1947); we know from their experiences that we perceive and respond rapidly to highly significant events (Vernon,

1962). The greater and more sudden the change in the environment, the more likely it attracts attention. Any situation, such as an extreme climate, which may be potentially dangerous (for example, drought and floods) will stimulate the observer. On the contrary, peaceful and slow changes are overlooked, for we do not maintain prolonged awareness of a relatively slow change. Our expectations and decisions are related to the probability of occurrence of events belonging to one of these two subjective categories. A great deal is to be done to make people more aware of the importance of some types of events (for example, climate change) and in some perceptual situations (for example, tourism in mountain environments).

PERCEPTION AND NATURAL HAZARDS

The disastrous effects of climate modification, such as floods or snow storms, were often called acts of God. Nowadays, since we have learned about hazards, the main question is how to evaluate the risks and to judge the acceptability of the risk. As W. Lowrance (1976) writes, 'a thing is safe if its risks are judged to be acceptable'.

This approach results from studies of natural hazards and human responses undertaken in the United States by geographers from Chicago and Boulder. The pioneer work was done by G. White on human adjustment to floods (1945), followed by studies of man and environment (1974a), and natural hazards – local, national and global (1974b). Recent geographic studies involve a broad range of people, working in different fields of investigation, from man-made to natural hazards.

The search for measurement of human responses to natural hazards, by Burton, Kates and White (1978) allows us to define an event according to seven dimensions: magnitude, frequency, duration, areal extent, speed of onset, spatial dispersion, and temporal spacing. To paraphrase these authors:

- Magnitude concerns occurrences exceeding some common level; for climate it could be the maximum temperature reached during a cold wave.
- Frequency asserts how often an event of a given magnitude may be expected to occur on average in the long term. Thus, a snowstorm of a given magnitude (depth of snow accumulation in cm) may occur only once in ten years (but not, of course, every ten years).
- Duration refers to the length of time over which a hazard persists; thus, an unexpected frost may cause much damage in a few hours.
- Areal extent refers to the space over which a hazard extends. A snow storm's areal extent may be a short and narrow swathe, while a drought may cover thousands of square kilometres.
- Speed of onset refers to the length of time between the first appearance

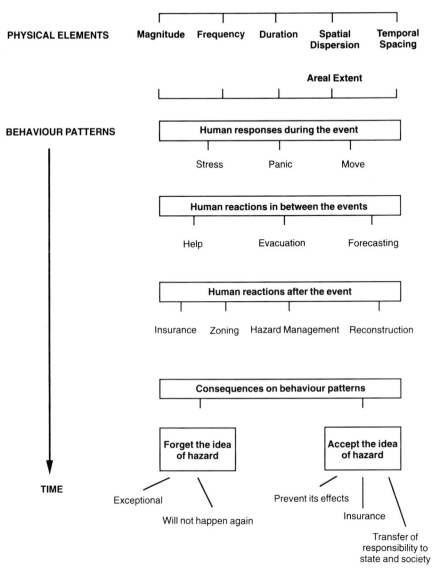

Figure 20.1 Human reactions to natural hazards.

of an event and its peak. Drought is a slow-onset hazard, whereas avalanches are fast-onset hazards.

- The first five of these are measures of the aggregate of separate events. Spatial dispersion and temporal spacing refer to the distribution of a population of events over space and time.

321

How do people react to these dimensions? Since it is rare that individuals have access to full information, their responses are related both to their perception of the event and to their awareness of opportunities to make adjustments. On a diagram drawn from Burton *et al.* (1978), we show the theoretical range of responses (Figure 20.1). These may range from immediate actions to long-term ones.

Most research on human reactions to climate change has been linked to violent hazards such as floods, tropical cyclones. Non-extreme events, which are not immediately hazardous to people, are not the object of a behavioural approach in the Alps. However, a basic bibliography of publications in the social and political sciences is given by Dürrenberger and Jaeger (1992). The authors show that the range of responses depends on cultural adaptation, the capacity to absorb information, and economic constraints. People imagine, in their cultural environment, not only probable outcomes, but develop subjective views of those probabilities.

Expected utility, outcome, and past experiences are basic to the understanding of the process of coping with risk in an uncertain environment; it is essentially a process of information and interpretation which leads to hazard appraisal and a perception of the alternatives.

Research related to climatic change, undertaken in mountainous environments such as the Alps, is linked more to planning and images of the mountains than to reaction to climatic change itself. However some useful studies from the Grenoble Alpine Institute should be mentioned (Guérin, 1984; Gumuchian, 1984; special issues of the *Revue de Géographie Alpine* on Homo Turisticus) as researches linked to the now-terminated National Climate Program (ProClim) in Switzerland (ProClim Workshop 4, 1989; Rebetez-Beniston, 1994). As a bibliographical base we can also use the 'man–environment' MAB publication (Messerli, 1989) and the perception of mountainous environments by Auchlin and Vietti-Violi (1983). Even if these publications do not deal precisely with climate change as such, they have to take into account indirectly the climatological environment such as cold spells, snow storms, avalanches, or floods. They are the basis for a better historical knowledge of human reactions to climate changes in the Alps.

RELATIONSHIPS BETWEEN INFORMATION ON CLIMATIC CHANGE AND HUMAN PRACTICES

Information on climate change, and predictions about climate are fundamental for tourism; for example remembering a previous Christmas without snow is going to orientate some people towards destinations other than alpine ski resorts; on the other hand, an early snowfall, even if the snow cover is not thick enough at low altitude to allow skiing, will reverse the trend. All this information, summarized in Figure 20.2, with other non-climate

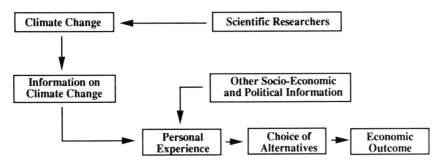

Figure 20.2 The relationship between climate change and human choices.

information (economic trends, world tensions, war, etc.), will lead to a choice of alternatives, with an economic outcome for tourist regions.

The cognitive process is linked to climate confirmation, which has to be in the most useful form if it is to be relevant for decision making. Only then will the return exceed the cost of diffusing the information. An example in Alpine regions is related to whether or not ski resorts should in future invest in ski lifts and snow-making machines, as an economic 'guarantee' for the future of these ski resorts in increasingly snowfree winter conditions.

The information has to give answers to practical questions, to clarify the causes of trends and the effects of climate change in mountain environments for predictive purposes. Tourism-climatology should emphasize the relation between climate change and leisure practices, especially since there is an evolution through time of the perception of climatic change.

The global environment is subject to drastic changes; what is still in doubt is the scale of change, especially in mountainous environments where all human activities (particularly activities linked to leisure outdoors) depend on these changes. Since the duration and persistence of these changes cannot be forecast, it is still difficult to ask people to adjust now to an uncertain situation; people living in the Alps are engaged in activities that may be viewed as a game with capricious Nature. Social scientists have to make people aware of the probable course of events and give them ways to cope with possible alterations by a new organization of society. These adjustments involve new agricultural and touristic activities that are more in line with nature, and more adapted to warming trends.

However most governments still do not deal with the social aspects of climatic change, except for punctual interventions in extreme situations. They now have to prepare our society for probable changes and rediscover the advantages of diversity and flexibility, a way to build a better environment adapted to climate change in the Alps. In this connection more behavioural studies have to be undertaken in order to gain a better knowledge of human reactions. A simple model by P. Bartelmus (1986) used

323

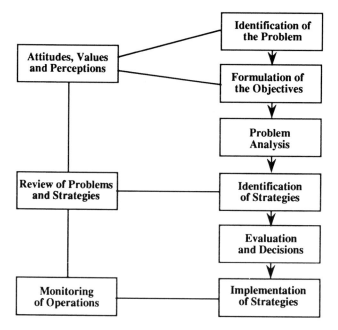

Figure 20.3 Major stages of the planning process.

for planning purposes provides an insight into environmental perception (Figure 20.3).

We presented in this paper one issue concerning environmental perception, but others could be studied to enable societies to face environmental realities. Some of these might be:

- erosion and soil degradation;
- contamination of the biosphere, air, water, soil, living bodies;
- deforestation, desertification;
- destruction of natural habitats;
- loss of biodiversity.

Science needs to be used not only to forecast future risks but also to inform and to influence human practices and public decisions.

The history of humanity shows that, for all of the tragedies and errors of human behaviour, people have an innate capacity to learn from and to evaluate experience. But this capacity is often insufficiently shared to avert disaster, and environmental deterioration can affect the welfare, health, and quality of life of many, maybe all, people.

NOTE

1   The term 'geography of representations' is widely used in francophone social science, but is not commonly employed in English. This concept refers, according to Piaget, to the manner in which human beings experience: how they presently perceive, and how their present perception is based upon an accumulated set of experiences and memories; how they represent, imagine and express value in verbal, written, mental, or graphical form.

REFERENCES

Auchlin, P. and Vietti-Violi, P. (1983) 'Perception du Pays-d'Enhaut par ses touristes (1980–1981)', *MAB-Schlussbericht* Nr 3.

Bailly, A. (1974) 'La perception des paysages urbains. Essai méthodologique', *L'Espace géographique*, III/3: 211–17.

—— (1984) 'Probabilités subjectives et géographie humaine. Fondements, méthodologie et applications', in *Géographes aujourd'hui; mélanges offerts en hommage à François Gay*, Université de Nice.

—— (1991) 'La géographie des risques naturels', in A. Bailly *et al.* (eds), *Les concepts de la géographie humaine*, 2nd edn, Masson, Paris, pp. 181–6.

Bailly A. and Debarbieux, B. (1991) 'Géographie et représentations spatiales', in A. Bailly *et al.* (eds), *Les concepts de la géographie humaine*, 2nd edn, Masson, Paris, pp. 153–60.

Bartelmus, P. (1986) *Environment and Development*, Allen & Unwin, London.

Bridel, L. (1986) 'Gestion des dangers naturels, théorie et étude de cas', postgraduate course, Planification et projet, nouvelles démarches, EPFL, Département d'architecture.

Brunet, R. (1974) 'Espace, perception et comportement', *L'Espace géographique*, III/3: 189–204.

Burton, I., Kates, R.W. and Snead, R.E. (1969) 'The human ecology of coastal flood hazard in Megalopolis', University of Chicago, Department of Geography, Research paper, 115.

Burton, I., Kates, R.W. and White, G.F. (1978) *The Environment as Hazard*, Oxford University Press, New York.

Caldwell, L. (1990) *Between Two Worlds: Science, the Environmental Movement and Policy Choice*, Cambridge University Press, Cambridge.

Claval, P. (1974) 'La géographie et la perception de l'espace', *L'Espace géographique*, III/3: 179–87.

Dürrenberger, G. and Jaeger, C. (1992) 'Klimaänderung und Risikodiskurs', NFP 31, *Sclussbericht Vorstudie 42*.

Eddy, A. *et al.* (1980) *The Economic Impact of Climate*, vol. I, Norman, Oklahoma.

Geipel, R. (1977) 'Friaul. Sozialgeographische Aspekte einer Erdbebenkatastrophe', *Münchener Geographische Hefte* Nr. 40.

Grunder, M. (1984) 'Ein Beitrag zur Beurteilung von Naturgefahren im Hinblick auf die Erstellung von mittelmassstäbigen Gefahrenhinweiskarten (mit Beispielen aus dem Berner Oberland und der Landschaft Davos)', *Geographica Berneensia*, 23.

Grunder, M. and Kienholz, H. (1986) 'Gefahrenkartierung', in O. Wildi and K. Ewald (eds), *Der Naturraum und dessen Nutzung im alpinen Tourismusgebiet von Davos*, Ergebnisse des MAB-Projektes Davos, Birmensdorf, EAFV, *Bericht*, 289, 67–86.

Guérin, J.P. (1984) *L'Aménagement de la montagne. Politiques, discours et productions d'espaces*, Ophrys, Gap.

Gumuchian, H. (1980) 'La montagne française vue par l'enfant', *Revue de Géographie Alpine*, special issue, pp. 35–68.

—— (1984) *Les Territoires de l'hiver ou la montagne française au quotidien*, Editions des Cahiers de l'Alpe, Grenoble.

—— (1988) *De l'espace au territoire. Représentations spatiales et aménagement*, Collection Grenoble Science, Université Joseph Fourier, Grenoble.

Kates, R.W. (1967) 'The perception of storm hazard on the shores of Megalopolis', in D. Lowenthal (ed.), *Environmental Perception and Behavior*. University of Chicago, Department of Geography, Research paper 109, 60–74.

—— (1975) *Risk Assessment of Environmental Hazard*, John Wiley & Sons, Chichester.

Knoepfel, P. (ed.) (1988) *Risiko und Risikomanagement*, SAGUF, Basel.

Lieberherr, B. (1981) 'Système écologique et conflits d'affectations In les régions de montagne; recherche méthodologique', *MAB-Fachbeitrag*, 8.

Litaudon, C. and Noth, N. (1990) 'Représentation des catastrophes naturelles: le cas de Nîmes', Université de Genève, Département de géographie.

Lowrance, W. (1976) *Of Acceptable Risk*, Kaufman, Los Altos.

Lynch, K. (1960) *The Image of the City*, MIT Press, Cambridge.

Maunder, W. (1986) *The Uncertainty Business: Risks and Opportunities in Weather and Climate*, Methuen, London.

Messerli, P. (1989) *Mensch und Natur im alpinen Lebenstaum. Risiken, Chancen, Perspektiven*, Zentrale Erkentnisse aus dem Schweizerischen MAB-Program. Haupt Verlage, Berne, Stuttgart.

Pellegrino, P. *et al.* (1983) 'Identité régionale et représentations collectives de l'espace', PNR '*Problèmes régionaux en Suisse*', 25.

Pfister, C. (1985a) 'Klimageschichte der Schweiz 1525–1860. Das Klima der Schweiz von 1525–1860 und seine Bedeutung in der Geschichte von Bevölkerung und Landwirtschaft', *Academica Helvetica* 6, Haupt, Berne.

—— (1985b) *Climhist. Banque de données pour l'histoire du climat*, Meteotest, Berne.

Piaget, J. and Inhelder, B. (1947) *La Représentation de l'espace chez l'enfant*, PUF, Paris.

Rebetez-Beniston, M. (1994) 'Perception du climat en Suisse romande', doctoral thesis, Lausanne.

Saarinen, T.F. and Sell, J.L. (1980) 'Environmental perception. Progress reports', *Progress in Human Geography*, 4: 525–48 and 5: 525–47.

Saarinen, T.F., Seamon, D. and Sell, J.L. (1984) *Environmental Perception and Behavior: An Inventory and Prospect*, Chicago University, Department of Geography, Research paper 209.

Schneider, S. (1990) *Global Warming: Are We Entering the Greenhouse Century?* Lutterworth Press, Cambridge.

Vernon, M.D. (1962) *The Psychology of Perception*, Penguin Books, Harmondsworth.

White, G. (1945) *Human Adjustment to Floods*, Research Paper 29, Department of Geography, University of Chicago.

—— (1974a) 'Comparative field observations on natural hazards', in M. Pecsi and P. Ferenc (eds), *Man and Environment*, Research Institute of Geography, Hungarian Academy of Sciences, Budapest.

—— (ed.) (1974b) *Natural Hazards: Local, National Global*, Oxford University Press, New York.

Whyte, A.V. and Burton, I. (1982) 'Perception of risks in Canada', in I. Burton, J. Fowle and B. McCullogh (eds), *Living with Risks: Environmental Risk Management in Canada*, Environmental monograph no. 3, Institute of Environmental Studies, Toronto, 39–69.

21

CLIMATE CHANGE AND WINTER TOURISM

Impact on transport companies in the Swiss canton of Graubünden

B. Abegg and R. Froesch

INTRODUCTION

Tourism in the Alps is highly dependent on winter sports. In the Canton of Graubünden (or Grisons, eastern Switzerland), 54 per cent of the overnight stays and about two thirds of the earnings are made in winter. Considering the dominance of winter sports in numerous touristic regions, the economy of these regions – while being dependent on tourism in general – can be considered as excessively monostructured, given the narrow case of winter tourism. This could prove to be highly problematic in view of the uncertainty about climatic change.

Climatic conditions are part of the environment in which outdoor activities take place. The most fundamental effects concern the presence, or absence, of direct climatic resources such as snow cover for winter sports. Skiing is very sensitive to climate and weather and has strict requirements for it to be possible or enjoyed. It follows that anything which alters the length of the skiing season is likely to have major repercussions for the skiing industry (McBoyle and Wall, 1987; Galloway 1988). Figure 21.1 provides an overview of the region considered in this study.

CLIMATE CHANGE

Assuming that the mean temperature rises by about 3°C in coming decades and that precipitation and evaporation varies in the present range of ± 10 per cent, the following consequences can be sketched for the Swiss Alps (Foehn, 1991):

- The snowline in winter would rise by 300 m in the Central Alps and by 500 m in the Prealps.

328

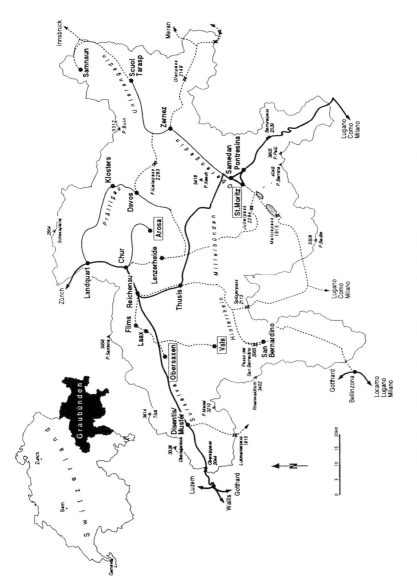

Figure 21.1 Map of Graubünden (Cartography by Martin Steinmann).

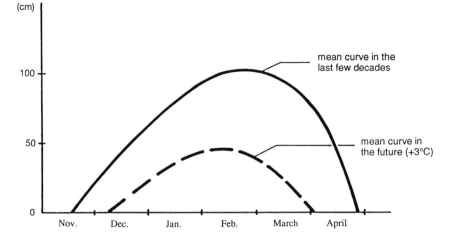

Figure 21.2 Snow depth and duration of snow cover at an altitude of 1,500 m in the last few decades and in the future (schematic illustration).

- Below an altitude of 1,200 m continuous snow cover would seldom be found.
- The first snowfall would be delayed, would melt prematurely, and the length of snow cover would be reduced by one month compared to the present. At the same time, the depth of snow would certainly be considerably reduced (Figure 21.2).

In order to judge the snow-reliability the so-called 100-day-rule has to be applied (Witmer, 1986). This rule states that to operate a skiing industry with profit, a snow cover sufficient for skiing (i.e. 30 cm) should last at least 100 days between the first of December and the end of April.

As a rule in most regions of the Swiss Alps, ski resorts above 1,200 m have enjoyed abundance of snow in the last few decades. However, if significant atmospheric warming were to take place, only stations above 1,500 m would be snow-reliable and locations exposed to sunny conditions would be confronted with insufficient snow conditions even up to 2,000 m (Figure 21.3).

THE EFFECTS OF SNOW-DEFICIENT WINTERS ON TOURISM

The three winters from 1987/88 to 1989/90 were generally too warm, and up to the end of January or mid-February were extremely snow-deficient. Snow deficiency in December is not rare, but three winters in a row with such precarious snow conditions are in fact most unusual.

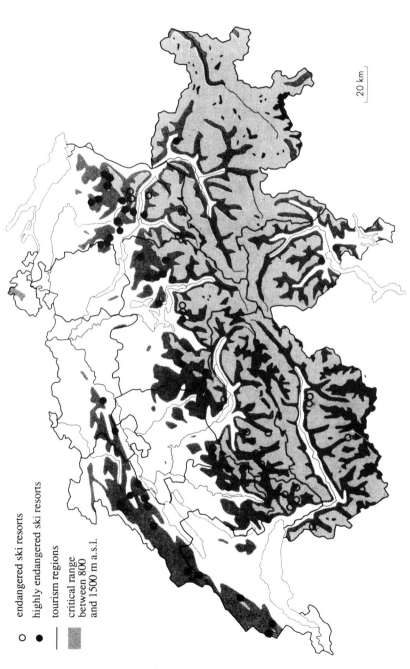

Figure 21.3 Snow deficiency and endangered ski resorts in Switzerland according to a warming of +3°C. It can be easily recognized that ski resorts in the Jura mountains and in the northern Prealps would be endangered. *Source:* Messerli, 1990: 26.

○   endangered ski resorts

●   highly endangered ski resorts

—   tourism regions

▨   critical range between 800 and 1500 m a.s.l.

20 km

In the following paragraphs we try to sketch the consequences of these snow-deficient winters – perhaps the harbingers of a global warming – on the tourism industry in the Canton of Graubünden. We will focus mainly on the most directly affected branch, the suppliers of transport services: the transport companies (cable cars, chair-lifts and ski-lifts). In order to achieve this goal, we carried out a survey for which the operating data of the transport companies for the period between 1984 and 1990 were analysed. The survey contains a period of 'normal' winters as a base for comparison and the winters with snow deficiency mentioned above.

Demand in ski tourism

The unfavourable snow conditions had direct consequences on the demand in ski tourism. In some regions and ski resorts, heavy drops in passenger numbers were reported. In the whole Canton the difference between the best and the worst year within the observed period was about 30 per cent. On a regional level even stronger differences could be noticed. The losses here are up to over 50 per cent (Hinterrhein), while other regions – with drops of about 20 per cent – were not hit as badly.

The situation in the snow-deficient winters can be illustrated with examples of a few ski resorts (Figures 21.4 and 21.5). To illustrate this, four companies of different size, situated in different regions and altitudes were chosen. (The reader is referred to Figure 21.1 for the location of these stations.) To sum up, two characteristic aspects for Graubünden can be observed:

1 There are big local and regional differences in the extent of the drop in demand. This can vary from 5 to over 60 per cent. As expected, there is a close connection between the depth of snow and the demand.

2 The change of demand over the years shows very different patterns as well. In some resorts we find a more or less continuous drop (e.g. Vals), in other resorts there are only certain years with extreme losses (e.g. Obersaxen, St Moritz). In most regions the last of the three winters with poor snow conditions, the winter 1989/90, was the worst one for business. The situation was different in the Engadine where the low was in the winter of 1988/89.

The data concerning snow conditions and the change of demand show no evidence of a clear spatial trend. Areas at higher altitudes tend to have an advantage, but altogether few conclusions can be drawn concerning favoured or unfavoured regions within the Canton.

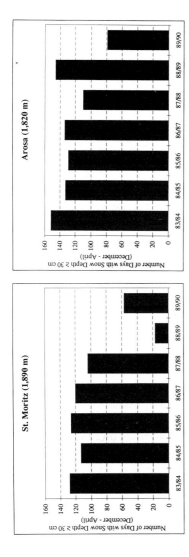

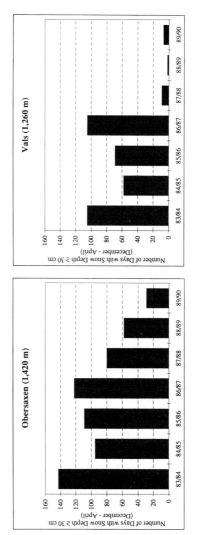

Figure 21.4 Snow conditions, according to the 100-day-rule; the number of days where snow depth exceeds 30 cm are indicated.

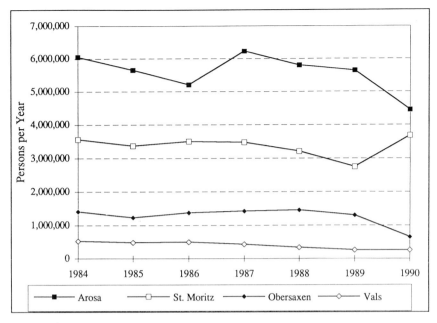

Figure 21.5 Number of persons transported (per year).

Economic consequences for the companies

The considerable reductions in persons transported obviously had an impact on the earnings and profits of the companies (Froesch, 1993). The summary of all business results (the yearly profits or losses respectively) shows that total profits have dropped. The overall result remains positive, but in 1990 this was only due to the profits of the transport companies in the Engadine.

Significant differences can be found in the situation of the individual transport companies. An increasing number of companies not only had to face reduced profits but even real losses. A survey of the number of transport companies with losses clearly shows this situation.

In the 'normal' years (1984–87) most companies worked with a profit. Only a few, mainly smaller ones made losses. In the following bad winters a large number of companies got into the 'loss-zone'. The number of companies with losses tripled by 1990 (Figure 21.6).

The selected examples show that the economic consequences were different for individual companies (Figure 21.7).

The figures of the yearly profits or losses are not sufficient for a complete understanding of the economic situation of the individual transport companies. They can only point to the economic problems the companies in question had to face during this period.

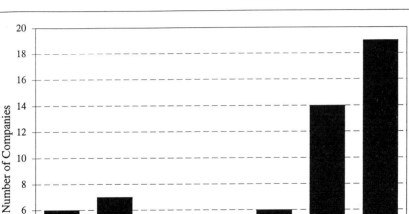

Figure 21.6 Number of companies with losses in Graubünden in 1990
(46 companies in total).

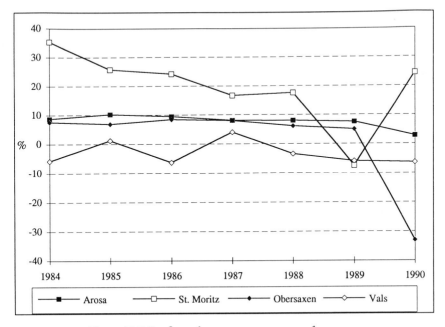

Figure 21.7 Profits or losses as a percentage of turnover.

335

Long-term impacts of the snow-deficient winters

As a result of the good snow conditions in the two following winters (1990/91 and 1991/92) and because of a significant rise in demand, the companies were able to cope with the losses. Therefore there were neither bankruptcies nor shut-downs in Graubünden – unlike other regions of Switzerland. Existing capacity could be maintained. The economic situation of the transport companies in the Canton of Graubünden is 'more favourable' than the Swiss average. But the results also show that longer-lasting periods with similarly unfavourable climatic conditions could have more serious consequences. The survival of a number of companies would then be open to question.

Effects on the tourism industry as a whole

The effects of the snow deficient winters differ among the different regions and sectors. The losses for the entire tourism industry – especially in the accommodation sector – were smaller than those of the transport companies. The overnight stays were less affected, the drops in Graubünden being of only a few per cent. A certain relation between the overnight stays and the snow conditions can be seen in the local distribution of the overnight stays, but the drops in the relative number of overnight stays were nowhere as serious as the drops the transport sector had to face. This means that the demand for holiday tourism as a whole did not react very strongly in the Canton of Graubünden. Altogether the loss of income for the entire tourism industry in Graubünden was estimated at 100 million Swiss francs (approx. $US 65 million) in total during the three bad winters.

In other regions of Switzerland the effects of the snow-deficient winters on ski resorts and on the entire tourism industry were more serious than in Graubünden. Ski resorts at low altitudes in particular suffered more from the lack of snow. As a result a number of transport companies were in financial difficulty and there were a few bankruptcies. Some of the companies which were not able to sustain their business with their own resources received financial aid from local and state governments (Kaspar, 1992: 219).

In other touristic regions stronger drops in the number of overnight stays could be observed as well, such as in the Bernese Oberland which was considerably affected. There the drop in the number of overnight hotel stays in the winter of 1989/90 was about 12 per cent compared with the preceding period (BIGA and BRP 1991: 6).

The tourism industry as a whole could cope with these losses because of the recovery of demand and the increase that resulted from the summer season. But it is a fact that in longer periods of unfavourable snow conditions tourists change their habits. In this situation, a big drop in the number of vacationers and therefore considerable losses for the entire tourism industry have to be expected.

FUTURE DEVELOPMENT AND STRUCTURAL CHANGE

The future development of the transport companies in the ski resorts is uncertain. Different scenarios seem plausible. On the one hand there are still signs for strong growth:

- The competition between the companies leads to a continuous improvement of the supply. In a market which has rather stagnated since the beginning of the eighties, capacity was nevertheless extended, in order to be able to satisfy clients' demand for modern and comfortable installations and for short waiting periods.
- With technically necessary replacement and modernization of the facilities an increase of transport capacity takes place.
- It is difficult to predict future development on the skiing market.

Some authors point to a saturation of the market, others count on a new growth in Swiss tourism. Many studies see a growing market potential for ski-tourism. The last two good winters have shown that the main problem is not the lack of demand but how to manage peaks in demand.

In this situation individual companies find it difficult not to compete. There is still a continuous pressure to grow. In many places extension projects have been studied or undertaken.

On the other hand, climatic change can call the scenarios of growth into question. A larger variability of climate with aperiodic rows of snow-deficient winters is enough to get many companies into serious financial trouble.

In spite of dynamic growth, structural problems in the industry have become apparent in the last few years, even without taking into consideration additional problems of snow deficiency. For the expansion of capacity and the technically necessary renovations of the installations, greater and greater financial means are necessary. The cycles of technical innovations and the periods of depreciation have become shorter as well. Therefore the financing of the investments has become more difficult for a lot of companies. The changing relation between self-earned capital and borrowed capital makes it clear that such investments can often only be made with additional outside funds (Figure 21.8).

This situation favours the position of the big, financially strong companies which are most likely to cope with the rising costs. For the whole of Switzerland, one third of the transport companies are estimated to be making losses, another third are just about able to break even and only one third are really working with a profit.

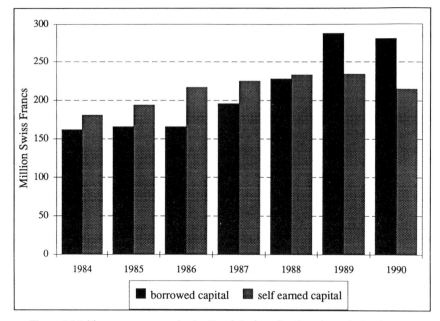

Figure 21.8 Transport companies in Graubünden: the changing relation between self-earned capital and borrowed capital.

Snow-making installations represent an additional financial constraint, when unfavourable snow conditions pushed the construction of these installations forward. The same competition and a similar obligation to adapt as in the extension of transport capacity takes place.

CONCLUSIONS

Tourism will experience a complex mix of winners and losers in most areas. A possible scenario is ruinous competition leading to a concentration on the most efficient companies in areas with good natural conditions for winter sports on one side, and an elimination of the economically weaker companies in areas with marginal conditions on the other side.

Such restructuring shows different problematic aspects:

- In the winner areas where the tourism industry concentrates, environmental impacts are likely to increase. A shortage of snow at ski resorts in lower parts of the mountain range may induce an increasing use of artificial snow and might initiate the building of new resorts and ski-runs at higher locations. This is likely to increase the pressure on what remains of the natural environment.
- Because of the probable bankruptcies in the loser areas, the regionally

balanced economic growth which tourism has provided is threatened. Furthermore, with the elimination of small businesses a part of the variety would get lost, a variety which is one of the strong points of Swiss tourism.

These are some of the factors which raise the question of increased subsidies for failing transport companies. The industry generally refuses subsidies for reasons of fair competition, but in exceptional cases where a transport company has crucial importance for a regional economy, subsidies might be considered.

The structural adjustment that can be expected in the transport facilities section does not necessarily have to result in ruinous competition, but can proceed in a weakened form. In fact, there are a number of strategies that can be taken up by the individual businesses.

Co-operation

Nowadays there are only small, local businesses. In the Canton of Graubünden there are almost fifty individual companies. By business co-operation or fusion of companies, rationalizing effects can be obtained and the financial base of the companies can be improved. There is a possibility of better co-ordination of investment strategies. The pressure of competition that forces the companies to continuous expansion can be moderated.

Diversification into other markets of the tourism industry

For bigger companies it is possible to diversify into other activities outside the transport business, for example as organizers in the entertainment, sports, and cultural sector (Smith, 1990: 177). Such activities can also help to improve summer business.

Niche-policy

A chance for small companies could be a policy aimed towards niches outside the big skiing regions; this implies a deliberate renunciation of expansion and a specialization in cheap rates and special offers.

The potential of such measures should not be overestimated. Even a successful diversification can only be an addition to skiing tourism. An increase of summer business that has been observed in the last few years cannot compensate for losses from the winter season. Snow remains the most important source of income for the transport business.

ACKNOWLEDGEMENTS

We wish to acknowledge the support we have received from the Swiss National Science Foundation.

REFERENCES

BIGA and BRP – Bundesamt für Industrie, Gewerbe und Arbeit and Bundesamt für Raumplanung (1991) *Beschneiungsanlagen – Neue Ausrichtung der Bundespolitik*, Swiss Federal Printing Office, Berne.

Foehn, P. (1991) *Was ist in Zukunft die Regel: Schneereiche oder schneearme Winter?*, Argumente der Forschung 3: 3–10.

Froesch, R. (1993) 'Sättigung im Tourismus – Probleme und Lösungsmöglichkeiten', Ph.D. dissertation, University of Zürich.

Galloway, R.W. (1988) 'The potential impact of climate changes on Australian ski fields', in G.I. Pearman (ed.), *Greenhouse: Planning for Climate Change*, CSIRO Publications, Melbourne.

Kaspar, C. (ed.) (1992) 'Jahrbuch der Schweizerischen Tourismuswirtschaft 1991/92', Internal Report, Institute for Tourism, University of St Gallen.

McBoyle, G. and Wall, G. (1987) 'The Impact of CO_2–induced warming on downhill skiing in the Laurentians', *Cahiers de Géographie du Québec* 31, 82: 39–50.

Messerli, P. (1990) 'Tourismusentwicklung in einer unsicheren Umwelt – Orientierungspunkte zur Entwicklung angemessener Strategien', *Die Volkswirtschaft* 12: 21–7.

Smith, K. (1990) 'Tourism and climate change', *Land Use Policy* 7 (2): 176–80.

Witmer, U. (1986) 'Erfassung, Bearbeitung und Kartierung von Schneedaten in der Schweiz', *Geographica Bernensia* G25, Berne.

22

LOCALIZING THE THREATS OF CLIMATE CHANGE IN MOUNTAIN ENVIRONMENTS

M. Breiling and P. Charamza

INTRODUCTION

Global climatic change has been an ongoing research topic for more than fifteen years. However, little has been done to bring adequate awareness to decision makers on a local level. Impacts in alpine regions will occur first on the local level resulting later on in supra-regional problems. Usually the local decision makers are not aware of how global climate change can affect local life and how to take it into account in various planning processes.

Alpine regions are particularly vulnerable to the impacts of global climatic change, both from an economic and environmental perspective. Over 40 years ago, most alpine regions made a living from subsistence agriculture. Nowadays, agricultural income is only a small proportion of the income in tourism. In Austria 8.4 per cent of the gross national income is directly earned by tourism (Österreich Werbung). In addition about 20 per cent of GNP is indirectly generated by tourism, approximately 80 per cent of which is in mountainous areas. Half of the income comes from winter tourism.

Economic development also became possible as a result of expensive safety constructions to reduce typical alpine environmental risks such as torrents, floods, and avalanches. As a result of global climate change, natural hazards could increase severely. Additional installation of safety constructions such as flood, torrent, and avalanche protections will become necessary. However, it is questionable whether most alpine communities will be able to afford them, because revenues from winter tourism are shrinking with the increase of temperature.

Global climate change models yield results on a large grid and are not directly applicable to local planning processes. However, if global climate change is a key issue in alpine regions, local decision makers should have a simulation tool to imagine the possible consequences of climate change. Our approach to this was to construct an interdisciplinary model of one Austrian mountainous district – the Hermagor district in the eastern Alps – and to test

341

the sensitivity of the climate-dependent parameters. Our aim was to develop three submodels of the same region describing the development in time over the same period (1951–1991) and to give long-term forecasts (2021) for each of them:

- a 'population-economic' submodel
- a 'land-use' submodel
- a 'hydrological' submodel.

Our goal is to find out the major relations among these three submodels and to construct a unifying model of the region. We hope that general considerations concerning our model construction are useful for other regions as well.

In this paper we mainly deal with the 'population-economic' submodel of the district. First we characterize the region of study. Second we mention other global climate change impacts not taken into account in the 'population-economic' submodel, mainly an expected increase of natural hazards. Third we introduce our 'population-economic' model. We give some basic theoretical considerations and also some simple examples of possible simulations with our model. Consequences of global climate change on the number of people working in different sectors are then estimated. In the final section we mention options for policy discussions that might help local decision makers to take appropriate actions against possible threats of global climate change.

CHARACTERISTICS OF HERMAGOR DISTRICT AND OBSERVED CLIMATE CHANGE

General

The Hermagor district covers 807 km² with a population of 20,000 people. This is about 1 per cent of the area of Austria and 0.25 per cent of the Austrian population. The district is divided into seven communities (Dellach, Gitschtal, Hermagor, Kirchbach, Kötschach, Lesachtal, St Stefan). It covers the valley of the river Gail (Lesach- and Gailtal), flowing west to east and the Gitschtal, north of the Gailtal. The lowest point is at 450 m and the highest point is 3,200 m above sea level. The timberline is at 1850 m altitude.

Originally agriculture was the main source of income. Nowadays it is tourism and the service sector. With 1.5 million guest-nights, more than 1 per cent of Austrian tourism numbering 130 million guest-nights in 1991 are in this area. The gross local product is US$ 234 million. The average income is less than 80 per cent of the Austrian average. There is a lack of work opportunities for people with higher education. At the same time there is a demand for seasonal work. Most farmers in the eastern part of the district

Location of Hermagor in Europe

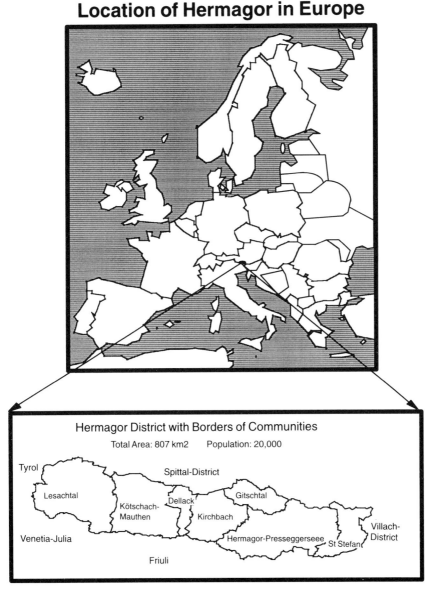

Figure 22.1 The Hermagor district in Europe and its division into communities.

having better access to the main economic centres Villach and Klagenfurt also have a second job. Tourism supports the local agriculture which in turn provides maintenance of the cultural landscape, which is an important resource for mountain tourism.

The main environmental problems are caused by the construction of ski-tracks and other infrastructure for winter sports (mainly at the Naßfeld skiing resort, where a snow-making plant was built recently), and the construction of forest roads. Both activities contribute to soil erosion and destabilization of the alpine landscape. Almost half of the district (47 per cent) is covered by a productive forest – a good natural protection against environmental hazards. Marginal agricultural land is afforested and contributes to a further stabilization of the landscape against natural hazards, but is at the same time diminishing the variety of the cultural landscape. The elimination of wetlands for agricultural purposes was leading to a loss in biodiversity. Air pollution and forest die-back were only registered in the east nearest to the industrialized centres. Water pollution and domestic waste problems connected to the increase of tourism have become a problem during the last decades. A sewage treatment plant is under construction.

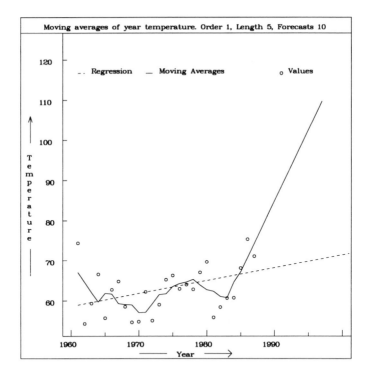

Figure 22.2 Surface temperature increase in Hermagor (Station Kornat, 1,050 m a.s.l.). *Source*: Austrian Meteorological Service (1992).

Table 22.1 Average number of days with mean under 0°C from 1851 to 1950

Altitude	100 year mean from eastern Alps	0.75 degree increase	1.5 degree increase	3 degree increase
400	77	68	60	27
600	90	84	77	60
800	101	95	90	77
1000	110	106	101	90
1200	120	115	110	101
1400	130	125	120	110
1600	144	137	130	120
1800	163	155	144	130
2000	178	170	163	144

Source: Aulitzky (1985) concerning the annual mean temperature value in the eastern Alps for 1851–1950 and WMO/UNEP (IPCC, 1990) for the expected increase of the frostline

Observed climate change

Compared to the global increase of surface temperature which is reported to be 0.3 to 0.6 degrees Celsius during the last 100 years (IPCC, 1992), the rate of change in Hermagor district between 1961 and 1991 was 0.9 degrees Celsius. This is much higher than the global average. However, there is only one continuous measurement station available for the whole district and we could not prove the rate of increase of the neighbouring stations.

In Figure 22.2 the plain line shows the five-year averaged mean temperature curve of Kornat, a station at 1,050 m altitude. The dotted line explains the trend. The temperature scale is 0.1 degree Celsius. Table 22.1 provides an estimate of present and future climate trends.

EXPECTED CONSEQUENCES FOR HERMAGOR DUE TO GLOBAL WARMING

Impacts on economy

Some major economic and environmental disturbances can take place. Their magnitude will depend on the rate of change. It might start with a decrease in tourist revenues in winter and additional costs for snow-making equipment and infrastructure. Later on, additional environmental safety constructions are needed, which will demand higher investments. During recent years, some US$ 40 million has already been spent on snow-making equipment at Naßfeld skiing resort, the main winter tourist resort of the district.

Impacts on ecosystems

Natural vegetation, in particular the high alpine protective forest, is endangered. The capacity for retaining soil moisture and soil erosion, cleaning of water, or the resilience against pests depends on the rate of warming. Some species can live within a relatively broad temperature range, while others are more sensitive to slow temperature change. First the variety of plants may decrease, and later on more robust tree and plant species (which might initially have increased in area and number) might be pushed out from their acceptable range of temperature, and large-scale destabilizations would then occur. Particularly in Hermagor, where we have a high percentage of forest and therefore good natural protection which can buffer non-benign environmental impacts, climate change would lead to more significant impacts than in areas with scarcer vegetation cover.

Extreme precipitation is expected to increase. Gordon *et al.* (1992) anticipate a tripling of extreme precipitation events and dry periods for Europe in the mid-latitudes. These are conditions under which vegetation suffers permanent environmental stress, leading to a decreased resilience. Simultaneously, extreme runoff will increase and alter the number and severity of catastrophes. Each bit of additional damage will increase the vulnerability of the entire ecosystem of the district. There might be a serious gap between diminishing funds available from local income (see next section) and the increased demand for safety construction (see Conclusions). There is a dilemma: the magnitude of observed damage does not yet justify major investment in additional safety construction, but at the time when damage will occur the entire economy of the district could be ruined. During the last 27 years (1965–1991) the annual damage due to catastrophic events was reported to include 21 episodes where damage was estimated to be less than US$ 1 million, 5 episodes with damage less than US$ 2 million and 1 episode in 1966 with damage reported at over US$ 7 million; at today's rates, this would correspond to damage of over US$ 25 million. In comparison to earnings the damage was in the worst case some 11 per cent of the district income, but usually under 1 per cent.

THE POPULATION-ECONOMIC SUBMODEL OF HERMAGOR

General

We constructed a baseline local development scenario for the 'population-economic' submodel, then undertook additional simulations with a change of the key parameters affected by warming. The difference of the simulations should then quantify the relative importance of global climatic change.

Similar approaches are planned for the 'land-use' submodel and the 'hydrological' submodel.

In order to develop an interdisciplinary model of Hermagor district we have to link these more specific submodels (which could themselves be split into even more disaggregated sub-submodels) and then try to unify them to the general model in the final stage. Further on we introduce some ideas and results concerning the population-economic model.

Population part

First we describe briefly the nonlinear population regression model. According to historical data (collected in 1951, 1961, 1971, 1981) and to general experience from the region, we use the following curves for different sectors:

1) logistic curve for relative number of people working in services
2) decreasing logistic curve for relative number of people working in agriculture
3) exponential function of negative quadratic function (Gaussian curve) to fit the relative number of people working in industry
4) linear trends for people working in business and others sectors.

The relative numbers (relative to number of inhabitants in the district) were considered. The software GAMS (Brooke *et al.*, 1988) was used for parameter estimates with the following results (by Time we mean (Year-1951)/10 + 1):

People (Agric. and Forest., Year) = $0.453 - 0.453/(1 + 6.582 \times 0.475^{\text{Time}})$
People (Industry, Year) = $\exp(-1.34 + 0.276 \times \text{Time} - 0.064 \times \text{Time}^2)$
People (Services, Year) = $0.303/(1 + 10.347 \times 0.452^{\text{Time}})$
People (Bus. and Trans., Year) = $0.062 + 0.021 \times \text{Time}$
People (Others, Year) = $0.166 + 0.017 \times \text{Time}.$

In order to get estimates for the absolute numbers of people working in different sectors we have to multiply right-hand sides of previous equations with the number of inhabitants in given Time. The following residuals for relative numbers were obtained for the years 1951–1981:

Table 22.2 Results of the GAMS model for Hermagor

	1951	1961	1971	1981
Agriculture and Forests	−0.002	0.006	−0.007	0.004
Industry	−0.003	0.008	−0.008	0.002
Services	0.003	−0.005	0.004	0.000
Business and Transport	0.003	−0.003	−0.002	0.002
Others	−0.004	0.009	−0.008	0.002

For the purpose of this paper we used only the data concerning the whole district of Hermagor. In fact, there are significant differences between smaller political units of the district communities. This was proven using ANOVA (ANalysis Of VAriance) which is summarized in the Appendix. Hence the strategy derived here for the 'global scale' of Hermagor district can be undertaken separately for the 'local scale', i.e., communities. This can give different results for different communities. From the 'global' point of view our results are, however, valid. Since the data for 1991 will not be available until October 1993, we had to give forecasts for this year instead of using the real data.

Joining population and economy

In this section we introduce a deterministic dynamic linear model which we use for the description of the economical and demographical behaviour of our region. We use the following notation:

GLP^t = vector, whose i-th coordinate is a value of GLP (Gross Local Product – i.e. Gross National Product produced in Hermagor district) produced in the i-th sector at time t, i {Agriculture, Industry, Services, Business, Others}.

INC^t = vector of average incomes per capita at time t. i-th coordinate is related to the i-th sector again.

POP^t = vector of number of inhabitants at time t dependent on the i-th sector.

P^t = square matrix (5 × 5 in our case). The ij element of this matrix is the fraction of GNP^t_j which an average person produces in the i-th sector.

Z^t = square matrix (5 × 5 in our case again). The ij element of this matrix is the amount of US$ produced by a man in the j-th sector which will be evaluated as the part of GLP^t_i.

Q^t = square matrix. On the ij position there is the amount of US$ produced by a man from the j-th sector which will become real income of an average man in the sector i.

The elements of defined matrices and vectors we will denote by small italics with sub indices, e.g. p^t_{ij} denotes the element in the i-th row and j-th column of the matrix P^t.

Under this notation we suppose the following dynamics in economy:

$$Z^t.POP^t = GLP^t$$
$$P^t.GLP^t = INC^t$$
$$POP^{t+1} = c.INC^t,$$

where c is some constant. Here the linear relation between POP^{t+1} and INC^t can be considered since the time difference in our case is ten years. If the time difference is smaller then a more general relation should be taken into account.

348

Generally, a change of *POP* reflects changes in *INC*. The reflection is almost zero initially and becomes linear after some time. A continuous case would need deeper considerations which are not of interest to this paper. In the model (1) the elements of matrices Z^t and P^t are of main importance. In fact, our model supposes a closed system with respect to income and output. If we would relate this model to geographically more complex areas, the matrices would become more dimensional matrices, etc. The structure of the matrices reflects general relations between different sectors of the studied region. Estimation of elements of these matrices is of course the major task. It depends on the data available and on our experience with the region. Local policy scenarios can be very substantial for the structure of these matrices. In the next paragraph we give a small example of how this model can be used for estimating the change of population development under global climate change. We will show how to measure this change, having very little information about the region at our disposal. We will assume the very special structure of the model matrices, which is not true in real life. However, our results can give rough estimates of the behaviour of the local economy in the region.

Simple example of use of previous model – estimate of effect of global climate change

According to values from Jeglitsch (ÖROK 1989), taking AS 10 as exchange rate for US$ 1 we estimate GLP for Hermagor district in 1991 as US$ 234 millions. Further on we suppose, according to standard economic development during recent years, the annual increase of Gross Local Product about $GLP = 0.01$.

Taking global warming into account we need to estimate the change of income for the region. In this paper we consider only economic change, and so we do not deal with possible consequences for land-use or hydrology, etc. According to the daily temperature data since 1961 (see Figure 22.2: station Kornat – 1,050 m high), we found a statistically significant increase of temperature in the Hermagor district. This increase is 0.00007 Celsius degrees per day (standard deviation of this estimate is 0.00001 degree). This gives an annual change of temperature of 0.0255. In this paper we give forecasts through to the year 2021 under three different situations. Under the 'optimistic' scenario that global warming will remain on the same level as during the last 40 years (for Hermagor district a nonlinear trend in warming was not proven statistically) we get 0.75 degrees warming in this time frame. Under the 'pessimistic' scenario (doubling of CO_2) we get 3 degrees warming, and under the 'middle' scenario we consider a 1.5 degree change of temperature for 2021. For our model the major effects of temperature change concern tourism, resulting from increases in the elevation of snow cover which can sustain skiing.

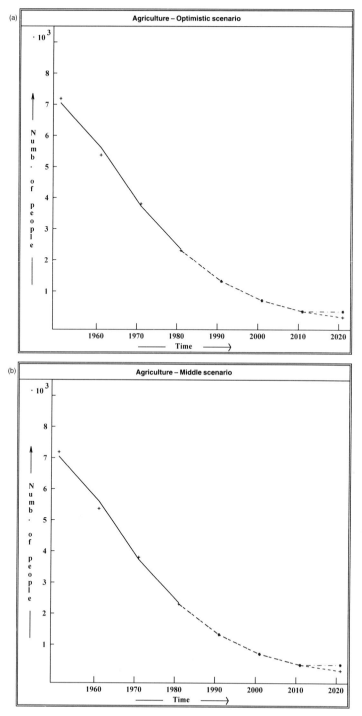

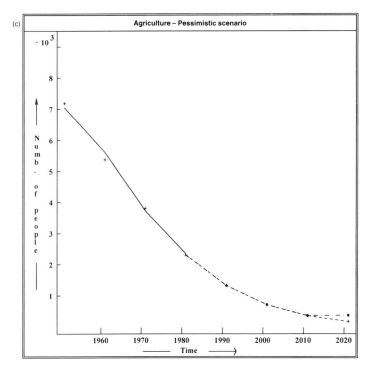

Figure 22.3 Forecasts for people living in the agricultural sector under the
'optimistic', 'middle', and 'pessimistic' warming scenarios in 2021. *Source*: Austrian
National Statistical Office. Forecasts using the FAMULUS software.

From Table 22.1 we obtain the average change of winter tourist nights as
the average of differences between the first and other columns. Hence we get
for the optimistic scenario 7.5 days difference, 14.7 days for the middle and
31.8 days for the double CO_2 scenario. We take average length of season as
the number of days with mean temperature below freezing point at the 800 m
altitude. Different definitions may be taken here according to the information from the district or community-related variable we denote by *LES*
(Length of Season). Under our definition we get about 7 per cent shortened
length of season under the optimistic scenario, 15 per cent in the case of the
middle scenario, and finally 31 per cent under the pessimistic scenario. We
assume income per winter tourist night to be US$ 70 and we have 530,000
winter tourist nights in the district. Hence the income from winter tourism
can be estimated as US$ 37,100,000 which will be shortened according to the
mentioned proportions to US$ 34,503,000, US$ 31,535,000 or US$
25,599,000 respectively.

We use this previous information for our model computations. Since we do
not have sufficiently detailed information at our disposal we have to assume

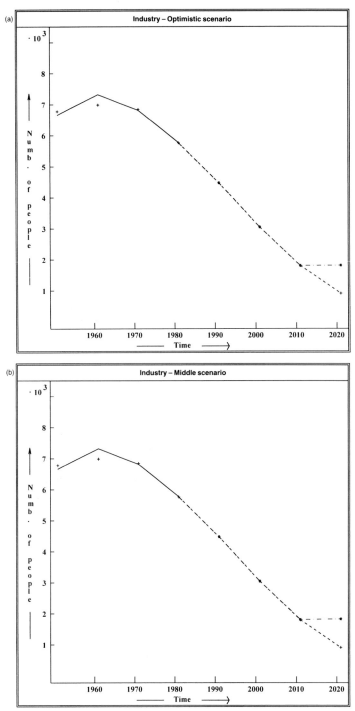

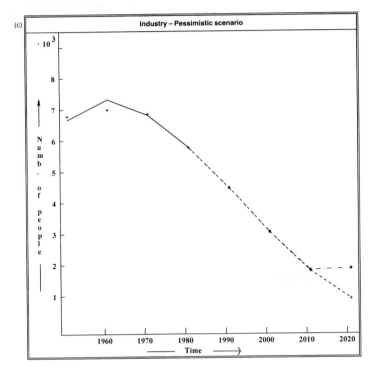

Figure 22.4 Forecasts for people living in the industry sector under the 'optimistic', 'middle', and 'pessimistic' warming scenarios in 2021. *Source*: Austrian National Statistical Office. Forecasts using the FAMULUS software.

a special structure of our matrices. In our simplification we assume that our economical sectors behave independently. This means all matrices Z^t and P^t we suppose to be diagonal. We consider the following values:

$$p^t_{ii} = \frac{c^t_1}{pop^t_i} \qquad i = 1, \ldots, 5.$$

$$z^t_{ii} = c^t_2 \qquad i = 1, \ldots, 5.$$

where $c^t_{1,2}$ are some real constants. It is relatively straightforward to interpret these values in that the GLP in some economic sector is linearly dependent on the number of people working in the sector (first equation of (1)) and the income of an average person in the i-th sector is some part of GLP produced in this sector which is proportional to $1/POP_i$. Now we take our estimates of population in different sectors gained from nonlinear regression analysis. We use these values taken at time $t+1$ to estimate GLP^t through the equation set above. This backward procedure helps us to extrapolate GLP^t for the

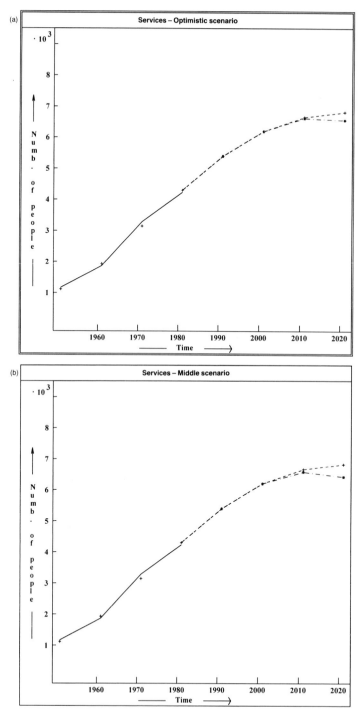

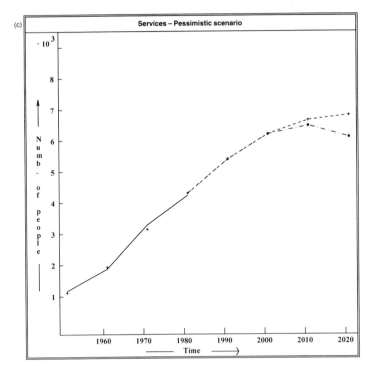

Figure 22.5 Forecasts for people living from services (mainly tourism) under the 'optimistic', 'middle', and 'pessimistic' warming scenarios in 2021. *Source*: Austrian National Statistical Office. Forecasts using the FAMULUS software.

years 1991, 2001, 2011. Now we subtract the estimated losses of GLP in services according to the considerations at the beginning of this section. These are our new data and the equations (1) must be recomputed now in the forward direction. We use the FAMULUS system for our computations. Its language is very close to PASCAL and it has attractive capabilities for graphics and graphic transfers. This program predicts the number of inhabitants working in different sectors in the next three decades under the three scenarios described above. The results are illustrated in fifteen figures. The numbers in any one column correspond to five given economical sectors for a given scenario. We denote by a solid line the fit of population in different sectors in the past (up to 1981). By --- we denote the situation without global climate change and by -.-.- the situation under global climate change.

The possible consequences are clear from the pictures. However, we must be careful in our considerations, in particular because of the relations between different sectors. Taking more detailed data, or possibly the expertise of local politicians, we can construct model matrices in more detail and hence our results would be more reliable.

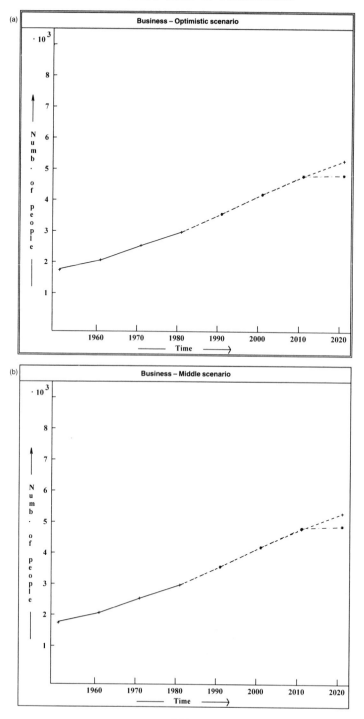

(a) Business – Optimistic scenario

(b) Business – Middle scenario

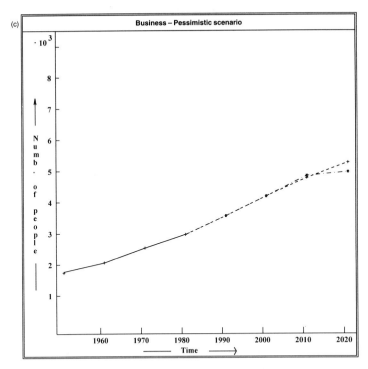

Figure 22.6 Forecasts of people living from trade under the 'optimistic', 'middle', and 'pessimistic' warming scenarios in 2021. *Source*: Austrian National Statistical Office. Forecasts using the FAMULUS software.

CONCLUSIONS AND POLICY DISCUSSION

As Hermagor district is highly dependent on winter tourism, climate change will have an impact on the employment structure. According to our calculations, tourism and service sectors will be subject to a decrease, while more people might be forced to be dependent on agriculture because of declining job opportunities elsewhere. In addition unemployment (included in 'Others' sector) might also rise in Hermagor.

Having to face potentially considerable losses, local politicians should be motivated to think about possible countermeasures. There are two main options: to reduce the production of greenhouse gases, or to adapt to modified environmental conditions. Both strategies imply considerable uncertainties and the numbers given reflect the present situation and not necessarily future conditions.

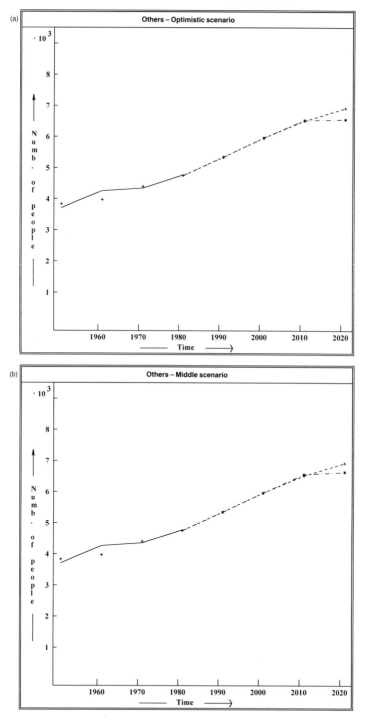

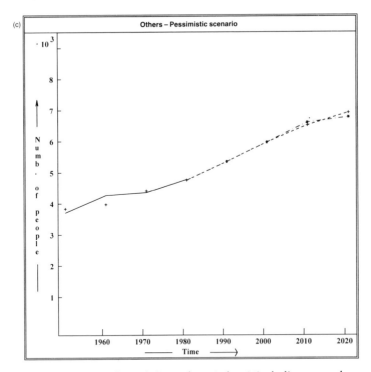

Figure 22.7 Forecasts of people living from 'others' (including unemployment) under the 'optimistic', 'middle', and 'pessimistic' warming scenarios in 2021. *Source*: Austrian National Statistical Office. Forecasts using the FAMULUS software.

(a) Strategies to counter emissions of greenhouse gases

Taking the Austrian average carbon emissions of 2t C per person and assuming an equal distribution of every person, Hermagor emits 40,000 t C. If a currently discussed carbon tax were to be introduced, some US$ 170 should be paid for one ton of emitted C in an OECD country (Messner and Strubegger, 1991). Some US$ 6,800,000 or 3 per cent of the GLP would therefore be due from the district. Another option currently under discussion is emission trading between countries with high costs for reducing carbon emissions, such as Austria, and countries with comparatively low costs, if both countries agree. This could decrease the cost of reducing CO_2 production to less than 1 per cent of the GLP.

(b) Strategies to adapt to a modified environment

As the number of extreme events (days with extreme precipitation and days with droughts) is expected to increase, it is estimated that current safety

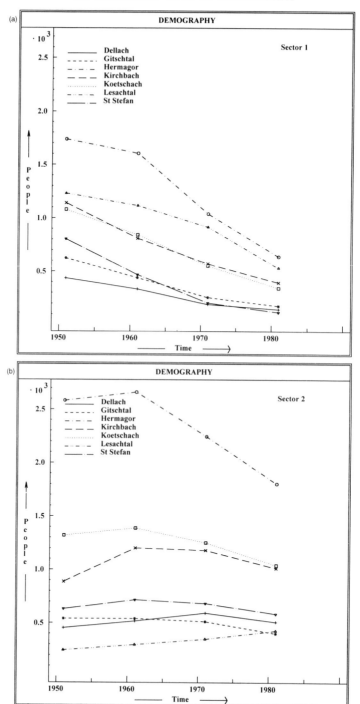

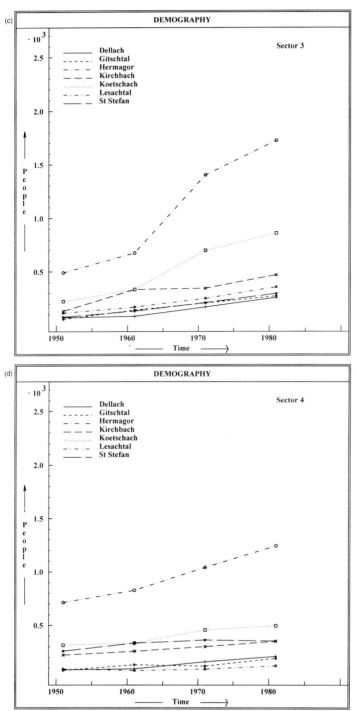

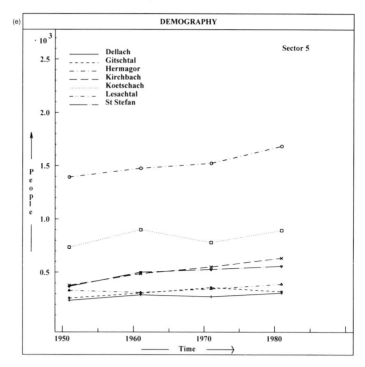

Figure 22.8 Differences of development in five economic sectors in the communities of the Hermagor district. *Source*: Data from Austrian National Statistical Office.

measures will not be sufficient in the future. Approximate estimates give us an area of 400 ha or 0.5 per cent of the total area additionally needed for safety construction and a price of US$ 600,000 to build over one hectare. This implies that some US$ 240 million are required for this purpose or some 5 per cent of GLP for 20 years. This option does not really seem practicable if the annual damage is on average less than 1 per cent of the GLP. However, in the long run average annual damages of up to 10 per cent of GLP cannot be excluded, and this option might become realistic at a more severe stage of warming.

(c) Policy recommendations for the current situation

According to the extent of losses predicted by the calculations presented here, the modified income of the Hermagor district would not justify compensation payments for greenhouse gas emissions within Austria. In addition, a greater increase of safety construction because of global climate change does not seem appropriate yet. The most desirable and cost efficient option is the investment in mitigation strategies outside of Austria. This option assumes an informed public ready to undertake measures against climate change at an early stage. Therefore it seems desirable for local decision makers to get more involved in the ongoing climate change debate and to begin to play a more active role.

REFERENCES

Aulitzky, H.E.A. (1985) *Grundlagen der Wildbach und Lawinenverbauung*, Universität für Bodenkultur, Vienna.

—— (1987) *Bioklimatologie II*, Universität für Bodenkultur, Vienna.

Breiling, M. (1990) 'The impact of mass tourism and the change of landscape in the Carnian Alps region', in L. Morbey (ed.), *Tourism and Landscape Management*, Conference Proceedings, Internal Report, University of Vienna.

—— (1992) 'Österreichs Tourismus der Zukunft – absehbare Umweltprobleme und deren mögliche Kosten', in K. Brodrick (ed.), *Umwelt-Tourismus-Verkehr*, Conference Proceedings, Hammer, Velden, Austria.

—— (1993) 'Die zukünftige Umwelt- und Wirtschaftsituation peripherer alpiner Gebiete', Ph.D. thesis, Inst. f. Landschaftsgestaltung, Universität für Bodenkultur, Vienna.

Breiling, M., Hetzendorf, I., Mattanovich, E., Schaffer, H. and Strasser, H. (1985) *Landschaftspflegeplan Naßfeld*, Institut für Landschaftsgestaltung, Universität für Bodenkultur, Vienna.

Brooke A., Kendrick, D. and Meeraus, A. (1988) *GAMS (General Algebraic Modeling System), A User's Guide*, The Scientific Press, Redwood City, California.

Embacher, H. (1992) *Urlaub am Bauernhof – das natürliche Angebot des österreichischen Tourismus*, Bundesverband Urlaub am Bauernhof in Österreich, Vienna.

Franz, H. and Holling, C.S. (1974) 'The Obergurgl Model. A microcosm of economic growth in relation to limited ecological resources', *Alpine Areas Workshop*, IIASA, International Institute of Applied System Analysis, Laxenburg, Austria.

Gordon, H.B., Whetton, P.H., Pittock, A.B., Fowler, A.M. and Hazlock, M.R. (1992) 'Simulated changes in rainfall intensity due to the enhanced greenhouse effect: implications for extreme rainfall events', *Climate Dynamics*, 7 (133): 1–20.

Hanousek, J. and Charamza, P. (1992) *Modern Statistical Methods* (in Czech), Internal Report, University of Prague.

Hanousek J., Charamza, P., Maly, M. and Zvára, K. (1993) *FamStat – Integrated Statistical Programming System* (in Czech), Internal Report, Charles University, Prague.

IPCC (1990a) *Climate Change: the IPCC Impacts Assessment*, Geneva.

—— (1990b) *Policymakers' Summary of the Scientific Assessment of Climate Change*, Intergovernmental Panel on Climate Change, Geneva.

—— (1992) *IPCC Supplement*, Geneva.

Jeglitsch, H. (1989) 'Volkswirtschftliche Gesamtrechnung nach Bezirken', *ÖROK* 72, Vienna.

MECCA (Morogoro Environmental Charter Consulting Agency) (1991) 'Global environmental problems: possible impacts for the Austrian environment and economy', in M. Breiling and W. Schopfhauser (eds), *Environment and Development*, University of Vienna Publications, p. 21.

Messner, S. and Strubegger, M. (1991) 'Potential effects of emission taxes on CO_2 emissions in OECD and LDCs', *Energy* 16 (11/12): 1379–95.

Neter, J. and Wasserman, W. *et al.* (1985) *Applied Linear Statistical Models*.

Tukey, W. (1977) *Exploratory Data Analysis*, John Wiley, New York.

APPENDIX – ANOVA RESULTS FOR POPULATION STUDY

In this appendix we describe how population behaves in different communities of the Hermagor region. The question of primary importance is whether there are significant differences between different communities. If there are any then the population–economic submodel for Hermagor should be taken for all the communities jointly but the relations between communities would have to be conserved. This means that instead of two-dimensional matrices we should use a three-dimensional one which will preserve the substantial economic structure of the whole region. More detailed data are then needed to study the possible development of the whole region in the general global context.

The positive correlations between time and number of people working in 'Services', 'Business and Transport', and also 'Others' sector can be observed. On the other hand negative correlations are quite clear between time and people working in 'Agriculture'. An umbrella-type dependency on time of people working in 'Industry' is also obvious. It is also clear from the pictures that the time dependencies of numbers of people working in some specific sector are almost the same for all the communities. The significance of different factors in the population–economic submodel can be studied using three-way ANOVA. A brief description of the model is given below. The general structure can be expressed as

$$Y = a + b + c + d + e + f + g + noise$$

where:

Y = number of people working in some economical sector
a = common effect (for all times, communities, economical sectors)
b = effect of economical sector
c = effect of time
d = effect of community
e = common effect (interactions) of time and economical sector
f = common effect (interactions) of economical sector and community
g = common effect (interactions) of time and community.

The numerical results of the ANOVA analysis are shown in Table 22.3.

After this first ANOVA analysis we exclude sequentially the most significant factors resulting in the final conclusion. It was proved that effect of time and common effect of time and community are really statistically nonsignificant for the given model. This means that time effects can be studied only in common with effects of the economic sectors. The non-significance of the time and community effect has already been mentioned. For any sector it explains the almost identical functional dependency on time of people working in the sector, independently of the region considered (see

Table 22.3 Application of ANOVA to Hermagor data

Analysis of variance of Table 22.1: all factors		
Factor	F-ratio	Prob > F
d (community)	153.96	0
c (time)	0.21	0.8909
g (time and community)	0.08	1
b (economical sector)	81.95	0
f (sector and community)	8.29	0
e (sector and time)	11.91	0

Figures 22.1–22.5c). Before imposing nonlinear regression we started by applying linear regression techniques; this enabled us to use the standard backward elimination technique to eliminate zero-valued parameters. For the sake of brevity, we mention here only certain results without entering into technical details (these can be found in Breiling, 1993).

First there are the overall decreasing tendencies in time in the numbers of people working in agriculture. The decrease is compensated by the increase of the relative number of people working in services and business. There is a strong increase of the relative number of people working in industry until the seventies. The trends were proved statistically to be linear with time, excluding the development of people working in industry and services. In these two economical sectors the quadratic term was significantly non-zero. Moreover, the analysis of the residuals of our model showed that although there was a rapid increase in the percentage of people working in the services sector, though nowadays there is a moderate downturn of this trend. It follows that a logistic curve could be used instead of a quadratic one for fitting the time tendency, which would lead to more precise forecasts. Also in agriculture the decreasing logistic curve would provide better results.

Second, there is a significant positive difference in the relative number of people working in 'business and transport' between the Hermagor community and other communities (excluding St Lorentzen). This can be explained by the fact that Hermagor is the local capital of the region. Moreover there is also a positive difference concerning the people working in the 'services' sector between Hermagor and other communities (excluding Koetschach-Mauthen). There is a larger relative number of people working in agriculture in Lesachtal and Kirchbach compared with other communities. Dellach and Gitschtal have the same employment structure of people working in different economical sectors.

These factors can be used to construct matrices in a general economic model according to the equation system described in the main text.

23

SENSITIVITY OF MOUNTAIN RUNOFF AND HYDRO-ELECTRICITY TO CHANGING CLIMATE

C.E. Garr and B.B. Fitzharris

INTRODUCTION

In mountain environments where snow and ice-cover influence runoff, climate variability can greatly alter the water resource. Runoff from many alpine areas of the world is used for hydro-electric generation, so that the question arises as to the sensitivity of mountain environments to climate change. In the future, a warmer and perhaps wetter greenhouse climate needs to be considered. This situation is illustrated with the example of New Zealand, where 80 per cent of the country's electricity is produced by hydro-electric generation, and almost half of the generating capacity (total 7,305 MW) uses runoff from the Southern Alps. In an average year, the hydro plant in these mountains produces 15,000 GWh. This chapter examines the main impacts of climate change and the sensitivity of a mountain-based hydro-electric system to changes in snow storage, temperature, and precipitation. Impacts on both supply and demand sectors of the electricity system are examined.

The impact of climate on water resources in alpine areas has previously been examined by Gleick (1986, 1987a, 1987b), Martinec and Rango (1989). Similar studies have related electricity demand to climate (Warren and LeDuc, 1981; Leffler *et al.*, 1985; Maunder, 1986; Downton *et al.*, 1987). However, few have attempted to integrate these impacts of climate change by considering both electricity supply and electricity consumption (Jager, 1983). In 1987, a New Zealand Climate Change Conference concluded with a call to determine the sensitivity of New Zealand's economic and social activities to changes in climate. An impact group was formed and two scenarios were presented for scientists to use as a basis for investigation. These two scenarios are described by the Ministry for the Environment (1990) and are entitled the 'most likely scenario' (scenario 1) and the 'alternative warm scenario' (scenario 2). Climatic conditions for the first scenario are based on the period

366

of maximum warmth 8,000 to 10,000 years ago, when westerlies were weaker over New Zealand and there was more airflow from the northwest (Ministry for the Environment, 1990). It assumes a temperature increase of 2°C and a precipitation increase of 10 per cent. Scenario 2 envisages a higher frequency of the positive phase of the Southern Oscillation (La Niña state). There are frequent incursions of tropical air from the north, and westerlies decrease, although they still prevail. It assumes a temperature increase of 3°C and a precipitation decrease of 10 per cent. These scenarios are not meant to be predictions, but rather represent two plausible climate changes in New Zealand with greenhouse warming. The sensitivity of electricity supply and demand is assessed with respect to these scenarios. In addition, the impact of a range of changes in precipitation and temperature on snow storage and electricity demand are considered.

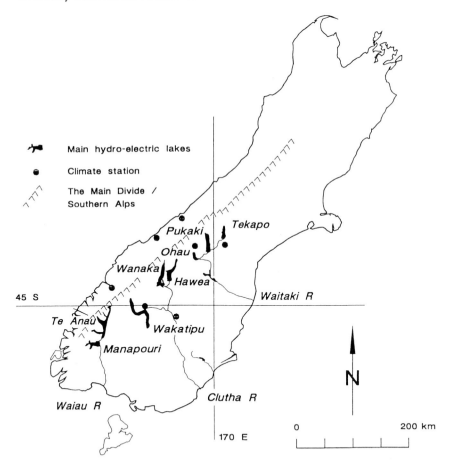

Figure 23.1 Location of main hydro-electric lakes in the South Alps of New Zealand.

HYDROLOGY OF SOUTH ISLAND HYDRO CATCHMENTS

The main hydro-electric storage lakes and catchments are located in the eastern part of the Southern Alps (Figure 23.1). The mountains are aligned southwest to northeast along the 800 km length of the South Island of New Zealand, and generally exceed 2,000 m in elevation, reaching over 3,000 m in the central part. The Southern Alps therefore provide a significant barrier to the Southern Hemisphere westerlies, which are consequently forced to rise. Enhanced uplift greatly increases precipitation, often by 3–4 times sea level values, so that it exceeds 10,000 mm/yr on the western flanks of the Southern Alps. The mountain ranges to the east receive much less, no more than

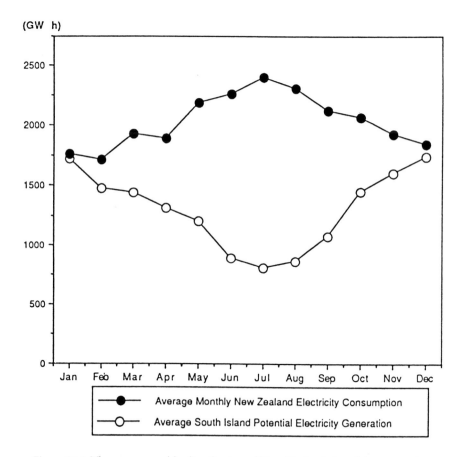

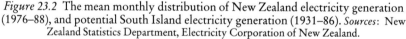

Figure 23.2 The mean monthly distribution of New Zealand electricity generation (1976–88), and potential South Island electricity generation (1931–86). *Sources:* New Zealand Statistics Department, Electricity Corporation of New Zealand.

3,000 mm/yr, and some of the intermontane basins record only 400 mm/yr Snow accumulation exceeds 4,000 mm near the Main Divide of the Southern Alps, but is usually less than 1,000 mm in eastern mountains.

The supply system has been examined using the nine large hydro catchments which rise in the Southern Alps, although the results for only one (Lake Wanaka catchment) are presented here. During winter, much of the precipitation falls as snow, and water is stored until it melts in spring and early summer. Consequently, river flows tend to be lowest in winter, and rise in spring and summer as snow and glacier melt make significant contributions (Figure 23.2). Long-term estimates of the water balance show an average precipitation of 2,820 mm/yr, with runoff at 2,270 mm/yr. Water accumulated as seasonal snow (snow storage), on average, amounts to 15 per cent of the total annual runoff. It is almost the same size as controlled lake storage, which is equivalent to 17 per cent of annual runoff. As shown in Figure 23.2, there is a mismatch between electricity demand, which is highest in the austral winter, and potential electricity generation (runoff), which is lowest at this time.

SENSITIVITY OF ELECTRICITY SUPPLY TO CLIMATE CHANGE

Runoff model

The method developed to examine the sensitivity of catchment runoff to changes in temperature and precipitation is based on a modified water balance model used by Fitzharris and Grimmond (1982), but incorporating a separate snow storage routine similar to that used by Gleick (1987a, 1987b) and Moussavi et al. (1989). Main inputs into the model are catchment hypsometry, runoff (R), and albedo, and mean monthly precipitation, temperatures, and solar radiation from nearby climate stations. Net radiation is calculated.

The model estimates annual and monthly components of catchment precipitation (P), evaporation (E) and storage (S). The snow storage routine uses a variant of the snow wedge model developed by Moussavi et al. (1989). It calculates the monthly mass balance of the snowpack at 1 m elevation bands throughout a catchment.

Precipitation in each elevation band is estimated using an exponential lapsing function similar to that used by Thompson and Adams (1979) and Moussavi et al. (1989). Snowmelt at each elevation band is obtained using a simple degree day model. The required temperatures are calculated by lapsing temperatures recorded at climate stations within the catchment, at a rate of 0.7°C per 100 m.

Once the storage is calculated, it is possible to estimate mean monthly runoff by:

$$R_{(mth)} = P_{(wb,mth)} - E_{(wb,mth)} \pm \Delta S_{(ss,mth)}$$

where the subscripts wb refer to results from the water balance model and ss refer to results from the snow storage routine.

The final stage of this analysis is to run the snowmelt model through a series of incremental temperature and precipitation changes. For each temperature change of 1°C (range –3°C to +5°C), precipitation is adjusted by 10 per cent (range –50 per cent to +60 per cent). These results are then plotted as a response surface. This is a two-dimensional representation of the response of mean annual runoff to changes in both precipitation and temperature.

Results – snow storage

Actual and modelled seasonal snow storages in the Wanaka catchment are shown in Figure 23.3. Snow begins to accumulate in May, with accumulation occurring until September. Average snow storage reaches a maximum of 198 mm. Snow melts from October onwards, disappearing by February, with most melt occurring in December (107 mm). Snow storage as predicted by the model, is slightly overestimated during the accumulation period, and slightly underestimated during the melt period. However, the model fit is good ($R^2 = 0.95$).

The sensitivity of snow storage to changes in precipitation and temperature is shown as a response surface for the Wanaka catchment in Figure 23.4. Snow storage declines as temperature increases and precipitation decreases, but the curved isolines indicate that the relationship is not linear. For example, with temperatures colder than present, snow storage becomes more sensitive to changes in precipitation. Climate warming envisaged by the scenarios produce much reduced snow storage: 60 per cent of present for scenario 1; and just 20 per cent of present for scenario 2.

Results – runoff

The modified water balance model is satisfactory, with estimated runoff closely matching the seasonal distribution of that observed (Figure 23.3). The general impact of scenario 1 in the Wanaka catchment is to increase annual runoff by 8 per cent. However, there are dramatic changes in the monthly distribution of runoff (Figure 23.5). During winter (June, July, August), runoff increases by 30 per cent. This is caused by increased precipitation, coupled with higher temperatures, so that a larger percentage falls as rain rather than snow. Higher temperatures also mean a higher rate of melt, and runoff during spring months (September, October, November) is enhanced by 23 per cent. The snowpack in scenario 1 is smaller than present, and is melted considerably sooner. Consequently, runoff is less than present-day

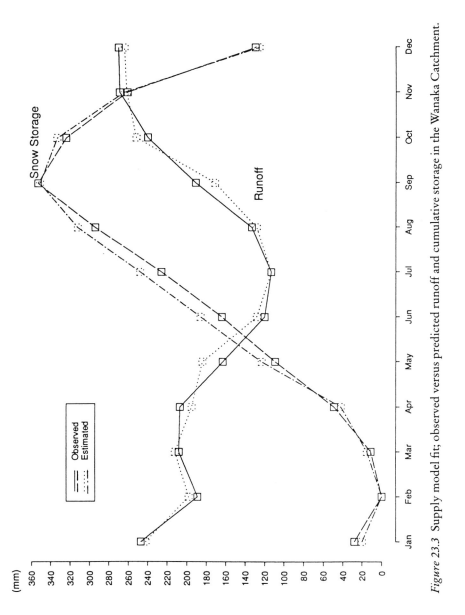

Figure 23.3 Supply model fit; observed versus predicted runoff and cumulative storage in the Wanaka Catchment.

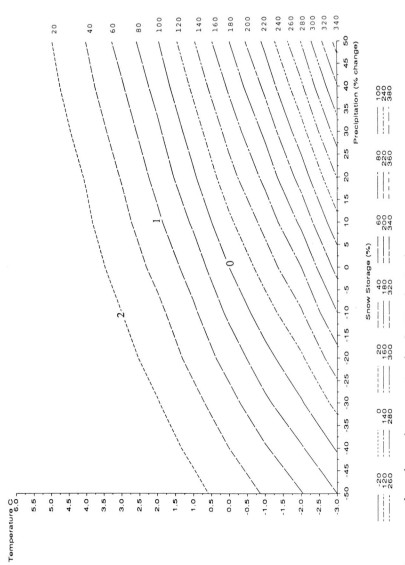

Figure 23.4 Response surface of annual snow storage for the Wanaka Catchment. *Note:* 0 indicates mean of present conditions, 1 indicates mean for climate scenario 1, and 2 indicates mean for climate scenario 2.

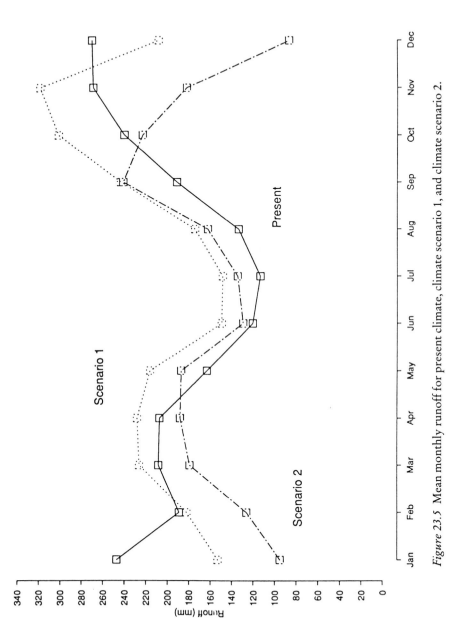

Figure 23.5 Mean monthly runoff for present climate, climate scenario 1, and climate scenario 2.

values during the latter part of the melt season (December, January, February). The greatest impact is on January runoff, which is reduced by 38 per cent.

The impact of scenario 2 is larger than for scenario 1. Annual runoff decreases by 18 per cent in the Wanaka catchment. Again, the monthly distribution of runoff changes dramatically. Average winter runoff is 18 per cent higher than present. However, during the remainder of the year runoff is much lower than present. Warmer winter temperatures and diminished precipitation reduce snowpack development, and it is melted by October. Summer runoff is not augmented by snowmelt, and higher temperatures lead to higher evaporation, so that summer runoff is much reduced by 68 per cent in December to less than present.

SENSITIVITY OF ELECTRICITY DEMAND TO CLIMATE CHANGE

Demand model

Daily and seasonal climate fluctuations have a significant influence on the amount of electricity consumed. A method is developed, using multiple regression, to analyse this impact on the national scale. Serial correlation is detected using the Durban-Watson d Test, and is treated by using the Yule-Walker procedure (SAS Institute, 1984). Two basic units measure the impact of climate on New Zealand electricity consumption. The first is the Heating Degree Day (HDD) which measures the electricity demand for heating. Heerdegen (1988) recommends that a base temperature of 15°C be used for New Zealand. The HDD index is summed over each month and weighted by the population in each region (Maunder, 1986). The second is the Cooling Degree Day (CDD) which measures the electricity demand for cooling. A base temperature of 20°C is used.

Model specification

The demand model specification is similar to that used by Leffler *et al.* (1985) to measure impact of weather on heating oil consumption. It determines sensitivity of electricity consumption to changes in temperature and is generally expressed as:

$$GWh_{(mth)} = F\{Yr_{(1977)} \ldots Yr_{(1988)}, Prdn_{(mth)}, WHDD15_{(Feb)} \ldots$$
$$WHDD15_{(Dec)}, WCDD15_{(Aug)} \ldots WCDD15_{(Jun)}, Mth_{(Feb)} \ldots Mth_{(Dec)}\}$$

where:

$GWh_{(mth)}$ = New Zealand monthly aggregated electricity consumption (GWh)

$Yr_{(1977)} \ldots Yr_{(1988)}$ = discrete dummy variables for each year
$Prdn_{(mth)}$ = number of production days per month
$WHDD15_{(mth)}$ = population weighted heating degree days per month for specified month
$WCDD15_{(mth)}$ = population weighted cooling degree days per month for specified month
$Mth_{(Feb)} \ldots Mth_{(Dec)}$ = discrete dummy variables for each month.

Long-term changes in electricity consumption caused by such features as increases in population, changes in the prices of electricity (and its substitutes), and changes in income, are represented as annual dummy variables. The impact of holidays and weekends is included with a production days variable. Intra-annual variation in electricity consumption is accounted for by using a temperature component and a non-temperature component. The non-temperature component varies from month to month, and is measured by monthly dummy variables. The model also allows for the sensitivity of electricity consumption to vary from month to month. For example, electricity demand may respond to a 1 degree temperature change in one season more than another.

Results

The linear form of the model is :

$$GWh_{(m)} = 1250 + 51Yr_{(1977)} + 83Yr_{(1978)} + 112Yr_{(1979)} + 134Yr_{(1980)} +$$
$$199Yr_{(1981)} + 277Yr_{(1982)} + 416Yr_{(1983)} + 531Yr_{(1984)} + 564Yr_{(1985)} +$$
$$617Yr_{(1986)} + 685Yr_{(1987)} + 681Yr_{(1988)} + 3.60WHDD15_{(Feb)} +$$
$$1.92WHDD15_{(Mar)} + 0.97WHDD15_{(Apr)} + 2.26WHDD15_{(May)} +$$
$$2.49WHDD15_{(Jun)} + 2.34WHDD15_{(Jul)} + 2.34WHDD15_{(Aug)} +$$
$$1.92WHDD15_{(Sep)} + 2.19WHDD15_{(Oct)} + 1.21WHDD15_{(Nov)} +$$
$$2.14WHDD15_{(Dec)} + 14Prdndays - 61Mth_{(Dec)} - 100Mth_{(Jan)} -$$
$$175Mth_{(Feb)}$$

Monthly dummy variables from March to November, HDD in January, and monthly weighted CDD variables are not significant and are excluded from the model. The model shows that New Zealand monthly electricity consumption is sensitive to changes in the population weighted HDD index from February to December. Changes in HDD have their greatest impact in February, June, and July, with April and November the months of least impact. The HDD index, alone, explains 10 per cent of annual aggregated New Zealand electricity consumption. Explanation is highest during the winter heating season (19 per cent) and lowest in summer and autumn (2 per cent).

Figure 23.6 compares observed electricity consumption with that predicted by the structural part of the model. Predicted consumption closely

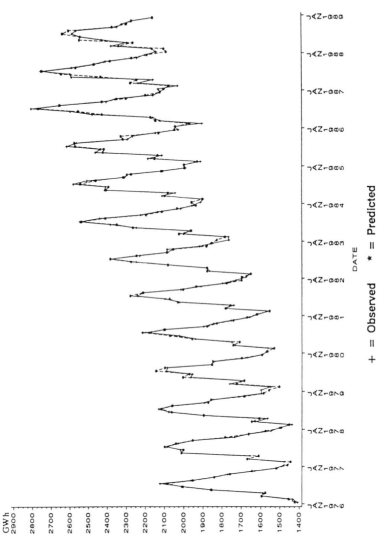

Figure 23.6 Demand model fit: observed versus expected New Zealand monthly electricity consumption (1976–88).

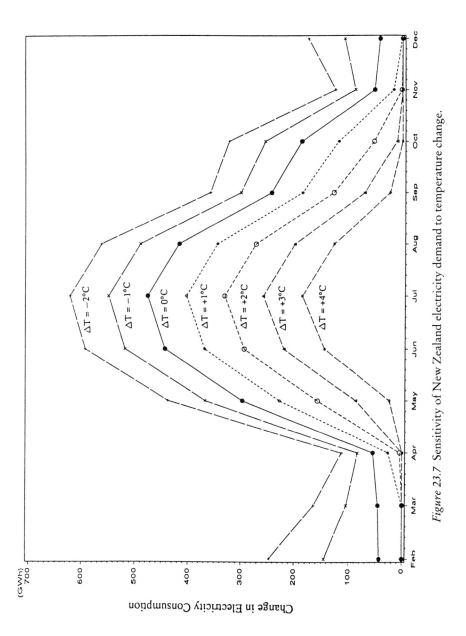

Figure 23.7 Sensitivity of New Zealand electricity demand to temperature change.

matches observed consumption. All variables are significant at the 95 per cent level. Model fit is high ($R^2 = 0.98$).

The sensitivity of monthly electricity consumption to temperature changes is represented by Figure 23.7. The warmest months of the year (December, January, February, March) exhibit the greatest change in electricity consumption as temperature decreases. Climate scenario 1 (2°C warming) leads to an annual decrease in consumption of 4.4 per cent. Change is greatest during winter, where electricity consumption decreases by up to 6.6 per cent. Scenario 2 (3°C warming) reduces electricity consumption by 6 per cent annually, and by up to 9.6 per cent in winter. Analysis of the CDD index suggests that large temperature increases are necessary before New Zealand electricity demand for cooling is significantly altered. This suggests that the peak of electricity demand will remain in winter and not shift to summer in a warmer climate.

A SYNTHESIS OF ELECTRICITY SUPPLY AND DEMAND

While electricity consumption models cannot predict actual electricity consumption forward to the middle of next century, they do describe sensitivity of the present system to climatic variability. Possible impacts of climate change scenarios can then be discussed on an *a priori* basis. Figure 23.8 shows changes in mean electricity consumption and mean potential electricity generation for the present, scenario 1 and scenario 2.

Scenario 1

A temperature increase of 2°C, and a precipitation increase of 10 per cent, raises annual potential hydro-electric generation by 12.1 per cent. Potential generation is higher than present in all months, other than January and February. Largest increases occur in August and September. New Zealand electricity consumption decreases, especially in winter.

An increase of 12 per cent in annual hydro lake inflows means a further 1,700 GWh of electricity could be produced (based on 1988 production). When combined with a suggested decrease in annual consumption of 1,060 GWh, a surplus of electricity of 2,800 GWh per year is produced. This is 12 per cent of present annual consumption.

There is higher generation potential during the winter months, when demand is greatest. The present demand mismatch between supply and demand is less severe. Water stored in hydro-electric lakes will be more effective and the system will have less reliance on alternative generation sources to meet peak demand.

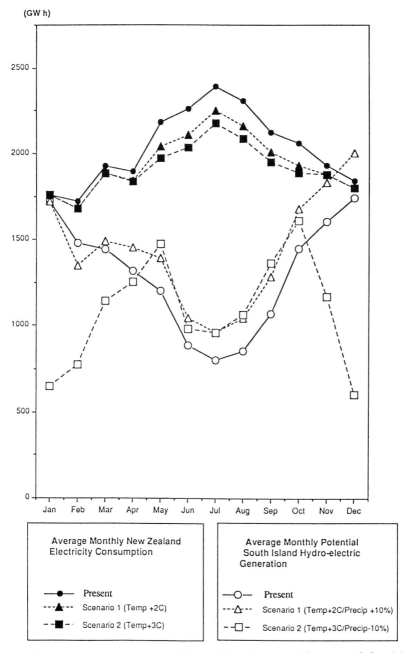

(GW h)

Figure 23.8 Synthesis of supply and demand: the mean monthly potential electricity generation from South Island catchments compared with the mean monthly aggregate New Zealand Electricity consumption for present climate, climate scenario 1 and climate scenario 2.

Scenario 2

A temperature increase of 3°C and a precipitation decrease of 10 per cent result in generation potential falling by 16 per cent annually. There are increases in potential generation of more than 25 per cent in winter, but decreases of up to 62 per cent in summer. Annual electricity consumption decreases by 6 per cent, mainly in winter. The integrated impact of a reduction in annual runoff of 16 per cent, or 2,500 GWh, and a decrease in annual consumption of 1,466 GWh produces a net deficit of 1,050 GWh compared with present. However, as with the first scenario, increased winter lake inflow raises potential generation at a time when demand is greatest. Thus the distribution of runoff through the year is again better matched to meet electricity demand, and is less reliant on storage than at present. However, the deficit between annual supply and demand compared with present would require further generation plant.

CONCLUSION

Mountain runoff (electricity supply) and electricity consumption (demand) are both sensitive to changes in precipitation and temperature. Long-term changes in future climate will have a significant impact on the seasonal distribution of snow storage, runoff from hydro-electric catchments and aggregated electricity consumption.

Figure 23.8 summarizes both potential generation and aggregate electricity consumption for present conditions, as well as for two climate change scenarios. It shows a convergence of electricity consumption and potential generation in New Zealand for two climate scenarios. In scenario 2, there is a fall in potential generation during spring and summer months.

The seasonal variation of electricity consumption is less pronounced than present, with the largest changes in winter, the time of peak heating requirements. There is also less seasonal variation in runoff and more opportunity to generate from existing hydro plant. In summary, the electricity system is made less vulnerable to climate variability in that water supply is increased, but demand is reduced. There is less need for new capital plant and water storage. On the whole, these are net benefits. The experience of the Southern Alps suggests that climate change will have important implications for hydro-electricity systems in other mountain areas.

ACKNOWLEDGEMENTS

This research was funded by the Electricity Corporation of New Zealand and the University of Otago. We appreciate the conference travel support accorded to us by the Convenor of the Davos Conference.

REFERENCES

Downton, M.W., Stewart, T.R. and Miller, K.A. (1988) 'Estimating historical heating and cooling needs: per capita degree days', *Journal of Applied Meteorology*, 27: 84–90.

Fitzharris, B.B. and Grimmond, C.S.B. (1982) 'Assessing snow storage and melt in a New Zealand mountain environment. Hydrological aspects of alpine and high mountain areas', *Proceedings of the Exeter Symposium*, July, IAHS Publ. No. 138.

Gleick, P.H. (1986) 'Methods for evaluating the regional hydrologic impacts of global climatic changes', *Journal of Hydrology*, 88: 97–116.

—— (1987a) 'Regional hydrologic consequences of increases in atmospheric CO_2 and other trace gases', *Climatic Change*, 10: 137–61.

—— (1987b) 'The development and testing of a water balance model for climate impact assessment: modelling the Sacramento Basin', *Water Resources Research*, 23 (6): 1049–61.

Heerdegen, R.G. (1988) 'An evaluation of the heating degree-day index', *Weather and Climate*, 8: 69–75.

Jaeger, J. (1983) *Climate and Energy Systems: A Review of their Interactions*, John Wiley & Sons, New York.

Leffler, R.J., Sullivan, J. and Warren, H.E. (1985) 'A weather index for international heating oil consumption', in W.F. Thompson and D.J. De Angelo (eds), *World Energy Markets, Stability or Cyclical Change, Proceedings of the 7th Annual North American, International Associations of Energy Economists*, Westview Press, Boulder and London, pp. 630–44.

Martinec, J. and Rango, A. (1989) 'Effects of climate change on snow melt runoff patterns', in *Remote Sensing and Large-Scale Global Processes, Proceedings of the IAHS Third Int. Assembly, Baltimore, Md, May 1989*, IAHS Publ. No. 186, pp. 31–8.

Maunder, W.J. (1986) *The Uncertainty Business*, Methuen & Co. Ltd, London.

Ministry for the Environment (1990) *Climatic Change, A Review of Impacts on New Zealand*, DSIR Publishing, Wellington, New Zealand.

Moussavi, M., Wyseure, G. and Feyen, J. (1989) 'Estimation of melt rate in seasonally snow-covered mountainous areas', *Journal of Hydrological Sciences*, 34: 249–63.

SAS Institute (1984) *SAS/ETS Users Guide, Version 5 Edition*, SAS Institute Inc., Cary, NC.

Thompson, S.M. and Adams, J.E. (1979) 'Suspended load in some major rivers of New Zealand', in D.L. Murray and P. Ackroyd (eds), *Physical Hydrology – New Zealand Experience*, New Zealand Hydrological Society, Wellington, pp. 213–29.

Warren, H.E. and LeDuc, S.K. (1981) 'Impact of climate on energy sector in economic analysis', *Journal of Applied Meteorology*, 20, 1431–9.

24

MOUNTAIN ENVIRONMENTS AND CLIMATE CHANGE IN THE HINDU KUSH-HIMALAYAS

S.R. Chalise

INTRODUCTION

The mountain environments of the Hindu Kush-Himalayas (HKH) have been undergoing rapid transformations in the last three decades at an unprecedented rate. Development imperatives have forced these transformations despite severely inadequate understanding of the biogeophysical processes in these high mountain environments. Development of infrastructures has increased accessibility into these otherwise isolated mountains. Increasing interactions with the outer world have made new demands on available resources as subsistence agriculture is being replaced by market-oriented economy. During the same period these inherently fragile and unstable environments have also witnessed a doubling of the population. Consequently pressure on natural resources has increased enormously, causing widespread degradation of the mountain environments and increasing poverty.

This situation has attracted worldwide attention since the early seventies and has been a major theme of global discussion (Eckholm, 1975, 1976; Myers, 1986; Ives and Messerli, 1989). It was this concern primarily which led to the establishment of the International Centre for Integrated Mountain Development, for the sustainable development of the Hindu Kush-Himalayas and the people living there (Glaser, 1984).

So far, the principal issues of major concern and debate with regard to the degradation of the Himalayan environment were not related so much to climate or climate change but more to overpopulation, deforestation, and soil erosion (Eckholm, 1975, 1976; Bajracharya, 1983; Mahat, 1985; Myers, 1986). However, it is well recognized that lack of reliable and long-term data on the extent and rate of deforestation, as well as on other environmental (including hydrometeorological) parameters for diverse ecosystems of the region, has given rise to serious uncertainty with respect to the degradation of

382

Himalayan environments and sustainable development (Thompson *et al.*, 1986; Ives and Messerli, 1989).

Any possible impact of global warming and climate change on the mountain environments of the Hindu Kush-Himalayas will further add to this uncertainty as regards the causes and consequences of such environmental degradations.

Significant research work on assessment of impacts of climate change has hardly started in the region. Subsequent discussions are based on the review of limited literature and data that were available at the time of writing. Conclusions on possible impacts of climate change in the region are tentative and drawn by extending the general conclusions of the IPCC (Houghton *et al.*, 1990; Tegart *et al.*, 1990) to this region.

GENERAL ASPECTS OF CLIMATE AND ECOLOGY OF THE HINDU KUSH-HIMALAYAS (HKH)

High diversity in climate and ecology are unique features of the HKH. Rising from a few metres above sea level to almost tropospheric heights, these highest mountain chains possess such enormous climatic and ecological diversity that within a span of less than 200 km they capture almost all types of climate that exist on earth and provide an unparalleled pool of genetic resources and biodiversity. Figure 24.1, which represents the cross-section for Central Nepal showing the dramatic rise in altitude and its diverse ecozones between India (south) and the Tibetan Autonomous Region of China (north) within a horizontal distance of 174 km, illustrates this point. Similar situations are found throughout the length (about 2,500 km) and breadth (about 200 km) of the HKH. These diversities in climate and ecology have also been the source of sustenance for the people and their rich cultural diversities in these mountains.

In the case of the HKH, generalization in terms of their ecology and climate is complicated because studies are sparse and relevant data are not easily available. However, it can be said that the ecology and climate of any particular area in this region are essentially characterized by altitude, topography, and the seasonality in precipitation induced by the monsoonal system. Thus this region abounds in extreme contrasts in terms of ecology and climate due, basically, to variations in these factors. For example, if Cherapunji in Meghalaya (northeast India) lying south of the main Himalaya, receives one of the highest rainfalls in the world (*c.* 11,000 mm per annum), then Leh, which lies in the rainshadow in the northwest Indian Himalaya, is one of the driest areas in the world (*c.* 75 mm per annum). Similar contrasts exist in vegetation, from the moist tropical and subtropical forests of *Shorea* in the southern plains, to the cold high alpine juniper forests and meadows, and the vegetation of arid deserts of Tibet and the trans-Himalayan areas of higher elevations in the north.

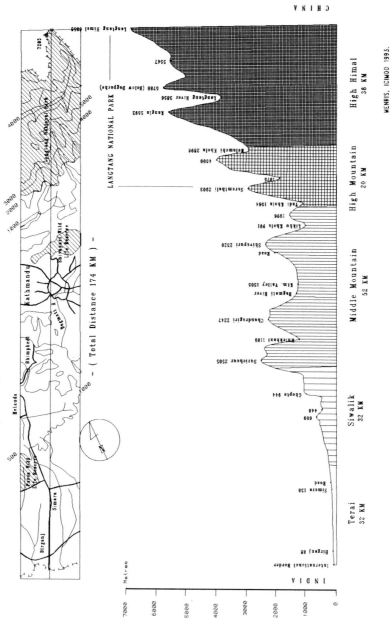

Figure 24.1 Cross-section of Central Nepal.

The climate of this region is essentially dominated by the southwest monsoon which provides most of the precipitation during the rainy summer months (June to September). However, the westerlies which predominate the rest of the year also bring snow and rain during winter and spring, most significantly in the western part of the region. These mountain chains block the northward advancement of the monsoon, which is about 6,000 m high (Mani, 1981: 11; Shamshad, 1988: 14), causing widespread and intense precipitation on the southern side of the Himalayas while, at the same time, making the Tibetan plateau and northern rainshadow areas among the driest in the world. They also intensify the precipitation processes in some preferred areas through orographic lifting, producing extreme examples as mentioned above. In fact, high and intense rainfall in the foothills of the northeastern Himalayas (e.g. Cherapunji) is a direct consequence of intensification of precipitation process by orographic lifting. The HKH mountains, therefore, act as an effective barrier between the lower and middle latitude climatic systems and can be looked upon as a huge climatic wall in the global atmospheric circulation system. The Himalayas, thus, are responsible for much moister summers and milder winters in South Asia than would otherwise be observed.

These mountain chains deflect the course of the summer monsoonal flow near the Bay of Bengal towards the northwest and there is a general decrease in monsoonal rainfall from east to west along the Himalayas. Thus, while the eastern Himalayas (Assam) have about eight months of rainy season (March–October) with fairly active pre-monsoon activities, the central Himalayas (Sikkim, Nepal, and Kumaun) have only four months (June–September) of rainy season. In the western Himalayas (Kashmir) the summer monsoon is active only for two months (July–August) (Mani, 1981: 9). Mean annual precipitation and monthly normals (1901–50) for some selected stations are given in Table 24.1 (Domroes, 1979), indicating the variation of rainfall from east to west as well as between trans-Himalayan and outer-Himalayan regions.

From Table 24.1 it is seen that whereas the outer Himalayas (e.g. Pasighat, Darjeeling) receive the highest rainfall, the trans-Himalayas (e.g. Kargil, Ladakh) is the region of lowest precipitation. The trans-Himalayan region gets its rain mostly from the western disturbances during winter. In general rainfall increases with altitude up to about 2,000 m south of the Himalayas but decreases sharply towards the north of the Himalayas. Similarly, rainfall generally increases from west (e.g. Simla: 1,542 mm) to east (e.g. Pasighat: 4,494 mm) in the HKH.

In general, the advent of the monsoon is characterized by the change in the direction of seasonal winds and the northward shift of the Inter Tropical Convergence Zone (ITCZ), from its normal position south of the Equator. The beginning of the monsoon season is not clearly defined in the mountains in contrast to the 'burst' of monsoon in the plains of India. Pre-monsoon

Table 24.1 Monthly rainfall normals, mm (1901–1950)

Station	J	F	M	A	M	J	J	A	S	O	N	D	Year
Kathmandu Nepal 1324 m	10	42	15	26	129	246	373	347	182	36	2	8	1416
Simla, Kumaun Himalaya 2200 m	65	70	64	46	60	149	416	419	182	33	10	2	1516
Nainital Kumaun 1953 m	70	73	53	38	84	391	769	750	363	61	13	2	2667
Pithoragarh Kumaun	44	56	40	28	73	183	300	287	149	33	7	22	1222
Pasighat/Tirap Frontier Tract, NE Assam	55	97	138	273	466	967	975	622	585	261	33	24	4496
Darjeeling, 2265 m	11	32	54	113	231	597	792	643	446	142	25	6	3092
Kyelong, N Punjab Himalaya	59	64	102	79	56	23	33	33	52	21	7	26	555
Srinagar, Kashmir Himalaya	70	74	94	90	60	34	56	63	40	28	13	37	659
Kargil, Ladakh 2682 m	37	38	60	42	25	7	7	10	10	6	3	21	266

Source: Domroes (1979)

activities gradually merge with the monsoon rains particularly on the eastern part, while in the western part its arrival is sudden with abrupt changes in cloudiness, temperature, winds, and rainfall (Mani, 1981).

Earlier attempts at generalization of climates in the Himalayas have been largely based on the relationship between climates and vegetation, and, to some extent, cultural practices at various altitudes (Schweinfurth, 1957, quoted by Ives and Messerli, 1989; Troll, 1967; Bagnouls and Meher-Homji, 1959; CNRS, 1981). Taking into consideration the horizontal and vertical variations in climate and vegetation, Troll (1967) has made a comprehensive classification of the Himalayas from west to east into the following divisions: the Indus Himalayas, the Punjab Himalayas, the West Nepal–Garhwal Himalayas, the East Nepal–Sikkim Himalayas, and the Assam Himalayas. More recently remote sensing techniques have been used to classify the Himalayas (Kawosa, 1988). Altitudinal zonation of climate and vegetation has also been the basis of a broad classification of the climates in the Nepal Himalayas (Dobremez, 1976; Hagen, 1980) and this has been the best approach to understand the general nature of climate and vegetation at different altitudes in the region. In their recent study, Ives and Messerli (1989) have provided excellent details on a new attempt undertaken by the Geographical Institute of the University of Berne on the regionalization of climates in the Hindu Kush-Himalayan region.

Difficulties in the generalization of climate in the Himalayas are largely due to the absence of long-term data and extremely limited number of meteorological stations particularly at elevations above 2,500 m. These difficulties have seriously hindered a better understanding of climates in the Himalayas (Barry, 1981; Mani, 1981; Das, 1983; Chalise, 1986).

LOCAL CLIMATES

As the Himalayas rise suddenly from the plains in a series of folds, they are the causes of several complexities in the climate of the region. A greater number of sub-climates and small-scale subdivisions exist in the region due to dramatic changes in orientation, altitude, and size of the mountains, slopes, valleys, and plateaux (Domroes, 1979). According to Flohn (1970), multiplicity of diverse local climates in these high mountains is caused by the interaction of different atmospheric processes – local heat budget, thermal circulation of varying magnitude, and convection and advection as a result of the synoptic processes occurring in the lower and upper troposphere.

An interesting feature of such local climates is that the valley bottoms in the HKH are generally characterized by dry, and the adjoining slopes and peaks by wet, climatic conditions. Flohn (1970) attributes this to a thermal circulation pattern, 'with valley breezes blowing up-valley throughout the day', through the influence of the Tibetan Plateau which acts as a heat source. This dry valley phenomenon is considered to be a unique feature of the Himalayas and particularly associated with the larger valley systems.

Strong winds are also characteristic features of the Himalayan valleys in Nepal (Hagen, 1980: 57) as evidenced by the deformed trees in the Kali Gandaki valley (Ohata and Higuchi, 1978). Possibilities of utilizing such winds for power generation in some potential sites in Nepal have drawn some attention, but systematic assessments are difficult as relevant data are not normally available (Chalise and Shrestha, 1982).

Another common feature is the wetness of the windward side and the dryness of the leeward side in these mountains. This is true at the local scale as well as at the macro scale. For example, Lumle (1,642 m) lying south of Annapurna range in the Nepal Himalayas receives about 5,000 mm of rain per annum, whereas Jomsom (2,750 m) lying north of it receives only about 250 mm per annum.

Despite the broad influence of monsoon in the summer and western disturbances in the winter on the precipitation pattern in the region, local variation in rainfall is significant. Orography influences the rainfall pattern very strongly at the local scale and contrasts in rainfall amount even in adjacent watersheds are quite common.

Studies of diurnal rainfall in Nepal (Dhar and Mandal, 1986) and India showed that its variation is small at places where precipitation is mainly the result of depression and cyclonic storms, as both phenomena act irrespective of the time of the day. The variation could however be significant where precipitation is caused by strong insolation. This can be seen outside the monsoon months, especially during the pre-monsoon period, where isolated heavy showers associated with convective clouds occur during the afternoon accompanied by thunder and lightning. According to Ueno and Yamada (1989), diurnal variation in precipitation in the Lamgtang Valley in Central Nepal is associated with three types of precipitation at different hours of the day. The mechanism of each type of precipitation is strongly influenced by the local circulation, topography, and the monsoon activity.

High intensity of rainfall is characteristic of the Himalayas (Domroes, 1979; Nayava, 1974). True measurements of rainfall in the Himalayas are problematic and of limited value not only because of the difficulty in installation, maintenance, and running of stations but also because of only local-scale validity of the observations obtained (Domroes, 1979). The daily rainfall in the Himalayas has not been given particular scientific attention so far, and detailed studies on this subject are necessary. The general rule that with increasing precipitation its variability decreases is of practical significance for flood control and for agriculture and soil management.

SEASONALITY AND TRADITIONAL KNOWLEDGE OF CLIMATES IN THE HKH

The climate of the Hindu Kush-Himalayas is essentially seasonal in character and seasonality has greatly influenced the agro-ecological practices here. Essentially, there are four main seasons in the region – winter (December to February); pre-monsoon or summer (March to mid-June); monsoon (mid-June to mid-September); and autumn or post-monsoon (mid-September to November) (Mani, 1981: 6).

The rhythm of seasonal changes in climate is interestingly quite uniform as is evidenced by the farming practices in the mountains. For example, the hill farmers of Nepal have developed their own local farming calendars over the centuries, based on intimate knowledge of local climates and the regularity in their seasonal changes. These calendars are still the basis of farming in the Nepalese mountains, using either the solar or the lunar calendar, and have been remarkably successful in helping the farmer to plan ahead and to choose crops suited to local climatic conditions. The unique rice planting calendar of Jumla (altitude *c.* 2,400 m) in western Nepal is one such example which is still being followed (HMGN, 1974). Similar traditional farming calendars are widely in use in other parts of the Hindu Kush-Himalayas as evidenced by rice cultivation at very high altitudes in the region. According to Uhlig (1985) the highest altitudes at which rice is cultivated is found in the HKH region.

A remarkable feature of these local calendars is their dependability from year to year for planning and managing agriculture and farming. This clearly indicates that through trial and error over the centuries farmers have obviously been successful in identifying the key climatic features of their local ecosystems, and have applied such knowledge in developing local

Table 24.2 Seasons and associated weather in Nepal (in terms of local months)

	Months	Associated Weather Type
1	Baisakh-Jeth (mid-April to mid-June)	Mainly hot and dry summer with rain and hail (due to thunderstorms and north westerlies)
2	Asar-Sawan (mid-June to mid-August)	Heavy monsoonal rains
3	Bhadra (mid-August to mid-September)	Hot and moist with scattered rainfall
4	Ahoj-Kartik – mid-Mangsir (mid-September to November)	Mild and fair with light rain or snow (due to Westerly disturbances)
5	Mid-Mangsir – Poush – Magh (December to mid-February)	Cold and dry with some rain or snow (due to Westerly disturbances)
6	Fagun-Chait (mid-February to mid-April)	Mild, fair and dry

Note: These local perceptions of associated weather with local calendar months show a remarkable degree of correlation with actual weather conditions during such periods and their annual cycle is quite regular. Baisakh is the first and Chait the last month of the Nepali calendar.

farming systems and farming calendars in terms of local months and calendars. Table 24.2 illustrates this for Nepal. Farmers across the HKH have surely no choice than to continue the use of such traditional calendars for a long time to come. Any significant and, particularly, abrupt change in such calendars as a result of climate change would cause disasters in local farming practices.

Over the centuries of isolation people across the HKH have learned to live with climatic disasters. Each year they have to survive either too much rain (causing landslides, floods, and debris flows) or too little (causing droughts). Hail is another much dreaded disaster which by its highly localized character is essentially unpredictable. Farmers have developed their own strategies to survive such disasters although such practices have no scientific justification. For example, in the middle hills of Nepal, local Buddhist priests (Lamas) are entrusted by villagers to ward off hail from their village (H. Gurung, personal communication, 1992). This is widely practised in Nepalese middle hills, probably because science has yet to reach there and the people still believe that climatic disasters are acts of God, and hence beyond human control and comprehension. Similarly it is widely believed in Nepal that climatic events and disasters have a 12-year cycle, which tends to agree with actual climatic events such as floods, in many cases. There are also traditional examples where cropping pattern, choice of diverse crops and a deliberate choice to have farm lands (big or small) in different localities and altitudes have helped the farmers to cope with climatic disasters.

Traditional knowledge of local climates and hydrologic events have not been well documented and it will be worthwhile to do so in selected ecosystems across the HKH. With further refinement, such knowledge could also provide useful and practical guidelines for improved agricultural practices and in developing strategies to face climatic disasters in future. As it is difficult to imagine that the network of hydro-meteorological stations could be expanded in the near future to generate meaningful data for the understanding of local climates and hydrology in the HKH mountains, it would be unwise to ignore such a useful and practical knowledge of local climates and hydrology possessed by local people in the region.

CLIMATIC FLUCTUATIONS IN THE HKH: USUAL OR UNUSUAL?

Climatic variability from year to year is more the rule than the exception in any locality in the HKH due to the diverse causes discussed in the preceding sections. Again, climatic fluctuations on different time scales and due to various natural causes have been known to occur during various geological periods (Kutzbach, 1976). As long-term climatological data for the mountainous areas of the Hindu Kush-Himalayas are lacking and the issues concerning global warming and impacts of climate change have started to receive some attention only very recently (Gupta and Pachauri, 1989), it is difficult to discuss quantitatively the impacts of global warming and climate change in the region. Some studies indicating fluctuations in atmospheric temperature in the region are discussed below, although they cannot be linked to the impacts of climate change without further studies and evidence.

Noontime temperature distribution in Kathmandu (1802–1803 and 1968–1990)

Hamilton's famous book on Nepal based on his fourteen months of stay during the years 1802 and 1803, also contains the oldest temperature records available so far for Kathmandu and its suburbs from 19 April 1802 to 16 March 1803 (Hamilton, 1990: 322–44). Unfortunately, it does not provide information on measurement, site, and equipment used, and temperature records are not available continuously for the subsequent years up to 1968. On the basis of Hamilton's records and the present availability of data (1968–1990) a comparison is made of the distribution of monthly means of noontime temperatures which is shown in Figure 24.2. Hamilton's original temperature records were on the Fahrenheit scale and they have been converted here to Celsius for convenience. Data for other hours of the day could not be used for comparison with the present records.

From Figure 24.2 it is seen that between the months of August and March mean monthly noontime temperature during 1968–1990 is higher than the monthly noontime temperature during 1802–1803. Between the months of April and July, Hamilton's records lie within the range of 1968–1990,

390

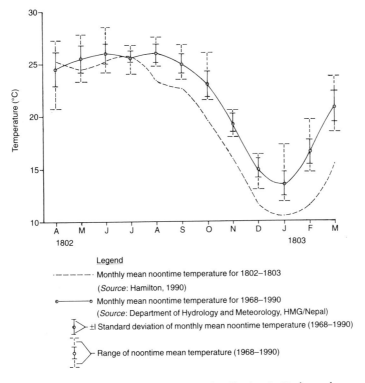

Figure 24.2 Noontime temperature distribution in Kathmandu (1802–3 and 1968–90). *Source*: S.R. Chalise 1993.

showing no significant trend. Although this preliminary analysis shows that the mean monthly noontime temperature has been increasing in recent years in Kathmandu, it needs further examination of factors, such as the impact of urbanization, which could influence such warming before definite conclusions are made in terms of its possible linkage with global warming.

Glacier fluctuations

Glacier studies in the Hindu Kush-Himalayas have received a lot of attention, particularly by scientists from outside the region. Most of the studies indicate that the glaciers are generally retreating. Mayewski and Jeschke (1979) have made local and regional syntheses of 112 records of such fluctuations in the HKH and their study shows that the glaciers in this region have been in a general state of retreat since AD 1850. Similar reports of glacier retreat and accelerated ablation are available from Nepal (Miller, 1989; Miller and Marston, 1989; T. Yamada, personal communication, 1991), although examples of both advancing and retreating glaciers have been reported from the Karakoram and Kunlun Mountains (Zhang, 1984; Wang *et al.*, 1984; Chen *et al.*, 1989; Zheng *et al.*, 1990). It is difficult to relate these fluctuations to climate change, although they indicate some warming of the atmosphere in the region.

IMPLICATIONS OF POTENTIAL IMPACTS OF CLIMATE CHANGE ON MOUNTAIN ENVIRONMENTS IN THE HKH

As relevant data on the concentration of greenhouse gases are not available for the region, it is difficult to discuss the possible impacts of global warming in the HKH due to sources from, or outside, the region. General deterioration in the quality of air is visible in many parts of the region, particularly in urban areas, and especially in urban areas located in the valleys (e.g. Kathmandu), which are significantly affected by growing urbanization, industrialization, and increased use of fossil fuels. Problems of air pollution are also exacerbated by climatic factors, such as low-level inversions, particularly during winter in the mountain valleys. The following discussion is based essentially on the extension of the IPCC's climate change impact scenarios (Houghton *et al.*, 1990; Tegart *et al.*, 1990) to the HKH.

Despite uncertainties involved in climate change scenarios, potential impacts of global warming (Houghton *et al.*, 1990; Tegart *et al.*, 1990) which might affect the HKH are: increased monsoon rainfall due to seasonal shift of ITCZs polewards; enhanced precipitation; general shift of agro-ecological zones towards higher elevations, and shrinking of the areas under snow and permafrost. The possible implications of such changes with respect to some basic issues concerning the mountain environments of the HKH are discussed below. These discussions deal mainly with those issues concerning emissions of CO_2 and CH_4, which are most relevant in the context of the mountain environments of the region.

Ecological issues: use and conservation of forest and biodiversity

The mountain environments of the HKH are already under stress through the combined pressures of human and livestock population on its natural resources, principally forest, land, and water. With a conservative estimate of 2 per cent growth rate, the present human population of 118 million (Sharma and Partap, 1993) will double within 35 years. Similarly the present livestock population of nearly 70 million (Bhatta, 1992) is also expected to grow at the same rate and double within that period. The total impact of additional demands for food, fodder, fuel, and fibre will be on the existing forests (Nautiyal and Babor, 1985) not only because additional lands will be needed to produce more food but also because of almost absolute dependence of traditional mountain farming system on forest in the HKH (Mahat, 1987).

The total area under forest in the HKH including pastures and protected areas is presently estimated as 227.714 million ha (Bhatta, 1992). The area under forest cover is estimated at around 86.3 million ha covering about 24 per cent of the land area of the HKH, and the rate of deforestation is estimated as between 141,000–354,000 ha/annum (Bhatta, 1992). Despite the 'uncertainty' that usually surrounds forest data in the region (Thompson *et al.*, 1986) the rate of deforestation is certainly going to exceed the above figures in the coming years, unless some drastic measures are taken to lessen the traditional dependence of farming on forests.

The role of forests and vegetative covers in stabilizing the slopes, controlling soil erosion, sedimentation, and mass wasting, modulating soil temperature extremes and the hydrologic regimes are well recognized. Equally important is the possible role of forests in absorbing carbon dioxide from the atmosphere and hence retarding or controlling the rate of global warming through increased afforestation on a large scale despite the enormity and complexity of the task involved (Melillo *et al.*, 1990). Although lack of relevant and reliable data on both climate and forest cover in the HKH does not permit us to draw any conclusion on the impact of one on the other, loss of forest cover does imply a net shrinking of the global carbon sink and hence increases in CO_2 and atmospheric warming. The burning of wood, principally as fuel for cooking and to a certain extent to keep warm during the cold period is another important issue connected to deforestation as well as increased CO_2 in the atmosphere. It is estimated that 80 per cent of the total wood used in developing countries is for burning as fuel (Jaeger, 1983). This is true for the HKH too.

It therefore appears that the deforestation and burning of wood which is occurring in the HKH is, to a certain extent, contributing towards increased CO_2 in the atmosphere. However, fuelwood adds much less CO_2 to the atmosphere than fossil fuels, and if forests are managed sustainably to supply fuelwood then much of the carbon released will be reabsorbed by forests (Houghton, 1989). This makes fuelwood a more benign energy source than is generally believed. Effective protection of existing forests and increased afforestation are, therefore, very important for the HKH. This could convert the HKH into a net carbon sink as part of a concerted effort towards reafforestation in the Indian Himalayas (Singh *et al.*, 1985). Cooperation among the countries of the region to increase the green cover in the HKH will be an important contribution towards achieving such objectives (Khoshoo, 1990a, 1990b, 1990c).

Similarly, encouraging people to plant trees in their marginal lands through a scheme which will guarantee the supply of food to mountain villages and an assured market for wood or wood products, in what may be called a 'Wood for Food' programme (Chalise and Joshy, 1983), could greatly help increase the forest cover in the mountains of the HKH.

Increase in non-combustive use of wood will not only reduce the build-up of carbon dioxide in the atmosphere, but will also provide economic opportunities to the poverty stricken people of these mountain environments. A concerted effort to increase such use of wood and wood products as much as possible, at the regional as well as at the global level, will provide incentives for better management of existing forests as well as encourage increased afforestation in the region on a sustainable basis. A global co-operative programme to develop renewable energy resources to provide alternatives to fuelwood in the region in the long term also needs to be considered.

Another important ecological issue which is again linked directly to deforestation or degradation of forests is the need for the conservation of biological diversity in which this region is exceptionally rich. Phyto-geographically the Himalayas as a whole are considered as an autonomous region, distinct from other regions such as the Sino-Japanese, Indian, and Central-Asiatic (Dobremez, 1976: 245). In terms of ecosystems, the Hindu Kush-Himalayan region is the meeting place of Palaeoarctic realm of the north and the Indo-Malayan realm of the south. This region is also at the crossroads of five major biogeographic sub-regions (Train, 1985: 9). These add to this region's unique richness in biological resources. For example, more than 100 species of mammals and 850 of birds have been recorded here and about 10,000 indigenous plants have been reported of which over 6,000 species are indigenous to Nepal only (Upreti, 1985: 19).

Loss of biodiversity is an issue of global concern. The Himalayas have, since ancient times, been considered as a reservoir of rare medicinal plants having recognized healing properties. *Taxus baccata*, a native species of Himalaya has recently attracted considerable interest in the western world as a potential drug against cancer. Several species of plants that are yet to be studied might be lost because of a poleward shift of climatic zones, estimated at 200–300 km per degree of warming. It is also estimated by WHO that 80 per cent of the people in developing countries depend on traditional medicines, of which 85 per cent involves the use of plant extracts (Tegart *et al.*, 1990: 3–25). Loss of useful species from the HKH will also be a global loss.

Conservation of such diverse ecosystems and *in situ* protection of the rich biodiversity of these mountains are therefore extremely important. Global and national attention has been drawn to this important issue and there is a good network of protected areas in the region. There are about 300 protected areas of all categories (as per IUCN's classification) covering approximately 101,000 sq km distributed over the entire region (Stone, 1992: 124–5). Although these areas do provide some protection to several endangered species of plants and animals in major ecosystems of the region, it may be useful to note that the existing network covers less than 2 per cent of these mountains in the eight countries (Afghanistan, Bangladesh, Bhutan,

Myanmar, China, India, Nepal, and Pakistan) of the Hindu Kush-Himalayas. The need for expanding the protected areas to include representative ecosystems of the region cannot be overemphasized. Again there is virtually no information available on partially protected areas in the region, although their role in the conservation of biodiversity and local hydrological balance is equally important.

Another potential impact of global warming on the loss of biodiversity could be due to the melting of permafrost and increased intensity of precipitation adding to greater sedimentation in rivers and streams, thus affecting aquatic life.

Understanding of the intimate relationship between climate and ecology in the HKH is extremely important not only to identify important species and processes but also to know more exactly how the change in the climate could affect the ecology of the region. Studies with such precise objectives are yet to be carried out in the HKH.

Agriculture

Another issue of fundamental concern is the impact of increased production of food and fodder to satisfy the needs of a rapidly growing human and livestock population, on these mountain environments. Intensive cultivation, as well as extension of the cultivated area at the cost of forest, pasture, or other common lands, has not only changed the land-use but also caused widespread degradation of the environment in the region.

For example, the study of two subwatersheds in Nepal has shown that during the last three decades 75 per cent of the total forest land was cleared and grazing/pasture land had almost disappeared (Shrestha, 1992). Similar conditions are not uncommon in other parts of the HKH, although great controversy exists as regards the decrease in forest areas in recent times (Ives and Messerli, 1989). Mountain agriculture has also become more unsustainable and traditional management systems are no longer able to cope with such high demands and intensive use of natural resources (Jodha, 1989 and 1992) as is going to be required in future. Hence, even if we disregard climate change, sustainability of the mountain agriculture and stability of the mountain environment are in question.

The influence of monsoon on agriculture in the HKH needs no elaboration. A major impact of global warming is the possible shift of the ITCZ polewards and consequent enhanced monsoon precipitation. It is not possible to predict for the HKH the potential impact of enhanced monsoon rainfall due to climate change, but it might increase availability of water as well as moisture and contribute towards increased agricultural production. However, intense rainfall might trigger more landslides and mass wasting, and consequently result in a net loss of agricultural land and soil nutrients, causing negative impacts on agricultural production. Atmospheric warming may cause an upward shift of agro-ecological zones and could make more land available for agriculture, although the soil condition may not be able to support crops. Again, local climates might be affected, and local calendars on the basis of which mountain people have been planning and managing their agriculture so far, could become no longer useful.

Greater availability of water might also encourage farmers to bring more land under rice cultivation with a consequent rise in methane production. Reliable studies on methane production by paddy rice in the region are yet to be carried out, and controversy about the estimated amount of paddy and animal methane production from the developing world already exists (Agarwal and Narain, 1991; WRI, 1991).

Enhanced monsoon rainfall might bring more water to the fields than could be handled by traditional farmer-managed irrigation systems which so far have made agriculture, particularly rice cultivation, possible in the HKH. Disruptions in local irrigation and water management systems would be disastrous for agriculture and slope-land management in these mountains.

Water and energy resources

The countries of the HKH region are passing through rapid economic transformations in order to meet the growing aspirations of the people for their overall development, and the demand for energy for diverse developmental needs has grown enormously in the region in recent years. Sole dependence on fuelwood for all the energy requirements in the past was the most critical factor for deforestation in the region. For the bulk of the population this dependence, for their own needs as well as for the needs of their livestock, still persists. However, wood is no longer considered the principal source of energy, and water is now seen to be the principal source for rapid economic development (Verghese, 1990).

Considering that the HKH is one of the richest regions on earth in terms of hydro-power potential, such hopes are not illogical. However, a proper and systematic assessment of the hydro-power potential (including mini- and micro-hydro) of all the big and small rivers of the region has yet to be made. For example, the hydro-power potential of Pakistan, which comes largely from the Indus system, is estimated to be 20,777 MW (Quershi, 1981 in Sharma, 1983: 262) and that of the Ganga-Brahmaputra-Barak basin on the order of 200,000–250,000 MW, of which half or more is considered to be viable for harnessing (Verghese, 1990: 169). Similarly, countries such as Nepal, which alone has a theoretical potential of 83,000 MW (Chalise, 1983), and Bhutan are extremely rich in hydro-power, with potentials to transform their economies dramatically if such power could be harnessed and shared in the region. Despite such a huge potential only a negligible percentage of it has been utilized so far by the countries of the region.

There are three major hurdles in harnessing this rich potential, namely economic, scientific/technical, and political. So far, the poverty that exists in the countries of the region appears to be the main reason for not utilizing this huge potential, but external input of capital will not be difficult if the projects are viable. Such examples already exist in many countries of the region. Thus the scientific and the political seem to pose more important problems than the availability of capital.

The basic constraint from the scientific or technical point of view is the lack of long-term historical data bases on relevant biogeophysical parameters for a proper understanding and modelling of the hydrological systems of these mountain basins, big or small, which are so different from each other in terms of geology, topography, altitude, vegetation, and land-use. Even smaller basins have not been systematically and adequately studied to understand the inter-relationships between precipitation, erosion, land-use changes, sedimentation, and streamflow. Hence the impacts of natural and anthropogenic processes are difficult to model or predict, and have given rise to much debate and uncertainty (Kattelmann, 1987). This problem is further complicated because the hydrology of high mountain areas is itself not very well developed and the peculiar combination of the extremely high altitudes, steep slopes, and intense seasonal monsoonal rainfall, limits and even precludes the application of hydrological techniques, principles, and models developed in temperate regions of Europe and America. These problems are the primary concerns of the proposed project, namely, Regional Network of Experimental Watersheds for Hydrological Studies (RENEWHS), which is being launched by UNESCO and ICIMOD jointly in close co-operation with the countries of the region (UNESCO/ICIMOD, 1990 and 1992).

The complexity of the problems associated with the development of water resources in the region could be further complicated by the potential impacts of enhanced monsoon rainfall and increase in rainfall intensity due to global warming. The implications of such an increase in the amount and intensity of rainfall, which are virtually impossible to quantify, on the highly energized, geologically active high mountain environments of the HKH are evident but uncertain. Similarly, the possible impact of shrinking areas under seasonal snow, ice, and permafrost could seriously alter the hydrologic characteristics of the river systems in the region.

Considering the serious inadequacy of present knowledge on mountain hydrology, future changes in the hydrologic regimes of the rivers because of climate change will add further to the uncertainty and risks in harnessing water resources in the region. Technical designs for large dams exceeding fifty years' duration will need to consider these uncertainties. The region faces a dilemma as to whether to ignore these warnings or face the prospect of bearing additional costs, and whether or not such costs are affordable. Either way mountain environments could be adversely affected.

There is the possibility of developing other renewable resources, particularly solar and wind, in the region. They are virtually untapped so far.

Relevant data bases for large-scale harnessing of solar and wind energy need to be developed. However, the potential impact of climate change due to increased availability of moisture and increased rainfall could result in more cloudy days, and this might adversely affect the prospects of solar energy utilization in the region.

Hazards and disasters

With the highest mountain chains which are geologically young, active, and still rising, the Hindu Kush-Himalayan environment is highly vulnerable to natural hazards and disasters. These mountain environments are commonly associated with earthquakes, landslides, mass-wasting, floods, and droughts. Most of the disasters commonly faced by the people are climate-induced and are generally considered as acts of God in the region. Sometimes these disasters become catastrophic such as the 1988 floods in Bangladesh (Rogers *et al.*, 1989), the 1992 floods in Pakistan, and the more recent floods, landslides, and debris flows in the southern part of central Nepal in July 1993. Failure in monsoon rain often causes serious droughts, which sometimes can also be calamitous, such as those in north India during 1985 and 1986. Hail is another climatic hazard which the people have to face quite frequently, as mentioned earlier.

A change in the climatic pattern in this region could further increase the incidence of such disasters. Increased monsoon rainfall, increase in rainfall intensity, melting of snow, ice, and permafrost with a consequent decrease in their surface areas (which in turn could diminish lean period discharge), all of which are envisaged as potential changes due to global warming, could singly or in combination with other factors induce more incidences of disasters in the region, although it is not possible to consider the magnitude or frequency of such disasters at this stage. Such potential impacts of climate change further add to the confusion and uncertainty particularly with respect to the impacts of deforestation on downstream flooding and increased sedimentation in the rivers. It is an issue which has the potential to spark off regional conflicts as, for example, the floods in Bangladesh are assumed to be linked to deforestation in the Himalayas despite the fact that there is no factual and scientific evidence of such linkages (Ives and Messerli, 1989; Bruijnzeel and Bremmer, 1989; Carson, 1985; Alford, 1992).

CONCLUSIONS AND RECOMMENDED ACTIONS

From the foregoing discussions it is obvious that the implications of potential impacts of global warming and climate change for the mountain environments of the Hindu Kush-Himalayas are uncertain although the impacts could be more adverse than benign. Obviously this is going to add further to the uncertainty that is attributed to these environments.

The only way to remove such uncertainties is to intensify research for better scientific understanding of the natural and anthropogenic factors, including global warming, that are impacting or are likely to impact on these mountain environments. As the origin of the uncertainties is associated with the absence of pertinent data and/or because they are unreliable, scanty, and even conflicting, a serious and sustained effort is needed to develop a comprehensive and reliable data base on pertinent environmental and hydroclimatic parameters of the region.

Better understanding of these processes in the mountains regions of the Hindu Kush-Himalayas is critical for the sustainable development of this region on all fronts. Without such an understanding and knowledge base, the rich potential natural energy resources (solar, wind, and forest), as well as biological resources (including use and conservation of its richness in biodiversity), cannot be fully and sustainably developed. Similarly, the modern infrastructure and agricultural revolution that is required in this region to eradicate endemic mass poverty cannot be implemented without a proper understanding of the climate and hydroclimatic processes. The July 1993 disaster unleashed by intense rainfall in the southern part of central Nepal, which damaged or destroyed bridges, roads, and dams, has once again demonstrated the futility of investment in the construction of infrastructures without a proper understanding of the hydroclimatic and other bio-geophysical processes in the mountain ecosystems. Hence, efforts to develop a minimum data base (Messerli, 1985) on such parameters should be given highest priority at both the national and regional levels.

Development of such a comprehensive data base will be a long-term programme demanding much human effort and financial resources. However, some initiatives are already underway in the region, and a number of recommendations can be formulated, as given below.

- Although establishment of a network of monitoring stations to match the diversity of climate in the region should receive high priority, the first step in this direction will be to make an inventory of climate-related programmes and available data on hydro-meteorology from the existing network of stations in the countries in the region. This will also contribute towards a global inventory of climate observation programmes which is considered as extremely important (Barry, 1992).
- Much could be undertaken for sustainable research on climate-related environmental issues in the region through a greater involvement of universities and academic institutions at various levels, if they were to be provided with adequate funds and if hydro-meteorological data were made accessible for research. Presently, hydrological data are neither easily accessible nor easily shared by researchers in the region. Major initiatives in this regard have to come from the countries of the region.
- Building a systematic knowledge and data base on mountain environ-

ments should also start immediately by recording and documenting the knowledge of the local people about their climate and natural resources (land, water, flora, and fauna). Traditional knowledge of the mountain people in diverse cultural and ecological settings must be documented and recorded before they are wiped out by the onslaught of rapid changes that are sweeping these mountains (Pei, 1991).

• Solutions to the problems of the HKH region will have to be found through work carried out in the region. Scientific and technical principles and approaches which might have been successful elsewhere in developing natural resources do not constitute a sufficient guarantee for their replication in this region. This is clearly exemplified by the frequent damage or destruction of vital infrastructures, constructed at huge cost, through floods, landslides, and debris flows, which are primarily caused by excessive and intense precipitation. As these damages often affect neighbouring countries also, governments in the region should co-operate and accord priority to monitor Himalayan watersheds for hydroclimatic and geo-ecological studies. Donors who often provide funds for the development of infrastructure can play a crucial role in initiating such research. Further delay will only perpetuate the waste of scarce national and international resources.

• Support for initiatives already taken by national or international institutions, such as the Regional Programme on Mountain Hydrology launched jointly by UNESCO and ICIMOD (UNESCO/ICIMOD, 1990, 1992) would be a good start. Similar regional programmes on mountain climate, and climate change studies, should be launched as soon as possible. Similarly, one of the regional centres of the International Geosphere Biosphere Programme (IGBP) could be devoted to problems of mountain environments (Barry, 1992) and located at ICIMOD for the HKH region. A concerted effort by all concerned with research and studies on climate and impacts of climate change, such as UNESCO, UNEP and WMO, to influence government donors through specific programmes for this region will probably be more effective. ICIMOD, being located in the region, should be encouraged to initiate and implement such programmes in close collaboration with national institutions of the countries of Hindu Kush-Himalayas.

REFERENCES

Agarwal, A. and Narain, S. (1991) *Global Warming in an Unequal World*, Centre for Science and Environment, New Delhi.

Alford, D. (1992) *Hydrological Aspects of the Himalayan Region*, Occasional Paper No. 18, International Centre for Integrated Mountain Development (ICIMOD), Kathmandu.

Bagnouls, F. and Meher-Homji, V.M. (1959) 'Types Bioclimatique du sud est asiatique', *Trav. Inst. Francias Pondicherry*, I (IV): 207–47.

Bajracharya, D. (1983) 'Fuel, food or forest? Dilemmas in a Nepali village', *World Development*, 11 (12): 1057–74.

Barry, R.G. (1981) *Mountain Weather and Climate*, Methuen, London.

—— (1992) 'Climate change in the mountains', in P.B. Stone (ed.), *The State of the World's Mountains*, Zed Books Ltd, London, pp. 359–74.

Bhatta, B.R. (1992) 'Forestry and pasture resources in the Hindu Kush-Himalaya region', forthcoming publication, ICIMOD, Kathmandu.

Bruijnzeel, L.A. with Bremmer, C.N. (1989) *Highland–Lowland Interactions in the Ganges Brahmaputra Basin: A Review of Published Literature*, Occasional Paper No. 11, ICIMOD, Kathmandu.

Carson, B. (1985) *Erosion and Sedimentation Processes in the Nepalese Himalaya*, Occasional Paper No. 1, ICIMOD, Kathmandu.

Chalise, S.R. (1983) 'Energy efficiency and conservation in Nepal', *Energy*, 8: 133–6.

—— (1986) *Ecology and Climate in the Mountain System: A Review*, Working Paper No. 12, ICIMOD, Kathmandu.

Chalise, S.R. and Joshy, D. (1983) 'Management of agroforestry and biomass programmes in Nepal and some prospects of regional cooperation', *Biomass Energy Management in Rural Areas: Proceedings of the Regional Workshop on Biomass Energy Management in Rural Areas for South Asia*, 27–29 December 1983, Hyderabad, India, pp. 14.1–14.7.

Chalise, S.R. and Shrestha, M.L. (1982) 'Preliminary consideration on the potential and constraints of utilisation of wind energy in Nepal', *Proceedings of the Seminar Workshop on Renewable Energy Resources in Nepal*, RECAST, Tribhuvan University, Kathmandu, pp. 251–65.

Chen, J., Wang, Y., Liu, L. and Gu, P. (1989) 'Surveying and mapping on Chongce Ice Cap in the West Kunlun Mountains', *Bulletin of Glacier Research* (Japan) 7: 1–5.

CNRS (1981) *Paléographie et biogéographie de l'Himalaya et du Sous-continent Indien – comptes rendus de la table ronde tenue a les 27–28 avril 1979*, Edition du CNRS, Paris.

Das. P.K. (1983) 'The climate of the Himalaya', in T.V. Singh and J. Kaur (eds), *Himalayas Mountain and Men*, Print House, Lucknow.

Dhar, O.N. and Mandal, B.N. (1986) 'A pocket of heavy rainfall in the Nepal Himalayas – a brief appraisal', in S.C. Joshi (ed.), *Nepal Himalaya: Geo-Ecological Perspectives*, Himalayan Research Group, Nainital, India, pp. 75–81.

Dobremez, J.E. (1976) *Le Nepal Ecologie et Biogeographie*, CNRS, Paris.

Domroes, M. (1979) 'Temporal and spatial variations of rainfall', *Journal of the Nepal Research Centre*, 2/3 (1978/79): 49–67

Eckholm, E. (1975) 'The deterioration of mountain environment', *Science*, 189: 764–70.

—— (1976) *Losing Ground*, World Watch Institute, W.W. Norton & Co., New York.

Flohn, H. (1970) 'Beitraege zur Meteorologie des Himalaya', *Khumbu Himal*, 7 (Innsbruck): 25–45. (English Translation 'Contributions on the meteorology of Himalaya', in *Studies on the Climatology and Phytogeography of the Himalaya*, published by Nepal Research Centre, Kathmandu, 1988, for internal circulation, pp. 21–58).

Glaser, G. (1984) 'The role of ICIMOD: a presentation of the Centre', *Mountain Development: Challenges and Opportunities*, ICIMOD, Kathmandu, pp. 59–63.

Gupta, S. and Pachauri, R.K. (eds) (1989) *Global Warming and Climate Change: Perspectives from Developing Countries*, Tata Energy Research Institute (TERI), New Delhi.

Hagen, T. (1980) *Nepal*, Oxford and IBH, New Delhi .

Hamilton, F.B. (1990) (second AES reprint; first published 1819) *An Account of the Kingdom of Nepal*, Asian Educational Services, New Delhi.

HMGN (1974) *Mechi Dekhi Mahakali* (in Nepali): Part IV, HMGN, Ministry of Communications, Department of Information, Kathmandu, pp. 477–8.

Houghton, J.T., Jenkins, G.J. and Ephraums, J.J. (eds) (1990) *Climate Change. The IPCC Scientific Assessment*, Cambridge University Press, Cambridge.

Houghton, R.A. (1989) 'The contribution of deforestation and reforestation to atmospheric carbon dioxide', in S. Gupta and R.K. Pachauri (eds), *Global Warming and Climate Change: Perspectives from Developing Countries*, TERI, New Delhi, pp. 145–55.

Ives, J. and Messerli, B. (1989) *The Himalayan Dilemma*, UNU/Routledge, London.

Jaeger, J. (1983) *Climate and Energy Systems*, John Wiley & Sons, Chichester.

Jodha, N.S. (1989) 'Global warming and climate change: impacts and adaptations in agriculture', in S. Gupta and R.K. Pachauri (eds), *Global Warming and Climate Change: Perspectives from Developing Countries*, TERI, New Delhi, pp. 67–84.

—— (1992) *Mountain Agriculture: Search for Sustainability*, Mountain Farming Systems Discussion Paper Series No. 2, ICIMOD, Kathmandu.

Kattelmann, R. (1987) 'Uncertainty in assessing Himalayan water resources', *Mountain Research and Development*, 7 (3): 279–86.

Kawosa, M.A. (1988) *Remote Sensing of the Himalaya*, Natraj Publishers, Dehra Dun, pp. 13–14.

Khoshoo, T.N. (1990a) 'Wanted: a regional ecology plan', *The Sunday Observer*, New Delhi, 7 January.

—— (1990b) 'Need for a regional environment agenda', *The Times of India*, New Delhi, 27 July.

—— (1990c) 'A "holy war" for ecology', *The Hindu*, 112 (183), New Delhi, 2 August.

Kutzbach, J.E. (1976) 'The nature of climate and climatic variations', *Quarternary Research*, 6: 471–80.

Mahat, T.B.S. (1985) 'Human impact of forests in the middle hills of Nepal', unpublished doctoral dissertation, 2 vols, submitted to Australian National University, Canberra.

—— (1987) *Forestry–Farming Linkages in the Mountains*, Occasional Paper No. 7, ICIMOD, Kathmandu.

Mani, A. (1981) 'The climate of the Himalaya', in J. Lall and S. Moddie (eds), *The Himalaya. Aspects of Change*, Oxford University Press, Delhi, pp. 2–15.

Mayewski, P.A. and Jeschke, P.A. (1979) 'Himalayan and trans-Himalayan glacier fluctuations since AD 1812', *Arctic and Alpine Research*, 11 (3): 267–87.

Melillo, J.M., Callaghan, T.V., Woodward, F.I., Salati, E. and Sinha, S.K. (1990) 'Effects on ecosystems', in J.T. Houghton, G.J. Jenkins and J.J. Ephramus (eds), *Climate Change*, IPCC, WMO/UNEP, Cambridge, pp. 283–310.

Messerli, B. (1985) 'Stability and instability of mountain ecosystems. An interdisciplinary approach', in T.V. Singh and J. Kaur (eds), *Integrated Mountain Development*, Himalayan Books, New Delhi, pp. 72–97.

Miller, M.M. (1989) 'Comparative accumulation regimes of Himalayan and Alaskan neves and the issue of global warming', *Environment and Society in the Manaslu-Ganesh Region of the Central Nepal Himalaya*, Foundation for Glacier and Environmental Research and University of Idaho, Idaho, pp. 89–96.

Miller, M.M. and Marston, R.A. (1989) 'Glacial response to climate change and epeirogency in the Nepalese Himalaya', *Environment and Society in the Manaslu-Ganesh Region of the Central Nepal Himalaya*, Foundation for Glacier Research and University of Idaho, Idaho, pp. 65–88.

Myers, N. (1986) 'Environmental repercussions of deforestation in the Himalayas', *Journal World Forest Resource Management*, 2: 63–72.

Nautiyal, J.C. and Babor, P.S. (1985) 'Forestry in the Himalayas. How to avert an environmental disaster', *Interdisciplinary Science Reviews*, 10 (1): 27–41.

Nayava, J. L. (1974) 'Climates of Nepal', *The Himalayan Review*, Journal of Nepal Geog. Soc. 11: 15–20,

Ohata, T. and Higuchi, K. (1978) 'Valley wind revealed by wind-shaped trees at Kali Gandaki Valley', *SEPPYO*, Japanese Society of Snow and Ice, Nagoya, 40: 37–41.

Pei, S. (1991) 'Ethnobiology: a potential contributor to understanding development processes', *Entwicklung + Laendlicher raum*, DSE/ZEL, GTZ and DLG, Frankfurt am Main: 2: 21–3.

Rogers, P., Lydon, P. and Seckler, D. (1989) *Eastern Waters Study*, ISPAN, Arlington, Virginia.

Shamshad, K.M. (1988) *The Meteorology of Pakistan*, Royal Book Company, Karachi.

Sharma, C.K. (1983) *Water and Energy Resource of the Himalayan Block*, Sangeeta Sharma 23/281 Bishalnagar, Kathmandu.

Sharma, P. and Partap, T. (1993) 'Population, poverty and development issues in the Hindu Kush-Himalayas', Paper for Presentation at the International Forum on Development of Poor Areas, 22–27 March 1993, Beijing.

Shrestha, S. (1992) *Mountain Agriculture: Indicators of Unsustainability and Options for Reversal*, Mountain Farming Systems Discussion Paper No. 32, ICIMOD, Kathmandu.

Singh, J.S., Tiwari, A. K. and Saxena, A.K. (1985) 'Himalayan forests: a net-source of carbon for the atmosphere', *Environmental Conservation*, 12 (1): 67–9.

Stone, P. (ed.) (1992) *The State of the World's Mountains*, Zed Books, London.

Tegart, W.J.McG., Sheldon, G.W. and Griffiths, D.C. (eds) (1990) *Climate Change. The IPCC Impacts Assessment*, Australian Government Publishing Service, Canberra.

Thompson, M., Warburton, M. and Hatley, T. (1986) *Uncertainty on a Himalayan Scale*, Milton, Ash Editions, London.

Train, R.E. (1985) 'Inaugural address of the international workshop on management of national parks and protected areas in the Hindu Kush-Himalaya', in J.A. McNeely, J.W. Thorsell and S.R. Chalise (eds), *People and Protected Areas in the Hindu Kush-Himalaya*, KMTNC/ICIMOD, Kathmandu, pp. 9–10.

Troll, C. (1967) 'Die Klimatishche und vegetations-geographische Gliederung des Himalaya Systems', in *Khumbu Himal*, Ergebnisse eines Forschungsunternehmen in Nepal Himalaya. Bd 1, 353–88, Springer, Munich. (English Translation 'The climatic and phytogeographical division of the Himalayan system', in *Studies on the Climatology and Phytogeography of the Himalaya*, published by Nepal Research Centre, Kathmandu, 1988, for internal circulation, pp. 97–151).

Ueno, K. and Yamada, T. (1989) 'Diurnal variation of precipitation in Lamgtang Valley', *Glacial Studies in Lamgtang Valley*, Report of the Glaciological Expedition of Nepal Himalayas 1987–88, pp. 47–58.

Uhlig, H. (1985) 'Geological controls on high altitude rice cultivation in the Himalayas and mountain regions of Southeast Asia', in T.V. Singh and J. Kaur (eds), *Integrated Mountain Development*, Himalayan Books, New Delhi, pp. 175–92.

UNESCO/ICIMOD (1990) *Mountain Hydrology in the Hindu Kush-Himalayan Region*, Report of the First Consultative Meeting of the Regional Working Group on Mountain Hydrology, 24–26 October 1990, UNESCO/ICIMOD, Kathmandu.

—— (1992) *Mountain Hydrology in the Hindu Kush-Himalayan Region*, Report of the Second Consultative Meeting of the Regional Working Group on Mountain Hydrology, 16–18 March 1992, UNESCO/ICIMOD, Kathmandu.

Upreti, B.N. (1985) 'The park–people interface in Nepal: problems and new directions', in J.A. McNeely, J.W. Thorsell and S.R. Chalise (eds), *People and Protected Areas in the Hindu Kush-Himalaya*, KMTNC/ICIMOD, Kathmandu, pp. 19–24.

Verghese, B.G. (1990) *Waters of Hope*, Oxford and IBH, New Delhi.

Wang, W., Huang, M. and Cheng, J. (1984) 'A surging advance of Balt Bare Glacier, Karakoram Mountains', in K.J. Miller (ed.), *The International Karakoram Project*, Cambridge University Press, Cambridge, pp. 76–83.

WRI (1991) *World Resources 1990/91*, World Resource Institute, New York.

Zhang, X. (1984) 'Recent variations of some glaciers in the Karakoram Mountains', in K.J. Miller (ed.), *The International Karakoram Project*, Cambridge University Press, Cambridge, pp. 39–50.

Zheng, B., Jiao, K.Q., Li, G. and Fushimi, H. (1990) 'The evolution of Quarternary glaciers and environmental change in the West Kunlun Mountains, Western China', *Bulletin of Glacier Research*, (Japan) 8: 61–72.

25

EVALUATING THE EFFECTS OF CLIMATIC CHANGE ON MARGINAL AGRICULTURE IN UPLAND AREAS

T.R. Carter and M.L. Parry

INTRODUCTION

Upland regions contribute a significant proportion of the world's agricultural production. However, while mountainous areas in the middle and high latitudes can present a bleak and inhospitable alternative for agricultural production compared with the warmer lowland areas, at lower latitudes the converse may be true, with the highland zone offering a cooler and more temperate climate than the hot, often dry lowlands. Therefore, while the term 'upland' can be applied loosely to any region characterized by high elevation, the suitability of high-altitude regions for agriculture depends fundamentally upon latitude, with important modifying factors including soil type, slope, aspect, and climate. In this paper, we focus on changes in the last of these factors, climate, and their possible effect on upland agriculture.

Upland regions are characterized by climatic gradients that can lead to rapid altitudinal changes in agricultural potential over comparatively short horizontal distances. Where elevations are high enough, a level will eventually be reached where agricultural production ceases to be profitable, or where production losses become unacceptably high. For the purposes of this paper, we shall adopt a definition of upland agriculture as 'the practice of crop cultivation or animal rearing at elevations close to the margins of viable production'. Viability can be interpreted in terms either of economic profit or of subsistence. In essence, therefore, our discussion will concentrate on marginal upland agriculture.

In an earlier paper (Parry, 1985), a research strategy was introduced for assessing the effects of climatic change on agriculture. We illustrate here the application of this strategy to upland agriculture, using examples from different parts of the world. After a brief outline of the research strategy, subsequent sections consider how successive steps in the strategy have been applied in regional studies of climate change and agriculture. Finally, we discuss the usefulness of the approach for evaluating the vulnerability of upland agriculture to climate change.

A STRATEGY FOR IDENTIFYING IMPACT AREAS

There are considerable difficulties involved in disentangling the effects of climate from effects of other factors on agriculture. Similar impacts (for example, changes in agricultural land-use, crop productivity, or quality) can occur as a result of quite different physical, economic and social factors, of which climate is only one.

In order to isolate the influence of climate, a research strategy has been proposed (Parry, 1985), which attempts to predict the location and type of impact on the basis of an understanding of the relationships between weather and agriculture, and then to test these predictions against historical reality. The approach, which focuses on crop production, has four steps:

1. To isolate the important climatic variables by modelling crop–climate relationships.

2. To establish critical levels of these variables by relating them to farming behaviour (such as through changes in the probability of reward or loss).

3. To resolve climatic fluctuations into fluctuations of the critical levels (such as probabilities).

4. To map these as a shift of isopleths to identify impact areas.

While this is a general assessment strategy, applicable to different regional types and circumstances, we will concentrate here on marginal upland regions.

ISOLATING THE CLIMATIC VARIABLES

The use of models

In order to establish which climatic variables are most important for marginal upland agriculture, a knowledge of the relationships between climate and agricultural production is required. Our discussion here concentrates on agricultural crops, but animal production can be treated in broadly the same way.

Most existing knowledge about crop responses to climate has been acquired through observations from field and laboratory experiments. The most convenient way of organizing and combining this information is in the form of mathematical models. These seek to simulate the growth, development, and/or production of crops as a function of climatic and other factors. The models can range in complexity from simple indices relating climate to crop suitability (for example, to the potential for crop maturation), through empirical statistical models, which seek to identify statistical associations between climatic variables and crop yield, to crop growth simulation models, which attempt to represent the basic mechanisms and processes of plant development and growth.

Each of these model types provides information on the relations between climate and crop production, at varying levels of sophistication. An important prerequisite for their use in climate impact assessment, is that they are adequately validated over the range of conditions experienced or anticipated in a study region. If this requirement is fulfilled, then models can provide the basic tools for identifying the critical tolerances of crops to climate, which are required in the second step of the approach.

Climatological data in upland regions

Another requirement for using models to study climatic effects on crop behaviour is that sufficient information be available as input to such models. In upland regions, however, data on climate, soils, and agricultural production are often very sparse, which can pose a formidable obstacle to analysis.

Climatological information is basic to the analysis, but upland areas have two disadvantages compared with lowland regions concerning data provision. First, they generally have a poorer coverage of meteorological stations, due to their sparse population, physical inaccessibility, or perceived lack of importance. Second, their need for data is arguably greater, as they tend to exhibit weather conditions which are spatially highly heterogeneous.

Uplands are characterized by altitudinal climatic gradients, which means that conditions can change rapidly within a small horizontal area. In describing the upland environment, climatologists are often compelled to make use of these altitudinal gradients, in order to interpolate climatic data between sparse observation sites.

SELECTING CRITICAL LEVELS OF CLIMATE OR OF CROP RESPONSE

The second step of the research strategy involves the selection of levels of climatic variables that can be regarded as critical for the crop (e.g., determining potential cultivation limits or productivity levels). Further, in order to make meaningful assessments of the impact of climatic variations on actual agricultural response, it is necessary to combine this information with knowledge about those factors that are critical to the farmer (e.g., acceptable levels of yield or profit).

For example, the maritime regions of northwest Europe are characterized by very steep altitudinal gradients of air temperature (negative) and precipitation (positive). In an attempt to map the climatically marginal areas for cereal cropping, generalized climatic limits have been determined based on the number of growing months with a mean temperature above 10°C and the amount of increase in the difference between precipitation and potential

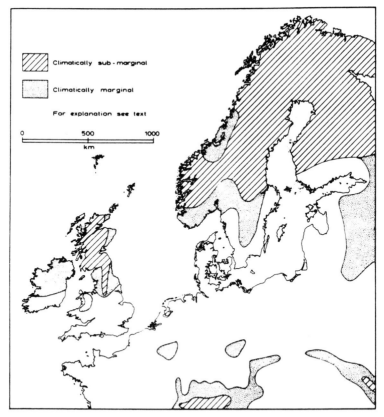

Legend:
- Climatically sub-marginal
- Climatically marginal
- For explanation see text

Scale: 0 — 500 — 1000 km

Figure 25.1 Climatically marginal land in northern Europe (after Parry, 1978).

evaporation (precipitation deficit) during the months July–September (Thran and Broekhuizen, 1965). The approximate limits of marginal land defined in such terms are (Parry, 1978):

a) upper limit: either, less than 5 months >10°C and no increase in precipitation deficit July–September, or, less than 3 months >10°C;

b) lower limit: either, less than 5 months >10°C and 0–50 mm increase in precipitation deficit July–September, or, 5–6 months >10°C and no increase in precipitation deficit July–September.

Upland areas defined as submarginal according to these criteria include the Scandinavian highlands, upland Scotland, and parts of northern England (Figure 25.1). These areas can be regarded as unlikely to be suitable for cultivation, at least without significant climatic improvement. Marginal areas are delimited as foothill areas in the United Kingdom, central and southern Norway, and central Sweden. In these areas, it can be hypothesized that cereal cultivation is prone to cool and excessively wet summer conditions, and only small variations in climate may determine the viability or otherwise of cereal cropping. Subsequent steps in the research strategy allow us to examine this hypothesis in more detail, below.

CLIMATIC VARIATIONS AND SHIFTS OF CROP POTENTIAL

Steps 3 and 4 of the research strategy consider how the climate in a region has varied in the past or is expected to change in the future, and how these variations might affect the pattern of crop potential. Climatic variations can be expressed as long-term average changes in a climatic variable, which are then compared with levels of the variable previously identified as critical for a crop. Alternatively, they can be evaluated as changes in the frequency of critical climatic anomalies that result in yield shortfall or crop failure. In this way, climatic changes are expressed as changes in the risk of impact.

If either of these approaches is applied across a network of sites in an upland region, maps can then be constructed showing the areas defined as suitable or otherwise for crop cultivation. In the case of changes of average climate, these maps would be of shifts in long-term crop potential. In contrast, if crop suitability is defined in terms of probabilities, then maps could be of changing levels of risk under changing climate. In both cases, the zones delimited by shifts in critical margins can be regarded as areas of potential impact from climatic variations. In the following section we can illustrate this with some examples.

DELIMITING AREAS OF POTENTIAL IMPACT: CASE STUDIES

We can demonstrate the application of the research strategy by reference to four studies conducted in upland regions: two from the UK, and one each from northern Japan and the Ecuadorian Andes.

Oats cultivation in southeast Scotland

Absolute and commercial limits to the cultivation of oats in the Lammermuir Hills, southern Scotland have been defined (Parry, 1978) as:

a) Absolute: accumulated temperature above a base of 4.4°C of 1,050 degree-days, potential water surplus (PWS, i.e., the excess of middle and late summer, up to 31 August, water surplus, when precipitation exceeds potential evaporation) over the characteristic early summer water deficit of 60 mm, and mean annual windspeed of 6.2 m/s.

b) Commercial: accumulated temperature of 1,200 degree-days, PWS of 20 mm and mean annual windspeed of 5.0 m/s.

When mapped (based on the most limiting factor), these represent long-term average climatic limits. However, the viability of crop cultivation on the

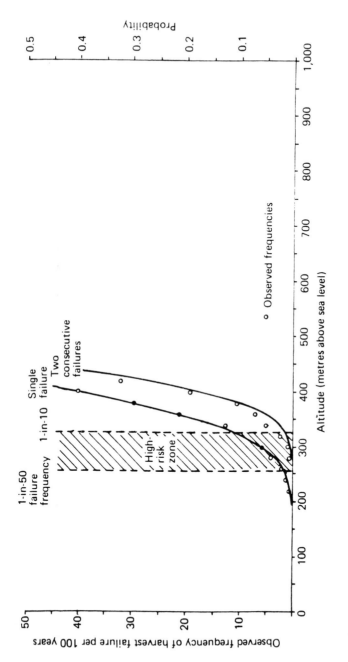

Figure 25.2 Actual and assumed frequencies of harvest failure (annual accumulated temperature below 970 degree-days) in southern Scotland, 1659–1981 (after Parry and Carter, 1985).

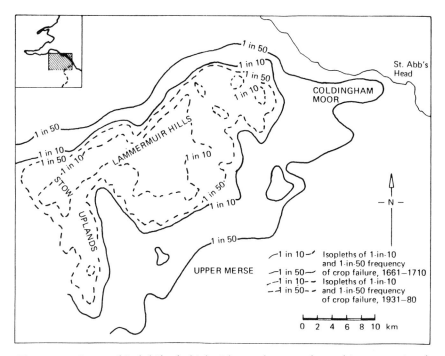

Figure 25.3 Geographical shift of a high-risk zone between the cool (1661–1710) and warm (1931–80) periods in southern Scotland. Risk is expressed as frequency of oats crop failure (after Parry and Carter, 1985).

ground is very much a function of the farmer's appreciation of the risk of harvest shortfall or outright failure. This requires an intuitive understanding of the underlying climatic variability, and probability of poor harvest years, which can themselves be quantified.

Following this line of inquiry, by looking at historical evidence for harvest failure, and the temperature conditions associated with this, an absolute lower limit of 970 degree-days was established as a critical threshold of summer warmth for successful oats ripening (Parry, 1978). At different elevations in the region, it was then possible to estimate the frequency of occurrence of hypothetical harvest failure. Using a temperature record of over 320 years, the average frequency of single and consecutive harvest failures was estimated for 10 m intervals (Parry and Carter, 1985). Both theoretical, assuming a normal distribution of annual accumulated warmth, and actual (counted) frequencies were evaluated (Figure 25.2).

The left-hand curve in the figure demonstrates the strongly nonlinear theoretical relationship between altitude and risk of a single crop failure. Actual frequencies (circles) follow this curve quite closely. The curve for consecutive failures exhibits even stronger nonlinearity. Moreover, this

411

theoretical relationship is further accentuated in reality through the tendency for 'poor' years to be clustered. Thus, at 340 m (approximately the limit of cultivation in the region) the observed frequency of consecutive failure is more than double that of the assumed frequency (Figure 25.2).

In order to study the predicted area of impact under long-term changes in temperature in the region, harvest failure probabilities of 1-in-10 and 1-in-50 were adopted as delimiting, respectively, the upper and lower edges of a 'high-risk' zone for oats cultivation. The 1-in-50 limit coincides approximately with the current upper level of cultivation or 'moorland edge'. Further information on contemporary local farming tolerances to crop failure would be required to fix these limits more accurately in an historical context. This high-risk zone was then mapped for the coldest and warmest 50-year periods in the 323-year temperature record (1661–1710 and 1931–1980, respectively). The shift of the zone is considerable (Figure 25.3), with a substantial part of the foothills (above 280 m) submarginal during the late seventeenth century, in contrast to a limit for the modern period of about 365 m, representing an altitudinal shift of some 85 m and a gain of about 150 km² of potentially cultivable land.

This result suggests that there is *prima facie* evidence for a much reduced agroclimatic resource base relative to present during the late seventeenth century. Earlier work suggested that this period marked the pessimum of a general decline in temperatures in northwest Europe beginning in the thirteenth century (Parry, 1978). If the hypothesis of a strong climatic influence on upland cereal cultivation is valid, some evidence of agricultural dislocation would be expected. This indeed proves to be the case in southern Scotland, where there is a close temporal and spatial fit between the distribution of permanently abandoned settlements and farmland and the fall of theoretical cultivation limits (Parry, 1978).

Modelling wheat suitability in northern England

In a separate study, a crop-climate simulation model was employed to compute yields of winter wheat in northern England (Carter, 1988). The model considered crop growth as the sum of processes of photosynthesis and respiration, which are themselves a function of climate. It was used to simulate the potential (well-fertilized) yields across a network of meteorological stations at different elevations. The model performance was tested against both experimental growth measurements and operational yields.

In order to demarcate the limits to viable wheat cropping, two criteria were specified. First, it was assumed that the occurrence of minimum temperatures below –2°C prior to harvest would lead to harvest failure. Second, a hypothetical yield below which the crop would cease to be profitable was defined, based on a survey of local farmers and advisers, and scaled-up to match simulated total above-ground dry matter (16.5 T/ha). Probabilities of

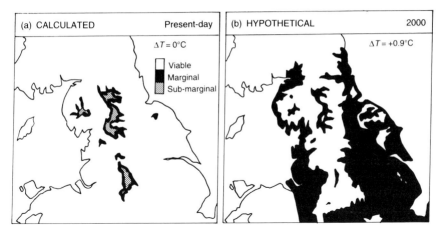

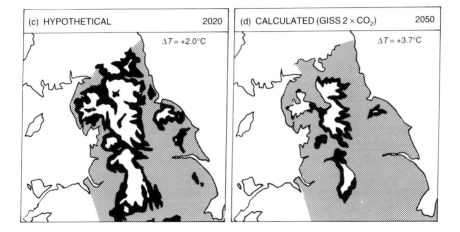

Figure 25.4 Estimated shifts of theoretical margins of winter wheat in northern England for a 30 September sowing date under: (a) baseline climate (1961–71) and progression to an equivalent $2 \times CO_2$ climate, through the years 2000 (b), 2020 (c), and 2050 (d) (after Carter, 1988).

hypothetical harvest failure were then mapped according to these criteria and a marginal zone defined as lying between probabilities of 0.1 and 0.5.

Under present-day (1961–1971) conditions, the lowland zone is delimited as viable for winter wheat cultivation, with a marginal zone occurring in the upland foothills, between about 390 and 480 m. Above this level, land is estimated to be unsuitable (Figure 25.4a). The wheat model was next run for changed temperatures (mean annual warming of 3.7°C) corresponding to climate model estimates for an equivalent doubling of carbon dioxide in the atmosphere. Crop model simulations assumed no change in management or in wheat variety from the present day. Probabilities of harvest failure were again computed, and the marginal zone mapped (Figure 25.4b).

The results indicate that a warming of less than 4°C could transform the potential suitability of this region for cultivation of current winter wheat varieties. Areas in the uplands that are submarginal today would become climatically viable for cultivation, whereas growing season temperatures in the present-day core lowland wheat growing area would reduce yields significantly by accelerating crop development and shortening the critical grain-filling phase of growth. The lowland area is thus delimited as submarginal.

Of course, these results are largely hypothetical. Other factors may militate against the cultivation of wheat in the high uplands (e.g., exposure, high precipitation, poor soils, etc.), and farmers in the lowlands would naturally adapt their farming practices to the changing conditions, for instance, by switching to more heat-tolerant varieties or crops and by altering the timing of operations. Nevertheless, the findings point to the potentially large changes in the agricultural resource base that can be induced by changes in climate.

Northern Japan (Yoshino *et al.*, 1988)

During the present century, there has been an expansion of paddy rice cultivation in Japan from the traditional rice growing areas both latitudinally into the northern island of Hokkaido, as well as upwards into higher altitudes. However, in these more marginal regions there is a much greater risk of crop losses due to inadequate summer warmth. Local varieties of rice require an effective accumulated temperature (i.e., accumulated temperatures above 0°C during the period when mean daily temperatures exceed 10°C) of at least 3,200 degree-days to complete the normal growth cycle in Tohoku district (northern Honshu) and 2,600 degree-days in Hokkaido, where early-maturing varieties are widespread. This information has enabled maps to be constructed of the current potentially cultivable area of rice production in northern Japan (Figure 25.5a).

To examine the effect of climatic variability on rice suitability, simple empirical indices have been developed relating rice yield to July–August mean air temperature in Tohoku and Hokkaido. In combination with information about regional lapse rates of temperature, estimates have been made of the effects of anomalous weather on yield and on potentially cultivable area. Estimates are summarized in Table 25.1 and Figure 25.5 for four situations: baseline (1951–1980) climate, a cool decade (1902–1911), a warm decade (1921–1930) and a possible future climate associated with greenhouse gas induced warming.

The results indicate that a July–August temperature 25 per cent departure

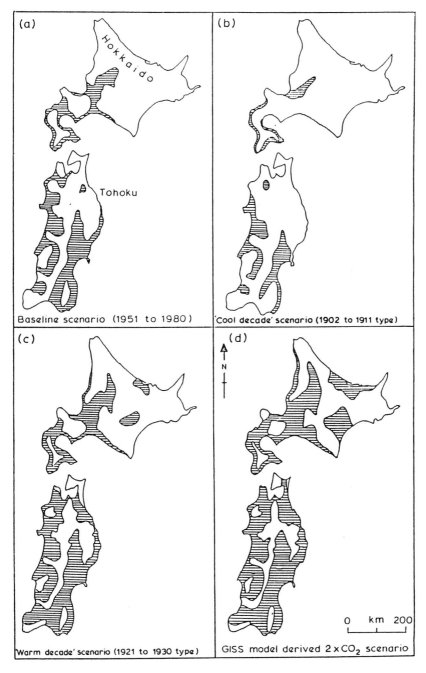

Figure 25.5 Safely cultivable rice area in northern Japan under four climatic scenarios (after Yoshino *et al.*, 1988).

Table 25.1 Effects of climatic scenarios on rice yields in Hokkaido and Tohoku districts, northern Japan

Climate scenario	Years	Mean July–August temp. (°C)	Anomaly (°C)	Yield index (% Baseline)
		HOKKAIDO		
Baseline	1951–1980	20.6	0.0	100
Cool decade	1902–1911	19.5	−1.1	75
Warm decade	1921–1930	21.0	+0.4	106
$2 \times CO_2$	Future	24.1	+3.5	105
		TOHOKU		
Baseline	1951–1980	22.8	0.0	100
Cool decade	1902–1911	21.8	−1.0	87
Warm decade	1921–1930	23.5	+0.7	106
$2 \times CO_2$	Future	26.0	+3.5	102

Source: After Yoshino *et al.* (1988)

of −1.1°C in Hokkaido would reduce yields by some 25 per cent relative to the average and the potentially cultivable area would contract to elevations below 100 m in a small area of southwestern Hokkaido, where the Sea of Japan acts to ameliorate the climate (Figure 25.5b). A comparable anomaly in Tohoku (−1.0°C) would produce a 13 per cent yield shortfall and the cultivable area would contract by some 20 per cent, to areas below about 200 m (Figure 25.5b).

Conversely, in warmer than average years, higher yields would occur in association with an upward shift in limits of potential cultivation (Table 25.1; Figure 25.5c). A temperature rise of more than 3°C, obtained under a scenario of future climate associated with an equivalent doubling of atmospheric carbon dioxide, would also lead to a yield increase, but less marked than in the warm decade (Table 25.1). This is because the analysis assumes that current varieties of rice are cultivated, which respond positively to small increases in temperature, but are less well adapted to large departures. It is likely, however, that later-maturing varieties would be adopted under such conditions, which could exploit the greater warmth. Under this scenario, most of the land below 500 m could become viable in Hokkaido, and land up to 600 m in Tohoku (Figure 25.5d).

Note that fluctuations in temperature above and below the mean would, of course, be expected by farmers as part of the inter-annual variability of climate. Such annual events alone would not affect the location of the average climatic limits to cultivation. However, the results shown above are for longer period anomalies, which are the types of events that can have a more lasting effect on the agroclimatic resource base and thus a more tangible impact in the more marginal regions.

The central Sierra of Ecuador (Bravo *et al.*, 1988)

In the central Sierra of the Ecuadorian Andes, close to the equator, subsistence agriculture is practised by indigenous communities up to high elevations (2,800–3,800 m). The valley interiors of this region display a variety of regional climates. Some are drought-prone whilst others are moist, depending on their disposition to prevailing rain-bearing winds. In the dry valleys, rainfall generally increases with elevation and farmers can minimize the effects of drought by cultivating crops higher up the valley sides. However, at these higher elevations, temperatures are lower and there is an increased risk of frost damage to crops. With these two dominant constraints in mind, farmers must choose which crops or crop mixes provide adequate returns in the prevailing climatic conditions while minimizing the risk of large-scale crop losses.

A detailed study has been made of the effects of climatic variations on agriculture in Chimborazo Province, a region of about 8,000 km² in the central Sierra. Here, elevations range from below 2,000 m in the valley floors to 6,675 m (the volcanic peak of Chimborazo). The upper limit of cultivation lies at about 3,800 m.

Maize is the major crop at lower elevations (below 3,300 m), because of its tolerance of high temperatures and drought. At higher elevations, potatoes, barley, and broad beans are cultivated along with other highland crops such as quinoa, oca, ullucu, and tarwi. The high grassland (*paramo*), above the cultivation limit, is used for cattle and sheep grazing, along with lower level land on large farms (*haciendas*) and on frost-prone flats.

Given the decline of temperatures with altitude, barley requires at least 10 months to mature at 3,200 m. Potato maturation increases from 5–6 months at 2,800 m to 9–10 months at 3,800 m. Planting times tend to be earliest at the highest elevations. Sowing may occur in August at 3,800 m, but as late as February below 3,000 m. Multiple planting times are also employed to help limit the effects of early season frosts.

The upper limit of intensive cultivation on flats (3,080 m) is lower than on slopes (3,800 m). It is probable that this limit is set by frost risk, since mean annual temperatures are higher than at the limit of cultivation on slopes. It is less clear what determines the upper limit of slopes cultivation. Yields of barley and potato do not appear to diminish markedly as mean temperature declines towards the limit. However, as crop maturation time extends, it appears likely that the susceptible growth phases are pushed into unfavourable (frost-prone) seasons.

On the basis of this hypothesis, estimates suggest that a decrease in late winter temperature minima of about 0.8°C, which is a plausible medium-term change in this region, could reduce the upper limit of cultivation by about 200 m. By mapping the area of cultivated land between 3,600 and 3,800 m, this suggests a potential loss of 8 per cent of arable land in the

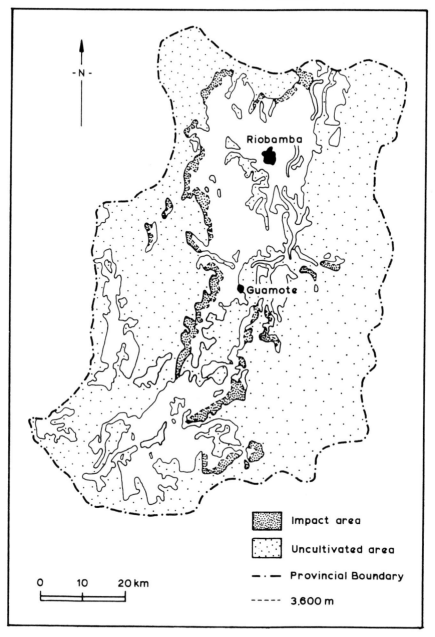

Figure 25.6 Effect of a 0.8°C cooling in late growing season temperature on the upper limit of cultivation in Chimborazo Province, Ecuador (after Bravo *et al.*, 1988).

Table 25.2 Potentially cultivable area (hectares) for barley and potatoes during average, anomalously dry, and anomalously wet years in Chimborazo Province, Ecuador

	Precipitation Scenarios		
	1-in-10 Dry	*Baseline*	*1-in-10 Wet*
Total area suitable for:			
Barley	340,460	346,120	163,570
Potatoes	70,140	274,015	128,000
Area with limitations:			
Frost	109,000	118,040	29,780
Drought	270,320	123,665	93,200
Water surplus	–	43,020	44,620
Total potentially cultivable area	340,460	397,630	207,331

Source: Adapted from Bravo *et al.* (1988)

province (Figure 25.6), possibly affecting some 4,500 mostly indigenous inhabitants. On flat land, a comparable cooling of the climate could lead to about 7 per cent of the currently cultivated area going out of production, affecting some 400 persons.

These results should be interpreted with caution, however, as many of the upper lands recently taken into cultivation are highly fertile, in spite of their frost risk. Moreover, temporary migration is one possible response to offset weather-related crop losses.

For a warming of the same magnitude (0.8°C), frost risk would decrease and the upper limit of slope cultivation could be raised by some 200 m to 4,000 m. A rise of only 120 m would increase by two-thirds the area of potentially fertile flat land in the province. Under the warming scenario, it is suggested that this level of expansion could be exceeded. Moreover, evidence of prehistoric raised fields at elevations up to 3,150 m in nearby Imbabura Province suggest that flats have been cultivated at higher elevations in the past.

In a further analysis, the potentially cultivable area for barley and potatoes has been defined according to their respective upper and lower annual precipitation and temperature requirements. These areas were then mapped for the present-day climate and for anomalously (1-in-10 probability) dry and wet years. The results are depicted in Table 25.2.

Under the dry year scenario (65 per cent of mean annual precipitation) total cultivable area declines by nearly 15 per cent, almost entirely due to a contraction of potato suitability in drought sensitive areas. In contrast, the areas that become too dry for barley are almost completely replaced by areas that would usually be too wet. Of course, this replacement is in *potentially* cultivable areas. Barley would not be grown in these areas on the basis of normal expectations.

A substantial area in the interior of Chimborazo, including most of the population centres, is located in the sensitive zone for drought according to these findings. Some of the population here have access to irrigation, or to land at higher levels. Other drought adaptation strategies are also important, such as supplemental off-farm employment, food for work schemes, and expansion of traditional irrigation networks.

The 1-in-10 wet year (134 per cent of mean annual precipitation) produces an even greater (48 per cent) decline in total cultivable area than the dry year (Table 25.2), with major losses occurring in both barley and potato potential. Moreover, the regions affected in wet years are quite different from the corresponding areas of impact in dry years. The vulnerable areas are located on higher level land, often away from major roads and settlements, and difficult to access during high rainfall seasons. In spite of their relative remoteness, perhaps one third of the province's farming population may be affected by this type of climatic anomaly.

CONCLUSIONS

It is clear from the foregoing examples that upland crop production, practised close to the margins of viable production, can be highly sensitive to variations in climate. However, the nature of that sensitivity varies according to the region, crop, and agricultural system of interest. In some cases, the limits to crop cultivation appear to be closely related to levels of economic return, through crop productivity. In these cases (e.g., wheat production in northern England), the average level of yield may change only gradually with changing altitude. However, yield variability often increases at higher elevation, and the effects of climate change may well be better interpreted as a change in risk of yield shortfall, rather than a change in mean yield. In many cases this accords well with farming perceptions in such regions, where farmers look to maximize their yields, while at the same time avoiding too frequent occurrences of yield shortfall.

In other cases (e.g., in the Ecuadorian highlands) the cultivation limit may not be related to levels of yield. In Ecuador, these do not appear to decline significantly towards the upper margin of cultivation. In such cases, we must look to other possible factors explaining the location of the limit. In subsistence systems, such as in Ecuador, or historically in northwest Europe, it appears that cultivation limits are determined more by risk of outright crop failure than by risk of shortfall *per se*. In the Ecuadorian example, late season frost is the most likely explanation for crop failure, and it is frost risk that appears to place the ultimate constraint on production.

This emphasis on climatically related risk, highlights the importance of extreme or anomalous events as possible determining factors in the location of upper limits to agriculture in the uplands. However, while the occurrence, and particularly the recurrence, of such events may *trigger* farming responses (e.g., farm abandonment, bankruptcy, etc.), it is still fair to assume that the main *underlying* control on cultivation limits is a function of the long-term mean climate.

Nor is it easy to separate the two factors, as long-term changes in the mean climate are usually accompanied by changes in the frequency of extremes. Since risk levels often increase exponentially with altitude, it can readily be appreciated that apparently small changes in the mean climate can induce large changes in agricultural risk (Parry and Carter, 1985). Application of the predictive strategy in Scotland shows that such changes have occurred in the past, with demonstrable impacts on crop cultivation in the marginal uplands. Predictions of substantial global warming due to the enhanced greenhouse effect indicate that shifts in agroclimatic potential may occur that are even greater and more rapid than those observed in the past. If so, farmers in the marginal uplands could be among the first and most vulnerable to change, whether it be beneficial or otherwise.

REFERENCES

Bravo, R.E., Cañadas Cruz, L., Estrada, W., Hodges, T., Knapp, G., Ravelo, A.C., Planchuelo Ravelo, A.M., Rovere, O., Salcedo Solis, T. and Yugcha, T. (1988) 'The effects of climatic variations on agriculture in the Central Sierra of Ecuador', in M.L. Parry, T.R. Carter and N.T. Konijn (eds), *The Impact of Climatic Variations on Agriculture. Vol 2: Assessments in Semi-Arid Regions*, Kluwer, Dordrecht, pp. 381–493.

Carter, T.R. (1988) *Climatic Change and Cropping Margins in Upland Britain*, unpublished Ph.D. thesis, University of Birmingham.

Parry, M.L. (1978) *Climatic Change, Agriculture and Settlement*, Dawson, Folkestone.

—— (1985) 'The impact of climatic variations on agricultural margins', in R.W. Kates, J.H. Ausubel and M. Berberian (eds), *Climate Impact Assessment: Studies of the Interaction of Climate and Society*, John Wiley & Sons, Chichester, pp. 351–67.

Parry, M.L. and Carter, T.R. (1985) 'The effect of climatic variations on agricultural risk', *Climatic Change*, 7: 95–110.

Thran, P. and Broekhuizen, S. (1965) *Agro-ecological Atlas of Cereal Growing in Europe: Volume I, Agro-climatic Atlas of Europe*, Amsterdam.

Yoshino, M.M., Horie, T., Seino, H., Tsujii, H., Uchijima, T. and Uchijima, Z. (1988) 'The effects of climatic variations on agriculture in Japan', in M.L. Parry, T.R. Carter and N.T. Konijn (eds), *The Impact of Climatic Variations on Agriculture. Vol. 1: Assessments in Cool Temperate and Cold Regions*, Kluwer, Dordrecht, pp. 723–868.

26

TOWARD A GLOBAL NETWORK OF MOUNTAIN PROTECTED AREAS

C.J. Martinka, J.D. Peine, and D.L. Peterson

INTRODUCTION

Representative examples of mountain landscapes receive some form of legal protection in many countries throughout the world. A recent tabulation of 442 prominent sites, those 10,000 ha or greater in size with 1,500 m or more relief, reveals global protection for more than 243 million ha of mountainous terrain (Thorsell and Harrison, 1992). These protected areas represent nature reserves, national parks, natural landmarks, and managed sanctuaries in each of the eight global biogeographic realms described by Udvardy (1975). Distribution of the 442 sites is skewed with more than half the number and nearly two thirds the land area located in the Nearctic and Palaeoarctic realms. When numerous smaller sites are considered, mountains appear to receive more extensive protection than most other biomes (Poore, 1992).

Mountain protected areas (MPA) sample a topographically complex environment that is important for the economic and spiritual sustenance of many human cultures. And as human populations continue to grow, use of this biologically diverse resource has accelerated, highlighting a strategic role for MPA that includes both conservation and scientific interests. From a global conservation perspective, protecting representative examples of mountain resources is a high priority goal. At the same time, scientists are more frequently using protected areas as controls for land-use research and as sites for long-term environmental monitoring. The prospect of global climate change has heightened the value of MPA as scientific resources (Bailey, 1991).

Mountains are biologically rich landscapes with complex regimes of climate and weather. Topographic gradients and atmospheric extremes are features that further enhance their usefulness for measuring climate change. We propose that protected sites represent a global sample of mountain environments that holds potential for enhancing the design of climate change research. To evaluate this idea and advance other values of MPA, we further propose that important site attributes be assembled as the informational foundation for a functional network of protected sites.

THE NETWORK CONCEPT

The possible causes and predicted effects of global climate change have precipitated an intense interest in the conservation of species and their habitats. Strategies to address conservation at the global scale require innovative approaches to the acquisition and use of knowledge. While the challenge includes the production of new knowledge, there is also unrealized opportunity in the extensive inventories of raw data provided by contemporary remote sensing technology.

During recent decades, the task of assembling information about species and their habitats has been undertaken by a variety of individuals and institutions. At the international level, inventories being developed by organizational coalitions are critical data sources for strategic conservation planning. The most comprehensive relating to protected areas is maintained by the World Conservation Monitoring Centre (WCMC) as a library of computer maps supported by structured data bases and paper files (WCMC, 1992). It represents a global overview of biological diversity that includes protected sites, threatened plant and animal species, and other areas of conservation concern. In this volume, P.N. Halpin provides a useful example of how these data can be used to understand the potential effects of climate change on nature reserves.

A geographically more limited endeavour is BRIM, the Euromab Biosphere Reserves Integrated Monitoring Programme (Gregg et al., 1993). Its focus is a survey of biological attributes, current knowledge, and science capabilities of biosphere reserves in Europe and North America. Strategically, the program is designed to facilitate the collection, reporting, and access of various resource data among biosphere reserves, sites designated by UNESCO for their conservation, education, and scientific values. However, it also serves as a useful model for other protected areas where research and monitoring are important functions. For MPA, a functional network requires the comprehensive global foundation provided by the WCMC as well as the kinds of resource data being assembled by BRIM. The objective would be documentation of attributes that foster a credible ecological evaluation of MPA at the global scale. In general, the requisite descriptors for each site would include direct or scaled measurements capable of supporting classification, comparison, selection, assessment, and other technical analyses of MPA. A hierarchical design provides the flexibility required for the information to be part of meta-data base networks now being considered by some organizations. Additionally, a hierarchy fosters maintenance of active files, data search procedures, and logical links to related data bases that include more comprehensive site data at individual MPA.

To begin the network construction process, a catalogue of information relating to the global network of MPA would serve an especially useful purpose. As a mechanism for marketing protected area values and identifying opportunities associated with their presence, its potential contributions would include a foundation for evaluating system status, identifying threats to its integrity, and developing standards for additional MPA. Moreover, an especially timely contribution would be elements of a framework for global research such as that described by Comanor and Gregg (1992) for the US National Park Service Global Change Research Program. And finally, a catalogue serves an educational purpose by defining a communication network and presenting knowledge of interest.

Designing an extended information base for MPA requires objectives formulated in response to a series of questions based on the descriptive power of the data base. What are the markets and uses for a catalogue of MPA? Which attributes contribute to the current ecological status of a protected site? Can the attributes be expressed as measurements suitable for statistical analyses? Are the attributes held in common by all protected sites? What are the most productive communication pathways and when is further data base development and management best handled at national, regional, or site levels?

Formulation of objectives is strategic and thus an essential prelude for several routine steps in developing a prototype network. These steps are sequential and include a thorough review of existing information systems, construction of a data base framework, and designing a format for data collection. In turn, the results become the resources for creating a working model that addresses objectives. To evaluate the prototype, we suggest that initial design include three replicates from each of the eight biogeographic realms for a total of 24 MPA. Criteria for site selection would include substantive contributions to existing data systems and willingness to be an active participant in evaluating and advancing the network concept.

PROTECTED AREA ATTRIBUTES

Determining the nature and number of descriptive site attributes is a critical step in the network proposal. A recent analysis for Glacier National Park in northwestern Montana revealed the potential complexity of the issue (Martinka, 1992). In that case, seven kinds of attributes that contributed to the conservation of biological diversity included enabling laws, economic factors, boundary characteristics, geographic location, management philosophy, environmental statutes, and public interest. Further analysis indicates that these attributes are conveniently grouped into three descriptive categories.

First are the primary physiographic features of a protected area, those which determine its basic potential for biological diversity. Geographic location, terrain characteristics, soil origin, and climatic features are especially important measures since they generally provide a basis for classifying MPA into various ecological categories. The usefulness of these basic attributes is illustrated by their role in regional ecosystem simulation models such as that described by Running *et al.* (1989).

Next are the ecological expressions of the basic features including species counts, primary productivity, energy flux, and similar measures of biological activity within a protected area. While useful as indicators of ecological conditions, they are also more difficult to document and interpret than physical features. None the less, expressive attributes add an essential dimension to the quantitative description of MPA.

A third category includes the legal and behavioural actions of humans which influence biological diversity in protected areas. Some with legal origins such as area size, boundary configuration, period of protection, and legislated mandates are relatively permanent and therefore important complements to physiographic features. Others such as regional setting, physical facilities, and organizational philosophy tend to modify the effects of physical and legal attributes; they too are important measures for a full description of MPA.

There is also a number of ancillary descriptors that are likely to enhance potential for a functional network. For example, the contents of a natural resources information system would provide for global research design based on existing knowledge. Or the presence of a library may provide local knowledge that is not available elsewhere. While these kinds of descriptors are important, they also add to the complexity of network design and should thus be selected with a clear view of the original objectives in mind.

NETWORK VALUES

Initially, the most important contribution of a network may be its ability to address questions about the nature of MPA. What are their global geographic attributes and how well do they represent major ecosystems? To what extent do MPA contribute to regional landscape conservation and the protection of biological diversity? Is size or replication sufficient to assure conservation of critical mountain resources? Does the system or representative components provide a sample for monitoring global environments?

In contrast to strategic analyses, there are also functional uses for the managers of protected sites. Are there sites with similar management philosophies elsewhere in the world that might provide information on an ecological issue? Can one design and implement an inventory or monitoring program that is exportable to other sites or be useful to global research design? How do various kinds of management potentially influence the ecological status of a new MPA? And in fact, the list of functional uses is limited chiefly by human imagination.

Strategic analyses and functional uses are valuable only to the extent that the data are accessible and accepted as credible knowledge. In turn, value is most effectively assured by applying current knowledge not just to new research, but also to education and management. With this in mind, climate change issues are likely to receive magnified benefits from a functional network of MPA. Poore (1992) suggests that mountain environments are sensitive indicators and that protected areas provide unequalled opportunities to detect and monitor changes in climate and air quality. He goes on to list a series of guidelines that stress their global role in understanding and conserving biological diversity.

CONCLUSIONS

When organized and presented in an appropriate form, knowledge improves the effectiveness of communications among scientific groups, conservation organizations, and managers of natural resources. The potential for a functional network of MPA is therefore enhanced by the assembly, presentation, and application of credible information about the hundreds of sites throughout the world. The concept is consistent with recent recommendations for new approaches to global environmental issues through the creation of international networks and centres (Carnegie Commission, 1992).

The design of a global network requires that its markets be fully explored to determine the role, functions, and applications of information to mountain environments. It seems reasonable to anticipate a diverse clientele, pointing to the need for relevant information that serves both the strategic and operational requirements of MPA. As such, the network becomes a contributor to the Sustainable Biosphere Initiative (Lubchenco et al., 1991) through international co-operation in science, education, and conservation.

ACKNOWLEDGEMENTS

The concept of a network of MPA emerged from an international workshop on protected areas in mountain environments sponsored in 1991 by the Environment and Policy Institute, East–West Center (EWC). The workshop emphasized the collaborative needs of networks and prompted preparation of this paper as one means of advancing the concept. We thank L. Hamilton, EWC, J. Thorsell, The World Conservation Union, and J. Harrison, WCMC, for encouraging us to continue with an exploration of the network concept. An anonymous referee provided suggestions that substantially improved the manuscript.

REFERENCES

Bailey, Robert G. (1991) 'Design of ecological networks for monitoring global change', *Environmental Conservation*, 18: 173–5.

Carnegie Commission (1992) *International Environmental Research and Assessment: Proposals for Better Organization and Decision Making*, Carnegie Commission on Science, Technology, and Government, New York.

Comanor, P.L. and Gregg, W.P. Jr (1992) 'Role of U.S. national parks in global change research', *The George Wright Forum*, 9: 67–74.

Gregg, W.P., Serabian E. and Ruggiero, M.A. (1993) 'Building resource inventories on a global scale', *The George Wright Forum*, 10 (1): 21–9.

Lubchenco, J., Olsen, A.M., Brubaker, L.B., Carpenter, S.R., Holland, M.M., Hubbell, S.P., Levin, S.A., MacMahon, J.A., Matson, P.A., Melillo, J.M., Mooney, H.A., Peterson, C.H., Pulliam, H.R., Real, L.A., Regal P.J. and Risser, P.J. (1991) 'The sustainable biosphere initiative: an ecological research initiative', *Ecology*, 72 (2): 371–412.

Martinka, C.J. (1992) 'Conserving the natural integrity of mountain parks: lessons from Glacier National Park, Montana', *Oecologia Montana*, 1: 41–8.

Poore, D. (ed.) (1992) *Guidelines for Mountain Protected Areas*, IUCN, Gland, Switzerland and Cambridge, UK.

Running, S.C., Nemani, R.R., Peterson, D.L., Band, L.E., Potts, D.F., Pierce, L.L. and Spanner, M.A. (1989) 'Mapping regional evapotranspiration and photosynthesis by coupling satellite data with ecosystem simulation', *Ecology* 70: 1090–101.

Thorsell, J. W. and Harrison, J. (1992) 'National parks and nature reserves in mountain environments and development', *GeoJournal*, 27: 113–26.

Udvardy, M.D.F. (1975) *A Classification of the Biogeographical Provinces of the World*, The World Conservation Union Occasional Paper No. 8.

WCMC (World Conservation Monitoring Centre) (1992) *The WCMC Biodiversity Map Library: Availibility and Distribution of GIS Datasets*, WCMC, Cambridge, UK.

Part IV

CONCLUSION

27

SHOULD MOUNTAIN COMMUNITIES BE CONCERNED ABOUT CLIMATE CHANGE?

M.F. Price

INTRODUCTION

Mountain regions are found on all of the world's continents. With the exception of Antarctica's mountains, all of these regions are inhabited to a greater or lesser extent. As Ives (1992: xiv) has noted, 'mountains and uplands comprise about a fifth of the world's terrestrial surface. Furthermore, they provide the direct life-support base for about a tenth of humankind and indirectly affect the lives of more than half.' Given these statistics, the question posed in the title is not merely academic: changes in mountain climates have the potential to affect hundreds of millions, if not billions, of people.

Few assessments of the impacts of climate change have been conducted in mountain regions, in contrast to other biomes such as tropical rainforests, coastal zones, and high-latitude, arid, and semi-arid areas. A variety of reasons can be stated, which are given below in no particular order.

First, mountain regions are economically and politically marginal. In individual countries, their direct economic importance to national economies is low, and they are often far from centres of political decision making. At the global scale, few mountainous countries have great political influence, while influential countries with many mountains tend not to encourage discussion about these marginal regions.

Second, the dominant feature of mountains – their relief – is so muted in most General Circulation Models (GCMs) that they are barely recognizable. Given the spatial resolution of current models, it is difficult to use the predominant GCM-based methodologies for assessing the potential impacts of climate change.

Third, the complexity of mountain systems presents major problems for assessing the potential impacts of climate change. This applies to assessments of changes in both biophysical systems (e.g., Rizzo and Wiken, 1992; Halpin,

1994) and societal systems, particularly because many of the most valuable 'products' of mountain regions are not easily quantified in monetary terms (Price, 1990).

Fourth, tourism, an increasingly important component of mountain economies around the world, is not an easily defined economic sector – in contrast, for example, to agriculture, forestry, or fishing – and, furthermore, is subject to substantial variations in demand and is closely tied into other aspects of mountain economies.

Finally, few mountain regions can be described comprehensively: data for almost all aspects (e.g., climatological, biological, socio-economic) are very heterogeneous in their length and frequency of record, spatial coverage, and availability. Any method of assessment (e.g., case study, analogue, modelling) requires considerable volumes of data (Kates *et al.*, 1985; Carter *et al.*, 1992; Riebsame, 1989).

While this paper is global in scope, the question posed in the title cannot be answered globally. Although the biological and physical processes characterizing mountain regions around the world may exhibit considerable similarities (Barry, 1992b; Gerrard, 1990), the human communities of these regions differ greatly and interact in very different ways with mountain climates: climate change will, for instance, have very different meanings for an Indian forester, a Peruvian farmer, or a Swiss ski instructor. Consequently, following this introduction, the first section of the paper provides a categorization of mountain regions in terms of their human populations. The second section briefly discusses mountain climates from the perspective of the climate impact assessment. The third section proposes some initial answers to the question posed in the title with respect to agriculture, forestry, water resources, tourism, energy, transport, and human health – topics that are all linked to a greater or lesser extent. The final section offers some conclusions for different mountain regions.

TYPES OF MOUNTAIN REGIONS

In most mountain regions, the dominant human population can trace their ancestry back some centuries, or even millennia into prehistory. In other regions – notably those settled by Europeans who displaced indigenous populations – the history of today's predominant population dates back only to recent centuries. Grötzbach (1988) has described these two broad categories as, respectively, 'old' and 'young' mountains. The latter, in North America, Australia, and New Zealand, are further characterized by sparse settlement, extensive market-oriented pastoral agriculture and forestry and, in the post-war period, the rise of tourism as a major economic force.

Old mountain regions, most of which are relatively densely settled, can be further divided into three broad categories. The first of these is characterized by the decline of traditional agriculture and forestry, which

432

are often kept going primarily through government subsidies. During the post-war period, tourism has grown rapidly, and may now have become the basis of the economy. Areas without tourism are often experiencing depopulation. These characteristics typify many of the mountains of western Europe and Japan.

The second category includes most of the mountains of developing nations. Traditional subsistence agriculture is largely intact, and there is a tendency toward over-population, ameliorated to some extent by short-term or seasonal migration for work. Most commonly, the inhabitants of these regions are mountain peasants, who may practise some transhumance (e.g., Andes; central America; much of the Himalaya-Karakorum-Hindu Kush; New Guinea; Southeast Asia; Sub-Sahelian Africa). In other regions, nomads occur in addition to these more settled people (e.g., High Atlas; western Hindu Kush-Himalaya; Middle East). This category also includes many of the mountains surrounding the Mediterranean, in countries which are otherwise industrialized to a greater or lesser extent. In all of these regions, tourism is becoming economically important in small, well-defined areas, often with considerable effects up to the national scale (Price, 1993). In some countries (e.g., Bhutan, Nepal, Rwanda), tourism is the primary source of foreign exchange. However, the initial or continued development of tourism may be limited by political instability or warfare (e.g., Afghanistan, Bosnia, Ethiopia, Peru).

The third category has, until recently, largely been characterized by collectivized or nationalized agriculture and forestry: the mountains of China and the former Soviet Union, and the Carpathians. With the fall of communist regimes in all of these countries except China, the social, economic, and political structures of these regions are now in a state of flux. In parts of the Carpathians and Caucasus, tourism based on the capitalist model is gradually replacing the former structure in which the supply of 'tourists' was guaranteed, so that these regions may move toward the first category. In other regions, tourism is developing in isolated pockets (e.g., southwestern China, other republics of the former Soviet Union). Thus, many of these regions are tending toward the second category; being far from centres of power, the degree of collectivization or nationalization was, in any case, often not as great as in more productive lowland regions.

The categories of mountain populations presented above are inevitably not clear-cut; the immense diversity of mountains rarely allows simplicity! However, they provide a framework for the discussion of reasons why mountain communities should be concerned about climate change, which is presented in the final sections of the paper.

MOUNTAIN CLIMATES AND HUMAN
POPULATIONS

Mountain climates are characterized by marked diurnal and seasonal cycles, with high variability at all spatial scales. In any mountain region, the diverse relief, aspect, and slope further increase both temporal and spatial variability (Barry, 1986; 1992b). Many of the resources on which mountain communities depend are climate-dependent, i.e., their availability may be affected in the short and long term by variability, extremes, and long-term changes in means of climatic parameters. In traditional mountain communities, these resources include water and both domesticated and wild plant, tree, and animal species. Many of these resources are also important in the many communities – in both developing and industrialized countries – where tourism has become economically important. Other climate-dependent resources in such communities include landscapes of natural and anthropogenic ecosystems, weather conditions that are suitable for specific activities and aid the dispersion of pollutants, and snow for skiing.

The various components of the hydrological cycle are particularly critical links between climate and the life of mountain communities. Water is essential for many activities, including agriculture, food preparation, power generation, sewage disposal, and snow-making. Natural snowfall must occur at the right time in relation to periods of heavy demand in the ski season. Snowmelt must also occur at the right time to allow crop or pasture growth and power generation. Similarly, the timing and volume of rainfall are critical constraints on many activities. Thus, intra-annual variations in precipitation are at least as important as inter-annual variations; snowmelt runoff is further influenced by air temperatures and solar radiation in the spring (Collins, 1989).

To understand the interactions of climate and human activities, particularly in highly variable environments such as mountains, requires detailed long-term knowledge of both sides of these very complicated relationships. However, the availability of long-term climatological data varies greatly both within and between mountain regions; and such data tend to be available more readily from settlements close to the mountains or in valleys, than from mountain sides and peaks where many activities take place. Relationships between climatic variables in mountains and neighbouring areas are highly complex (Barry, 1990, 1992b). Thus, even when data are available, they often do not provide a good overall description of the climate of an entire mountain area; extrapolations between locations with reliable data can be highly questionable.

Some mountain areas contain quite dense networks of meteorological stations at many altitudes. However, such networks often only record data for part of the year (e.g., winter, providing information useful for predicting avalanches). Equally, measurement at individual sites may not have taken place for long enough to provide a statistically valid – or useful – description of the local climate. Very small changes in the location of measurement sites can result in significant changes in the data collected (Linacre, 1992).

The varying levels of data availability and understanding of climates in mountain regions around the world have recently been described by Barry (1992a: 362–3). These are summarized below, from best to worst.

- The European Alps include many summit observatories with records for over 100 years, with dense data-collection networks. Both regional and smaller-scale research has been undertaken. Extensive data are also available for parts of the Carpathians.
- In the Pacific, detailed information is available for the island of Hawaii and for Mount Fuji, Japan.
- The mountains of western North America, Scandinavia, Britain, and the Caucasus have quite good station networks, but very few permanent mountain observatories. Considerable field research has been conducted into regional and smaller-scale phenomena.
- The northern Andes possess a reasonable network of stations, some at high altitudes, and microclimatological research provides additional information. However, large-scale climatic controls remain ill-defined.
- The Nepalese Himalaya has many stations in valleys, and expeditions have provided supplementary short-term data for high altitudes. The literature for the Tibetan Plateau is increasing.
- Other areas, including New Guinea, Ethiopia, Mount Kenya, and the mountains of the part of the former Soviet Union in Asia, are characterized by occasional research and sparse station networks.

In summary, there are considerable difficulties in describing most mountain climates except at the broadest scale. As noted by Barry (1992a: 374), no convenient inventory of mountain climate systems, much less a catalogue of data, exists; and most mountain areas have very few, if any, permanent stations above valleys. Theoretical studies of mountain meteorology have concentrated on synoptic-scale effects of mountain terrain and mesoscale features, while field research has emphasized mountain-valley winds, topographic effects on radiation, and weather modification. Many mesoscale modelling methods are well-developed, and 'nested' models (Giorgi and Mearns, 1991) may provide a way to use GCMs to develop climate change scenarios. However, in nearly all regions, modelling would be constrained by the availability of appropriate information and, even if this exists, the outputs of existing types of models may not be suitable for impact assessment or decision making in mountain communities.

CLIMATE CHANGE AND MOUNTAIN COMMUNITIES

As discussed above, the various mountain regions of the world differ greatly in their human history and economic bases, and the availability of information. This section attempts to tie together these themes and relate them to the existing literature on the potential impacts of climate change, which has been brought together particularly by Working Group II of the Intergovernmental Panel on Climate Change (IPCC) (Tegart *et al.*, 1990; 1993). For the sake of clarity, a number of topics are reviewed individually, although it must be recognized that, because of their interactions, none should be considered in isolation.

Agriculture

In 1985, in a global review of the sensitivity of agricultural production to climate change, Oram (1985: 150–1) noted that:

> Four marginal producer groups at particular risk can be identified .... farmers at high altitudes ... have received little attention until recently: they live in a wide range of conditions at varying altitudes and latitudes, and have an equally diverse range of production systems. In some regions altitude may modify harsh lowland climates favorably; in others, as in West Asia, it may compound summer heat with winter cold.... It is extremely difficult to predict how such diverse and complex situations would be affected by climatic change.

Seven years later, mountain farmers have still been paid little attention: for instance, a citation of Oram's statement is the only direct reference to mountain agriculture in the proceedings of the Second World Climate Conference (Swaminathan, 1991).

In both industrialized and developing countries, predictions have been made that climatic limits to cultivation will rise in the Alps (Balteanu *et al.*, 1987), Japan (Yoshino *et al.*, 1988), New Zealand (Salinger *et al.*, 1989), and Kenya (Downing, 1992). In developing countries, in-depth studies of the effects of climatic variations on agriculture have been undertaken in Ecuador's Central Sierra (Bravo *et al.*, 1988) and Papua New Guinea (Allen *et al.*, 1989). These show that crop growth and yield are controlled by complex interactions between solar radiation, temperature, precipitation, and frost, and that specific methods of cultivation – particularly those that protect against frost – may permit crop survival in sites whose microclimates would otherwise be unsuitable. Such specific details cannot be included in GCM-based impact assessments, which have suggested both positive and negative impacts, such as decreasing frost risks in the Mexican highlands (Liverman and O'Brien, 1991) and less productive upland agriculture in Indonesia and

Thailand, where impacts would depend on various factors, particularly types of cultivars and the availability of irrigation (Parry *et al.*, 1992).

In many mountain regions, a great variety of cultivars of many crops has been developed over generations. When planted and harvested according to detailed local knowledge in a wide range of microsites, these provide harvests that are adequate for food supplies (and sometimes surpluses), under diverse climatic conditions. In some regions, local social movements are emphasizing the cultivation of native crops in development strategies (e.g., Zimmerer, 1992). However, many food crops, in spite of their adaptability to mountain environments, high nutritional value, and tastiness, remain under-utilized even in their native regions (National Research Council, 1989); and others have been displaced by varieties that only give high yields under a narrow range of climatic conditions, often with high inputs of fertilizer, or by crops grown for urban or export markets. The maintenance and breeding of cultivars adapted to a wide variety of conditions seems to be a critical element in maintaining food systems in a changing climate (Erskine and Muehlbauer, 1990).

Given the wide range of microclimates that already exist in mountain areas and have been exploited by planting diverse crops, direct negative effects of climate change on crop yields may not be too great, and may be compensated for by increases in upper limits for cultivation where soils are suitable. Crop yields may increase if moisture is not limiting. This optimistic assessment might be upset, particularly by great increases in cloudiness or numbers of extreme events, or decreases in available moisture. In addition, increases in both crop and animal yields may be negated by increased populations of pests and disease-causing organisms, many of which have distributions which are climatically controlled. Interspecific interactions between pests and their predators and parasites may also change significantly (Parry *et al.*, 1990; Rosenzweig *et al.*, 1993). Thus, sedentary farmers and nomads might have to change seasonal patterns of pasturage and the composition of herds of domesticated animals.

Returning to the question at the start of the paper, the answer would appear to be 'yes,' but with a greater emphasis in 'old' regions with substantial population growth. In western Europe, the function of mountain agriculture is often as much to maintain rural populations in landscapes for tourism as to produce food; and specialized research institutions will, in all probability, be able to maintain supplies of suitable cultivars. Agricultural production in other 'old' mountains, however, is vital for vast and often rapidly increasing numbers of people (Stone, 1992). Considerable work still needs to be done with regard to optimum methods of cultivation and seed and food storage; the maintenance of adaptable genotypes; and under-standing interactions between domesticated plants and animals and pests and diseases. Thus, climate change provides an impetus to the evident need for greater involvement of mountain farmers in all aspects of land management and for the development and strengthening of development strategies emphasizing the cultivation and marketing of indigenous crops (Ives and Messerli, 1989; Zimmerer, 1992).

Forestry

In traditional communities in 'old' mountains, forestry and agriculture have long been inseparable, almost indistinguishable, parts of the economy. Trees provide fodder for livestock, shade for crops and livestock, and wood for a multitude of agricultural and domestic purposes. Particularly vital among these, and often dominant in terms of volumes harvested, is the use of wood for fuel. This pattern persists in most of these regions, except in Europe where wood harvests have declined greatly since the 1950s, with the availability of fossil fuels, the use of imported wood for construction, and decreasing employment in agriculture. This European pattern also applies in many 'young' mountain regions, although large-scale wood harvests continue where this is economically and environmentally permissible. Thus, in mountains now mainly inhabited by people of European ancestry, forests tend to be more important for regulating the quality and quantity of water flows, providing protection against 'natural hazards' such as avalanches, and as part of the landscape for tourism (Price, 1990).

Most assessments of the potential impacts of climate change on forests have used GCMs, concentrating on forests managed for wood production (Parry et al., 1990). Such assessments have usually been at the national scale, combining forests in mountain and lowland regions. However, regional-scale studies, both empirical and model-based, have been undertaken for eastern North America (Botkin and Nisbet, 1992; Cook and Cole, 1991; Overpeck et al., 1990), western North America (Winjum and Neilson, 1989), and the Swiss Alps (Brzeziecki et al., 1994; Kienast, 1991). All of these provide preliminary information about influences of climate on tree growth and survival.

A frequently expressed concern is that, because the rate of change of climate is likely to be faster than the ability of tree species to migrate, large areas of forest will die and, in the absence of significant human intervention, not be replaced (El-Lakany et al., 1991). This concern, however, derives particularly from the two-dimensional view of earth of GCMs. The verticality of mountain regions means that the horizontal distances over which species may have to move to reproduce and survive in response to changing climates will be much smaller than in other regions – though site conditions, particularly soils, would also have to be suitable (Price and Haslett, 1993). Furthermore, wood production might increase: various recent studies suggest that growth rates of trees in high-altitude forests have increased in recent decades in response to higher summer temperatures, the most important determinant of growth in these environments (Innes, 1991).

438

However, probabilities of frost damage in late winter might be greater because of higher early winter temperatures (Parry *et al.*, 1990).

As with crops and domesticated animals, some of the greatest and least understood concerns with regard to the potential impacts of climate change in forests relate to increases in susceptibility to, and outbreaks of, insects and diseases (Fanta, 1992). Another critical possibility is that, with increased evapotranspiration under warmer conditions, fire frequencies may increase (Fosberg, 1989; Franklin *et al.*, 1992). Thus, in many mountain areas in western North America and Europe, climate change could exacerbate the existing problem of increased fuel loading resulting from decades of fire suppression and decreasing harvests. In many mountain regions in eastern North America and Europe, the loss of forest cover through forest die-back ('*Waldsterben*') could also be accelerated by climate change.

Mountain communities around the world should evidently be concerned about climate change with regard to the forests which, in many ways, are essential for their survival. In the mountains of developing countries, considerable attention has been focused in recent years on decreases in forest cover and density, in spite of limited and often contradictory data (Hamilton, 1992). Climate change is yet another factor to be considered in these complex and ill-defined relationships. As with crops, there is a clear need for the maintenance of genotypes adapted to a wide variety of environmental conditions (Hattemer and Gregorius, 1990), and the development and use of appropriate methods for planting, maintaining and harvesting trees in both forests and other landscapes with varying densities of trees and crops (i.e., agroforestry).

In industrialized nations, where forest cover and density have often been increasing over recent decades, substantial synergetic negative effects may result from the addition of climate change to other sources of change: e.g., air pollution, fire, pests, and diseases. In general, the concern of mountain communities about forests in a period of climate change might be expected to grow in proportion to the density of human population and infrastructure in a given region. However, variations in the comparative importance and interactions of these diverse environmental and societal factors greatly limit possibilities for even suggesting likely impacts at any scale.

In summary, considerable work is required in all mountain regions to develop both better understanding – necessary to predict the behaviour of forest ecosystems, species, pests, and diseases in changing environments – and appropriate management approaches, including gene conservation (Tippets, 1992). Given the importance of forests to all mountain communities – whatever the current and anticipated mix of fuel, timber, protection against hazards, landscape, and other 'products' – such work is vital not only for mountain communities but to the billions of people who live downstream from them.

Water resources

Forests are an integral component of the hydrological cycle, influencing the discharge and sediment content of rivers through the interception, storage, and evapotranspiration of water. The significance of these processes for hydrological systems and the communities – both mountain and lowland – which depend on them varies with the spatial scale of analysis (Hamilton, 1992). Thus, as human activities change the area and structure of forests, they also affect the quality and quantity of water supplies which, in turn, affect and are influenced by other aspects of human life from agriculture to health and recreation.

In spite of limitations in the quality of historical data sets, discussed above, and inconsistencies in projections between GCMs, particularly for precipitation (Houghton *et al.*, 1990), assessments of the potential impacts of climate change on water resources, including snowfall and storage, have been conducted at a variety of spatial scales for most 'young' mountain regions and western Europe (Aguado *et al.*, 1992; Bultot *et al.*, 1992; Garr and Fitzharris, this volume; Lins *et al.*, 1990; Martin, 1992; Nash and Gleick, 1991; Oerlemans, 1989; Rupke and Boer, 1989; Slaymaker, 1990; Street and Melnikov, 1990). Lack of data, computing facilities, and personnel may explain the lack of assessments in other 'old' mountain regions.

Slaymaker (1990: 171) concludes that:

> three sources of uncertainty (the spatial variability of mountain systems, the range of predicted climate change scenarios, and the variable lag times of environmental systems to climate change) lead to qualitative estimates and predictions of tendencies rather than confident assertions about the geomorphic impacts of climate change.

The Working Group on Hydrology and Water Resources of IPCC Working Group II (Lins *et al.*, 1990: 4–25; Stakhiv *et al.*, 1993: 79) came to similarly qualitative conclusions about other impacts. In addition to decreasing uncertainties with regard to the latter two issues identified by Slaymaker (1990), they identified particular needs for:

- a uniform approach to the analysis of sensitivities of hydrological systems to climate change;
- assessments of water resource sensitivities in developing countries, especially in arid and semi-arid regions;
- better methods of water management under climate uncertainty, especially in relation to the greater variability of floods and droughts; and
- increased knowledge of potential effects of climate change on water quality.

The quality and quantity of water supplies are already constraining factors for mountain communities around the world, especially in winter when many supplies are frozen. Climate change may be characterized by changes in seasonal or annual precipitation, proportions of solid to liquid precipitation, or frequencies of extreme events. Whatever the directions and magnitudes of change, mountain communities and those downstream need to be prepared to implement flexible water management strategies which do not assume that recent patterns will continue. Events in recent history may provide useful guidelines for developing such strategies (Glantz, 1988).

Tourism

In recent decades, tourism has developed rapidly in mountain regions across the world. While widespread, tourism is not omnipresent in these regions: at any spatial scale, the degree of its development is highly variable and, over time, its importance may increase and then wane. Even in the Alps, whose economy is dominated by tourism, traditional agriculture and forestry remain the foundation of some communities. Similarly, in other well-known destinations, such as Nepal, large areas are not directly affected by tourism (Price, 1993). Climate change is likely to have both direct and indirect impacts on tourism in mountain areas. Direct changes refer to changes in the atmospheric resources necessary for specific activities. Indirect changes may result from both changes in mountain landscapes – the 'capital' of tourism (Krippendorf, 1984) – and wider-scale socio-economic changes: for example, in patterns of demand for specific activities or destinations and for fuel prices.

The marked seasonality of the climates of mountain regions means that their attractions for tourists vary greatly through the year. Various methodologies have been developed to assess the suitability of regions for specific activities in different seasons and over the course of the year (Besancenot, 1990). While such approaches are based on long-term averages, others have been developed to assess the economic implications of historical climate variability over short periods, mainly for the skiing industry (Lynch *et al.*, 1981; Perry, 1971).

Using scenarios derived from GCMs, these methods have been extended to examine the possible implications of climate change for skiing in Australia, eastern Canada, and Switzerland (Abegg and Froesch, this volume; Galloway, 1988; McBoyle and Wall, 1987; Lamothe et Périard, 1988). These studies show that, as the length of the skiing season is sensitive to quite small climatic changes, these could lead to considerable socio-economic disruption in communities that have invested heavily in the skiing industry. To some extent, such impacts might be offset by new opportunities in the summer season and also by investment in new technology, such as snow-making equipment, as long as climatic conditions remain within appropriate bounds. Such investments, following seasons with little snow, have provided some 'insurance' in mountain regions throughout North America. However, the introduction of snow-making has been far less widespread in Europe (Broggi and Willi, 1989), although recent experience has shown that seasons with little snow, especially at critical periods, such as the Christmas/New Year holiday, can be economically devastating to mountain communities.

The economic impacts on ski resorts of changing patterns of snowfall during climate change might appear to have little relevance to communities in the mountains of developing countries. Yet ski resorts are also found in the Andes (Fuentes and Castro, 1982; Solbrig, 1984) and the Himalayas, and changes in the length of the snow-free season would be of critical importance for most mountain communities. In South Asia, another important potential change for communities which increasingly depend on tourism concerns the monsoon, whose timing may well change (Houghton et al., 1990). This could have substantial effects on countries, such as Nepal and Bhutan, for whom tourists are the principal source of foreign exchange (Richter, 1989).

In addition to these potential direct impacts of climate change on tourism, a critical indirect impact should be noted. One of the most likely types of policy response to climate change will be the imposition of 'carbon taxes' on fossil fuels (Bryner, 1991; Pearce et al., 1991). These will increase the costs of these fuels, a major component of the cost of tourism, particularly to mountain regions, which tend not to be easily accessible. Other indirect impacts might include decreasing attractiveness of landscapes, new competition from other tourist locations as climates change (particularly at the seasonal scale and in relation to holiday periods), and concerns about the health risks of high levels of ultraviolet radiation at high altitudes.

Today, very few communities can rely on tourism as a reliable source of year-round employment and income. This is a particular problem in the many ski resorts where off-season work is limited because they are not based on pre-existing settlements. Seasonality is important not only in terms of employment, but also because facilities built for tourism represent investments that must be maintained year-round and paid off (Barker, 1982). As noted by Watson and Watson (1982), tourism is an 'industry of fashion', and while unexpected booms in demand are not unusual, predicted new demands, for which facilities have been constructed, do not necessarily materialize.

In both developing and industrialized countries, tourism has led to considerable changes in patterns of agricultural production, forestry and water resources management, and employment (Price, 1993). Thus, the implications of climate change for tourism must be regarded in conjunction with the implications for other resources (Breiling and Charamza, this volume). Returning to the question in the title, the answer clearly depends on the extent to which tourism has influenced the economy and life of a mountain community and the landscape surrounding it. Numerous examples of the decline of tourism in mountain communities exist. In some cases, the reversion to a traditional economy based on agriculture and forestry has been successful (Messerli, 1989). However, the ability to accomplish this depends on many factors, including the availability of an appropriate labour force, the extent to which essential resources – such as irrigation and terracing systems – have deteriorated, and climate.

In most mountain communities, the development of tourism has been characterized by rapid decisions based on far more limited information than decisions relating to other activities on which these communities have relied. Mountain people in both developing and industrialized countries recognize that they need to maintain or recover control over the forces of the tourist economy (Kariel and Kariel 1982; Lieberherr-Gardiol and Stucki, 1987; Moser and Moser 1986; Norberg-Hodge, 1991; Puntenney, 1990). A clearer understanding of these forces, including the role that climate change may play in them, is essential as a basis for future decisions about a diffuse economic sector for which long-term prognoses are hard to make.

Energy, transport, and human health

In addition to the topics considered above, IPCC Working Group II considered a number of others of central interest to mountain communities, particularly energy, transport, and human health (Ando *et al.*, 1993; Rouviere *et al.*, 1990). A few preliminary statements concerning the likely effects of climate change on mountain communities with respect to these topics can be made.

Energy supplies have been considered tangentially above with respect to both fuelwood and energy pricing policies. Higher costs for fossil fuels inevitably affect all economic activities. While warmer winter temperatures would decrease energy requirements for domestic heating, warmer summer temperatures could cause energy demands for cooling to rise. Climate change may lead to increased demands for, and probably development of, hydro-electric power as other sources become more expensive or less available for other reasons. Hydro-power developments vary greatly in scale from micro-generators for individual villages to massive projects, frequently to supply lowland cities and industries.

The scope for future large-scale hydro-electric projects in many western European mountains is limited because most available dam sites have already been developed, while environmental protection policies greatly limit future development in North America (Stone, 1992). In contrast, many large projects are under development or planned in Asia (Badenkov, 1992; Bandyopadhyay, 1992; Dhakal, 1990). As shown by research in the Southern Alps of New Zealand (Garr and Fitzharris, this volume), hydrological changes resulting from climate change would undoubtedly influence the viability of these projects, whose impacts on mountain communities may vary from the positive (e.g., availability of irrigation water and power) to the negative (e.g., loss of agricultural land, forests, settlements, and transport infrastructure).

Climate change may influence transport in mountain regions both indirectly, by altering the availability of energy sources as discussed above, and directly, especially through increases in the frequency of 'natural hazards,' such as avalanches, floods, and rockslides. Consequently, the maintenance of the stability of both forests and agricultural land is critical; but the likelihood of increased numbers of extreme precipitation events (Gates *et al.*, 1992; Lins *et al.*, 1990) means that communications – as well as many other activities – could be greatly disrupted. At the same time, a longer snow-free season would ease transport and decrease costs of snow removal.

Human health could be affected by climate change through changes in temperatures, air quality, and hydrology (Ewan *et al.*, 1990). Extreme temperatures, both hot and cold, could cause increased mortality from cardio-vascular and respiratory stress. Higher temperatures could also allow populations of disease-carrying organisms to increase significantly and to move to higher altitudes. Drier conditions could limit the availability of water for drinking and sewage disposal. Understanding of such relationships is at a very early stage of development. However, as both permanent and tourist populations increase, the maintenance of adequate health standards is a growing problem in many mountain regions. Climate change is likely to exacerbate this problem.

CONCLUSIONS

Clearly, mountain communities should be concerned about climate change. Their degree of concern will vary considerably in relation to a vast number of interacting factors, and it is important to stress that climate change should be seen in terms of not only negative impacts but also positive opportunities (Glantz *et al.*, 1990). Given the complexity of mountain environments at all scales and the lack of detailed scientific knowledge of these environments and how mountain populations utilize them, definitive conclusions are hard to make. The word 'scientific' is important; members of mountain communities whose history in a given region stretches back many generations are likely to have site-specific knowledge that is at least as relevant for the long-term future of their neighbours and their descendants as information collected according to objective criteria by scientists and their instruments.

Given the high variability of mountain climates, communities in old mountain regions with lifestyles that continue to be based, to some extent, on the opportunities and constraints of diverse micro-environments may be able to respond or adapt to changes in climate with relatively little difficulty – unless rates of change are very rapid. However, mere knowledge of these complex environments will not be adequate; suitable societal structures must be in place or be developed. The extensive networks of community and mutual co-operation that characterize many mountain communities (Beaver and Purrington, 1984; Guillet, 1983; Viazzo, 1989) may already be under stress as populations grow and change in structure, and increasing numbers of tourists arrive. However, such networks are likely to be increasingly essential, particularly in communities that are geographically or culturally distant from centres of power and/or in countries whose governments are not able to provide resources for their maintenance. This probably applies to the majority of communities of old mountain regions except western Europe.

In western Europe, tourism's rise to dominance has required massive investments in infrastructure and has been associated with great changes in agriculture and forestry. Similar patterns seem likely in other parts of Europe. Yet, as discussed above, tourism tends to be an unreliable long-term foundation for mountain communities, to an extent that is likely to increase in an era of climate change. Even in areas for which considerable scientific knowledge is available, increased efforts should be made to maintain and optimize the use of traditional knowledge and practices. These are generally declining but, linked to information derived from scientific research and monitoring, could be used to limit the negative effects of climate change and benefit from its positive aspects.

Young mountain regions tend to suffer from a dearth of both traditional knowledge and long-term climate records, although other aspects of their environments may be well documented. Densities of settlement tend to be quite low, except in tourist areas, and even in these, high population densities often occur only for relatively short periods. Transportation networks are also relatively sparse. As in other regions, communities depending largely on tourism may have to consider other alternatives as demand decreases or climatic conditions become unsuitable. One problem which may increase in severity is the loss of forest cover through epidemics, die-back, or fire, creating unstable and unattractive environments from which emigration appears the best alternative. Substantial loss of forest cover would also be a critical issue for lowland populations through effects on their water supplies; although this is true worldwide.

Whatever the course of climate change, the future of mountain communities and the environments which surround them is an issue which requires worldwide attention. Mountains are emerging in the fora of sustainable development, notably through the 1992 UN Conference on Environment and Development, whose Agenda 21, an action programme for the coming years, includes a chapter on mountains (Stone, 1992). Although sessions of previous meetings have considered the issue of climate change and mountains (Badenkov et al., 1990/91; Rupke and Boer, 1989), the conference at which this paper was first presented was the first devoted to this issue. It is to be hoped that the synergistic interactions between work on 'sustainable development' and climate in mountain regions, involving mountain people, scientists, and policy makers, will continue to grow from today's limited beginnings.

REFERENCES

Aguado, E., Cayan, D., Riddle, L. and Roos, M. (1992) 'Changes in the timing of runoff from West Coast streams and their relationships to climatic influences', *Swiss Climate Abstracts* (Special Issue: International Conference on Mountain Environments in Changing Climates): 3.

Allen, B., Brookfield, H. and Byron, Y. (eds) (1989) 'Frost and drought in the highlands of Papua New Guinea', *Mountain Research and Development*, 9(3): 199–334.

Ando, M., Hanaki, K., Harasawa, H., Karosawa, H., Masuda, K., Nishinomiya, S., Okita, T., Ball, R.H., Breed, W., Hobbie, D., Topping, J. and Nazarov, I. (1993) 'Energy; human settlement; transport and industrial sectors; human health; air quality; effects of ultraviolet-B radiation', in W.J.McG. Tegart, G.W. Sheldon and J.H. Hellyer (eds), *Climate Change 1992: The Supplementary Report to the IPCC Impacts Assessment*, Australian Government Publishing Service, Canberra, pp. 29–42.

Badenkov, Y. (1992) 'Mountains of the former Soviet Union: value, diversity, uncertainty', in P.B. Stone (ed.), *The State of the World's Mountains*, Zed Books, London, pp. 257–97.

Badenkov, Y., Hamilton, L.S., Ives, J.D., Ives, P., Galtseva, O. and Messerli, B. (eds) (1990/91) 'Proceedings of the Conference on the transformation of mountain environments', *Mountain Research and Development*, 10 (2) and 11 (1).

Balteanu, D., Ozenda, P., Huhn, M., Kerschner, H., Tranquillini, W. and Borten-schlager, S. (1987) 'Impact analysis of climatic change in the central European mountain ranges', in *European Workshop on Interrelated Bioclimatic and Land Use Changes*, vol. G, Noordwijkerhout.

Bandyopadhyay, J. (1992) 'The Himalaya: prospects for and constraints on sustainable development', in P.B. Stone (ed.) *The State of the World's Mountains*, Zed Books, London, pp. 93–126.

Barker, M.L. (1982) 'Traditional landscape and mass tourism in the Alps', *Geographical Review*, 72: 395–415.

Barry, R.G. (1986) 'Mountain climate data for long-term ecological research', in *Proceedings, International Symposium on the Qinghai-Xizang Plateau and Mountain Meteorology*, American Meteorological Society, Boston, pp. 170–87.

—— (1990) 'Changes in mountain climate and glacio-hydrological responses', *Mountain Research and Development*, 10: 161–70.

—— (1992a) 'Climate change in the mountains', in P.B. Stone (ed.), *The State of the World's Mountains*, Zed Books, London, pp. 359–80.

—— (1992b) *Mountain Weather and Climate*, 2nd edition, Routledge, London.

Beaver, P.D. and Purrington, B.L. (eds) (1984) *Cultural Adaptation to Mountain Environments*, University of Georgia Press, Athens.

Besancenot, J.-P. (1990) *Climat et tourisme*, Masson, Paris.

Botkin, D.B. and Nisbet, R.A. (1992) 'Forest response to climatic change: effects of parameter estimation and choice of weather patterns on the reliability of projections', *Climatic Change*, 20: 87–111.

Bravo, R.E., Cañadas Cruz, L., Estrada, W., Hodges, T., Knapp, G., Ravelo, A.C., Planchuelo Ravelo, A.M., Rovere, O., Salcedo Solis, T. and Yugcha, T. (1988) 'The effects of climatic variations on agriculture in the Central Sierra of Ecuador', in M.L. Parry, T.R. Carter and N.T. Konijn (eds), *The Impact of Climate Variations on Agriculture: Volume 2, Assessments in Semi-arid Regions*, Kluwer, Dordrecht, pp. 381–493.

Broggi, M.F. and Willi, G. (1989) *Beschneiungsanlagen im Widerstreit der Interessen*, CIPRA, Vaduz.

Bryner, G. (1991) 'Implementing global environmental agreements', *Policy Studies Journal*, 19 (2): 103–14.

Brzeziecki, B., Kienast, F. and Wildi, O. (1994) 'Simulating the distribution of forest communities in Switzerland using GIS technology', in M.F. Price and D.I. Heywood (eds), *Mountain Environments and Geographic Information Systems*, in press, Taylor and Francis, London.

Bultot, F., Gellens, G., Spreafico, M. and Schädler, B. (1992) 'Repercussions of a CO_2 doubling on the water balance – a case study in Switzerland', *Journal of Hydrology*, 137: 199–208.

Carter, T.R., Parry, M.L., Nishioka, S. and Harasawa, H. (1992) *Preliminary Guidelines for Assessing Impacts of Climate Change*, Environmental Change Unit, Oxford.

Collins, D.N. (1989) 'Hydrometeorological conditions, mass balance and runoff from alpine glaciers', in J. Oerlemans (ed.), *Glacier Fluctuations and Climate Change*, Kluwer, Dordrecht, pp. 305–23.

Cook, E.R. and Cole, J. (1991) 'On predicting the response of forests in eastern North America to future climatic change', *Climatic Change*, 19: 271–82.

Dhakal, D.N.S. (1990) 'Hydropower in Bhutan: a long-term development perspective', *Mountain Research and Development*, 10: 291–300.

Downing, T.E. (1992) *Climate Change and Vulnerable Places: Global Food Security and Country Studies in Zimbabwe, Kenya, Senegal, and Chile*, Environmental Change Unit, Oxford.

El-Lakany, H., Ramakrishna, K. and Salati, E. (1991) 'Task group 8: forests', in J. Jaeger and H.L. Ferguson (eds), *Climate Change: Science, Impacts and Policy*, Cambridge University Press, Cambridge, pp. 467–70.

Erskine, W. and Muehlbauer, F.J. (1990) 'Effects of climatic variations on crop genetic resources and plant breeding aims in West Asia and North Africa', in M.T. Jackson, B.V. Ford-Lloyd and M.L. Parry (eds), *Climatic Change and Plant Genetic Resources*, Belhaven, London, pp. 148–57.

Ewan, C., Bryant, E. and Calvert, D. (eds) (1990) *Health Implications of Long Term Climate Change*, University of Wollongong, Wollongong.

Fanta, J. (1992) 'Possible impact of climatic change on forested landscapes in central Europe: a review', *Catena* supplement 22: 133–51.

Fosberg, M.A. (1989) 'Climate change and forest fires', in J.C. Topping (ed.), *Coping with Climate Change*, Climate Institute, Washington DC, pp. 292–6.

Franklin, J.F., Swanson, F.J., Harmon, M.E., Perry, D.A., Spies, T.A., Dale, V.H., McKee, A., Ferrell, W.K., Means, J.E., Gregory, S.V., Lattin, J.D., Schowatter, T.D. and Larsen, D. (1992) 'Effects of global climatic change on forests of northwestern North America', in R.L. Peters and T.J. Lovejoy (eds), *Global Warming and Biological Diversity*, Yale University Press, New Haven, pp. 244–57.

Fuentes, E.R. and Castro, M. (1982) 'Problems of resource management and land use in two mountain regions of Chile', in F. di Castri, G. Baker and M. Hadley (eds), *Ecology in Practice*, vol. 2, Tycooly, Dublin, pp. 315–30.

Galloway, R.W. (1988) 'The potential impact of climate changes on Australian ski fields' in G.I. Pearman (ed.), *Greenhouse: Planning for Climate Change*, CSIRO, East Melbourne, pp. 428–37.

Gates, W.L., Mitchell, J.F.B., Boer, G.J., Cubasch, U. and Meleshko, V.P. (1992) 'Climate modelling, climate prediction and model validation', in J.T. Houghton, G.J. Jenkins and J.J. Ephraums (eds), *Climate Change 1992: The Supplementary Report to the IPCC Scientific Assessment*, Cambridge University Press, Cambridge, pp. 97–134.

Gerrard, A.J. (1990) *Mountain Environments*, Belhaven, London.

Giorgi, F. and Mearns, L.O. (1991) 'Approaches to the simulation of regional climate change: a review', *Reviews in Geophysics*, 29: 191–216.

Glantz, M.H. (ed.) (1988) *Societal Responses to Regional Climatic Change*, Westview, Boulder.

Glantz, M.H., Price, M.F. and Krenz, M.E. (eds) (1990) *On Assessing Winners and Losers in the Context of Global Warming*, National Center for Atmospheric Research, Boulder.

Grötzbach, E.F. (1988) 'High mountains as human habitat', in N.J.R. Allan, G.W. Knapp and C. Stadel (eds), *Human Impact on Mountains*, Rowman & Littlefield, Totowa, New Jersey, pp. 24–35.

Guillet, D. (1983) 'Toward a cultural ecology of mountains: the Central Andes and the Himalaya compared', *Current Anthropology*, 24: 561–74.

Halpin, P.N. (1994) 'A GIS analysis of potential impacts of climate change on mountain ecosystems and protected areas', in M.F. Price and D.I. Heywood (eds), *Mountain Environments and Geographic Information Systems*, in press, Taylor & Francis, London.

Hamilton, L.S. (1992) 'The protective role of mountain forests', *GeoJournal*, 27: 13–22.

Hattemer, H.H. and Gregorius, H.-R. (1990) 'Is gene conservation under global climate change meaningful?', in M.T. Jackson, B.V. Ford-Lloyd and M.L. Parry (eds), *Climatic Change and Plant Genetic Resources*, Belhaven, London, pp. 158–66.

Houghton, J.T., Jenkins, G.J. and Ephraums, J.J. (eds) (1990) *Climate Change: The IPCC Scientific Assessment*, Cambridge University Press, Cambridge.

Innes, J.L. (1991) 'High-altitude and high-latitude tree growth in relation to past, present and future global climate change', *The Holocene*, 1: 168–73.

Ives, J.D. (1992) 'Preface', in P. Stone (ed.), *The State of the World's Mountains*, Zed Books, London, pp. xiii–xvi.

Ives, J.D. and Messerli, B. (1989) *The Himalayan Dilemma*, Routledge, London.

Jaeger, J. and Ferguson, H.L. (eds) (1991) *Climate Change: Science, Impacts and Policy*, Cambridge University Press, Cambridge.

Kariel, H.G. and Kariel, P.E. (1982) 'Socio-cultural impacts of tourism: an example from the Austrian Alps', *Geografiska Annaler B*, 64: 1–16.

Kates, R.W., Ausubel, J.H. and Berberian, M. (eds) (1985) *Climate Impact Assessment*, John Wiley & Sons, Chichester.

Kienast, F. (1991) 'Simulated effects of increasing atmospheric CO_2 and changing climate on the successional characteristics of Alpine forest ecosystems', *Landscape Ecology*, 5: 228–38.

Krippendorf, J. (1984) 'The capital of tourism in danger', in E.A. Brugger, G. Furrer, B. Messerli and P. Messerli (eds), *The Transformation of Swiss mountain Regions*, Haupt, Berne, pp. 427–50.

Lamothe et Périard (1988) *Implications of Climate Change for Downhill Skiing in Quebec*, Climate Change Digest, 3, Atmospheric Environment Service, Downsview.

Lettenmaier, D.P. (1990) 'Hydrologic sensitivities of the Sacramento–San Joaquin river basin, California, to global warming', *Water Resources Research*, 26: 69–86.

Lieberherr-Gardiol, F. and Stucki, E. (1987) *Sur nos monts quand la nature...*, CERME, Château d'Oex.

Linacre, E. (1992) *Climate Data and Resources*, Routledge, London.

Lins, H., Shiklomanov, I. and Stakhiv, E. (1990) 'Hydrology and water resources', in W.J. McG. Tegart, G.W. Sheldon and D.C. Griffiths (eds), *Climate Change: The IPCC Impacts Assessment*, ch. 4, Australian Government Publishing Service, Canberra.

Liverman, D.M. and O'Brien, K.L. (1991) 'Global warming and climate change in Mexico', *Global Environmental Change: Human and Policy Dimensions*, 1: 351–64.

Lynch, P., McBoyle, G.R. and Wall, G. (1981) 'A ski season without snow', in D.W. Phillips and G.A. McKay (eds), *Canadian Climate in Review – 1980*, Atmospheric Environment Service, Toronto, pp. 42–50.

McBoyle, G.R. and Wall, G. (1987) 'The impact of CO_2-induced warming on downhill skiing in the Laurentians', *Cahiers de Géographie de Québec*, 31 (82): 39–50.

Martin, E. (1992) 'Sensitivity of the French Alps' snow cover to the variation of climatic parameters', *Swiss Climate Abstracts* (Special Issue: International Conference on Mountain Environments in Changing Climates): 50–1.

Messerli, P. (1989) *Mensch und Natur im alpinen Lebensraum: Risiken, Chancen, Perspektiven*, Haupt Verlag, Berne, Stuttgart.

Moser, P. and Moser, W. (1986) 'Reflections on the MAB–6 Obergurgl project and tourism in an Alpine environment', *Mountain Research and Development*, 6: 101–18.

Nash, L.L. and Gleick, P.H. (1991) 'Sensitivity of streamflow in the Colorado Basin to cliamtic changes', *Hydrology*, 125: 221–41.

National Research Council (1989) *Lost Crops of the Incas*, National Academy Press, Washington DC.

Norberg-Hodge, H. (1991) *Ancient Futures: Learning from Ladakh*, Rider, London.

Oerlemans, J. (ed.) (1989) *Glacier Fluctuations and Climate Change*, Kluwer, Dordrecht.

Oram, P.A. (1985) 'Sensitivity of Agricultural Production to Climatic Change', *Climatic Change*, 7: 129–52.

Overpeck, J.T., Rind, D. and Goldberg, R. (1990) 'Climate-induced changes in forest disturbance and vegetation', *Nature*, 343 (6253): 51–3.

Parry, M.L., Duinker, P.N., Morison, J.I.L., Porter, J.H., Reilly, J. and Wright, L.J. (1990) 'Agriculture and forestry', in W.J.McG. Tegart, G.W. Sheldon and D.C. Griffiths (eds), *Climate Change: The IPCC Impacts Assessment*, ch. 2, Australian Government Publishing Service, Canberra.

Parry, M.L., de Rozari, M.B., Chong, A.L. and Parvich, S. (1992) *The Potential Socio-economic Effects of Climate Change in South-east Asia*, United Nations Environment Programme, Nairobi.

Pearce, D., Barbier, E., Markandya, A., Barrett, S., Turner, R.K. and Swanson, P. (eds) (1991) *Blueprint 2: Greening the World Economy*, Earthscan, London.

Perry, A.H. (1971) 'Climatic influences on the Scottish ski industry', *Scottish Geographical Magazine*, 87: 197–201.

Price, M.F. (1990) 'Temperate mountain forests: common-pool resources with changing, multiple outputs for changing communities', *Natural Resources Journal*, 30: 685–707.

—— (1994) 'Patterns of the development of tourism in mountain communities', in N.J.R. Allan (ed.), *Mountains at Risk: Current Issues in Environmental Studies*, in press, Kluwer, Dordrecht.

Price, M.F. and Haslett, J.R. (1994) 'Climate change and mountain ecosystems', in N.J.R. Allan (ed.), *Mountains at Risk: Current Issues in Environmental Studies*, in press, Kluwer, Dordrecht.

Puntenney, P.J. (1990) 'Defining solutions: The Annapurna experience', *Cultural Survival Quarterly*, 14 (2): 9–14.

Richter, L.K. (1989) *The Politics of Tourism in Asia*, University of Hawaii Press, Honolulu.

Riebsame, W.E. (1989) *Assessing the Social Implications of Climate Fluctuations*, United Nations Environment Programme, Nairobi.

Rizzo, B. and Wiken, E. (1992) 'Assessing the sensitivity of Canada's ecosystems to climatic change', *Climatic Change*, 21: 37–55.

Rosenzweig, C., MacIver, D., Hall, P., Parry, M.L., Sirotenko, O. and Burgof, J. (1993) 'Agriculture and forestry', in W.J.McG. Tegart, G.W. Sheldon and J.H. Hellyer (eds), *Climate Change 1992: The Supplementary Report to the IPCC Impacts Assessment*, Australian Government Publishing Service, Canberra, pp. 44–63.

Rouviere, C., Williams, T., Ball, R., Shinyak, Y., Topping, J., Nishioka, S., Ando, M. and Okita, T. (1990) 'Human settlement; the energy, transport and industrial sectors; human health; air quality; and changes in ultraviolet-B radiation', in W.J.McG. Tegart, G.W. Sheldon and D.C. Griffiths (eds), *Climate Change: The IPCC Impacts Assessment*, ch. 5, Australian Government Publishing Service, Canberra.

Rupke, J. and Boer, M.M. (eds) (1989) *Landscape Ecological Impact of Climatic Change on Alpine Regions, with Emphasis on the Alps*, Discussion Report prepared for European conference on landscape ecological impact of climatic change, Agricultural University of Wageningen and Universities of Utrecht and Amsterdam.

Salinger, M.J., Williams, J.M. and Williams, W.M. (1989) *CO_2 and Climate Change: Impacts on Agriculture*, New Zealand Meteorological Service, Wellington.

Slaymaker, O. (1990) 'Climate change and erosion processes in mountain regions of western Canada', *Mountain Research and Development*, 10: 171–82.

Solbrig, O. (1984) 'Tourism', *Mountain Research and Development*, 4: 181–5.

Stakhiv, E., Lins, H. and Shiklomanov, I. (1993) 'Hydrology and water resources', in W.J.McG. Tegart, G.W. Sheldon and J.H. Hellyer (eds), *Climate Change 1992: The Supplementary Report to the IPCC Impacts Assessment*, Australian Government Publishing Service, Canberra, pp. 71–83.

Stone, P. (ed.) (1992) *The State of the World's Mountains*, Zed Books, London.

Street, R.B. and Melnikov, P.I. (1990) 'Seasonal snow, cover, ice and permafrost', in W.J.McG. Tegart, G.W. Sheldon and D.C. Griffiths (eds), *Climate Change: The IPCC Impacts Assessment*, ch. 7, Australian Government Publishing Service, Canberra.

Swaminathan, M.S. (1991) 'Agriculture and food systems', in J. Jaeger and H.L. Ferguson (eds), *Climate Change: Science, Impacts and Policy*, Cambridge University Press, Cambridge, pp. 265–77.

Tegart, W.J.McG., Sheldon, G.W. and Griffiths, D.C. (eds) (1990) *Climate Change: The IPCC Impacts Assessment*, Australian Government Publishing Service, Canberra.

Tegart, W.J.McG., Sheldon, G.W. and Hellyer, J.H. (eds) (1993) *Climate Change 1992: The Supplementary Report to the IPCC Impacts Assessment*, Australian Government Publishing Service, Canberra.

Tippets, D. (1992) 'Genes for surviving global climate change', *Forestry Research West* (April): 8–12.

Viazzo, P.P. (1989) *Upland Communities: Environment, Population and Social Structure in the Alps since the Sixteenth Century*, Cambridge University Press, Cambridge.

Watson, A. and Watson, R.D. (1982) *The Swiss Approach and its Relevance to Scotland*, Grampian Regional Council, Cambridge.

Winjum, J.K. and Neilson, R.P. (1989) 'Forests', in J.B. Smith and D. Tirpak (eds), *The Potential Effects of Global Climate Change on the United States*,

US Environmental Protection Agency, Washington DC, pp. 71–92.

Yoshino, M.M., Horie, T., Seino, H., Tsuji, H., Uchijima, T. and Uchijima, Z. (1988) 'The effects of climatic variations on agriculture in Japan', in M.L. Parry, T.R. Carter and N.T. Konijn (eds), *The Impact of Climatic Variations on Agriculture: Volume 1, Assessments in Cool, Temperate and Cold Regions*, Kluwer, Dordrecht, pp. 723–868.

Zimmerer, K.S. (1992) 'Biological diversity in local development: 'popping beans' of the central Andes', *Mountain Research and Development*, 12: 47–61.

INDEX